T0320338

Hands-On Data Analysis in R for Finance

The subject of this textbook is to act as an introduction to data science/data analysis applied to finance, using R and its most recent and freely available extension libraries. The targeted academic level is undergrad students with a major in data science and/or finance and graduate students, and of course practitioners/professionals who need a desk reference.

- Assumes no prior knowledge of R;
- The content has been tested in actual university classes;
- Makes the reader proficient in advanced methods such as machine learning, time series analysis, principal component analysis and more;
- Gives comprehensive and detailed explanations on how to use the most recent and free resources, such as financial and statistics libraries or open database on the internet.

Hands-On Data Analysis in R for Finance

Jean-François Collard

CRC Press is an imprint of the
Taylor & Francis Group, an **informa** business
A CHAPMAN & HALL BOOK

First edition published 2023
by CRC Press
6000 Broken Sound Parkway NW, Suite 300, Boca Raton, FL 33487-2742

and by CRC Press
4 Park Square, Milton Park, Abingdon, Oxon, OX14 4RN

CRC Press is an imprint of Taylor & Francis Group, LLC

Library of Congress Cataloging-in-Publication Data

Names: Collard, Jean-François, author.
Title: Hands-on data analysis in R for finance / Jean-François Collard.
Description: 1 Edition. | Boca Raton, FL : Taylor and Francis, 2023. |
Includes index. | Identifiers: LCCN 2022022739 (print) | LCCN 2022022740 (ebook) | ISBN 9781032340975 (hardback) | ISBN 9781032340982 (paperback) | ISBN 9781003320555 (ebook)
Subjects: LCSH: Finance--Study and teaching (Higher) | Finance--Databases. | R (Computer program language)
Classification: LCC HG152 .C65 2023 (print) | LCC HG152 (ebook) | DDC 332.076--dc23/eng/20220812
LC record available at https://lccn.loc.gov/2022022739
LC ebook record available at https://lccn.loc.gov/2022022740

ISBN: 978-1-032-34097-5 (hbk)
ISBN: 978-1-032-34098-2 (pbk)
ISBN: 978-1-003-32055-5 (ebk)

DOI: 10.1201/9781003320555

Typeset in Latin Modern
by KnowledgeWorks Global Ltd.

Publishers Note: This book has been prepared from camera-ready copy provided by the authors

Access the Support Material: (https://www.routledge.com/Hands-On-Data-Analysis-in-R-for-Finance/Collard/p/book/9781032340975)

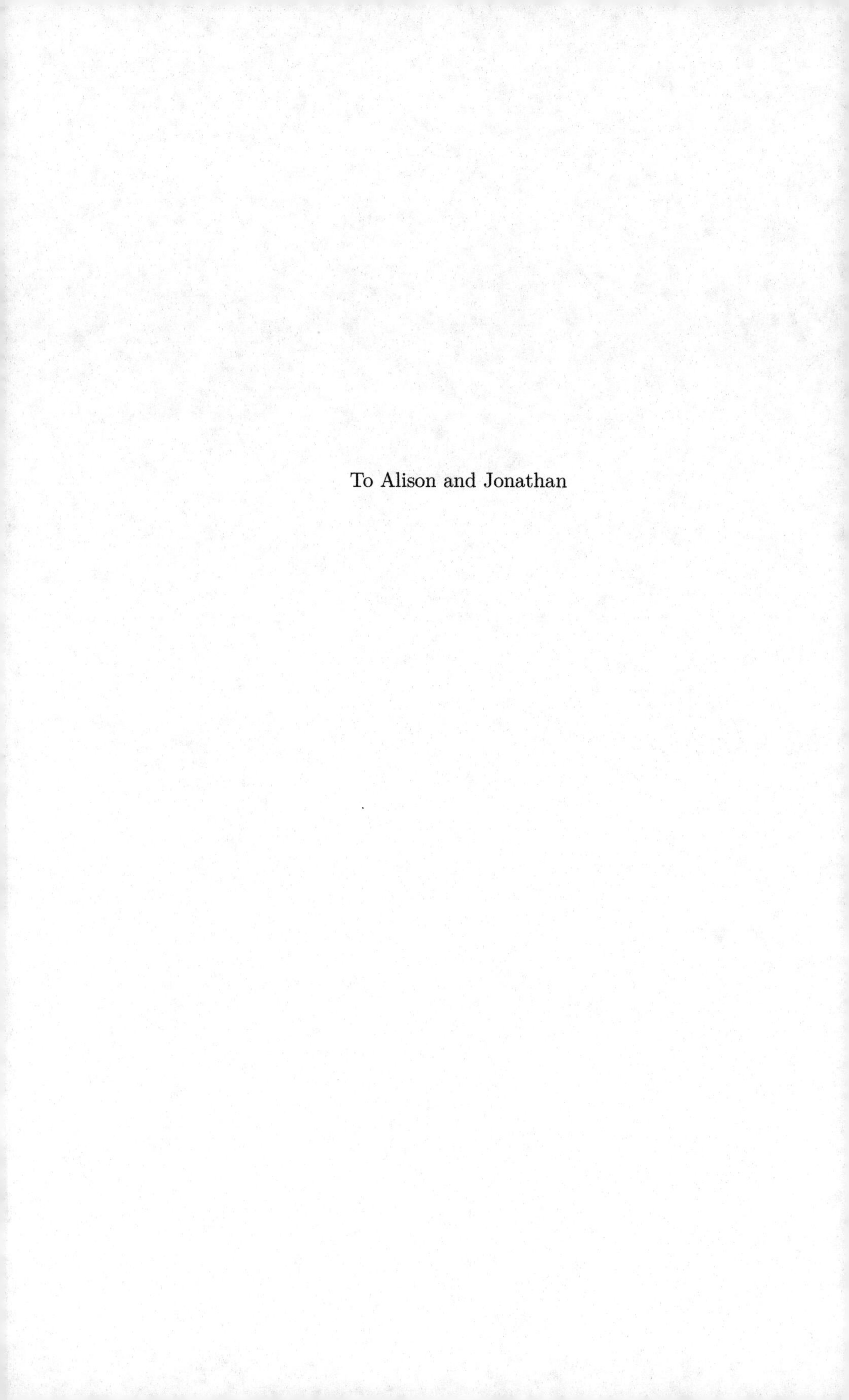

To Alison and Jonathan

Contents

List of Figures

Preface

This textbook evolved out of the lecture notes of two graduate classes taught at the Zicklin School of Business of Baruch College, City University of New York: STA9713 (Financial Statistics) and STA9750/OPR9750 (Introduction to Data Analysis in R). Its intended audience, however, also includes undergraduate students with a major in data science or finance, and assumes no prior knowledge of data analysis, the R programming language, or even finance. Another intended audience includes practitioners who need a reference book to quickly apply the tools that R and its numerous packages offer. Therefore, the guiding principle of this textbook is to provide as many real-life and concrete examples as possible, not only to explore the power and ease-of-use of R and packages that have been developed over the years but also to serve as a reference "cookbook" to practitioners.

Teaching about technology is very rewarding to students and instructors alike: to students, because they can immediately apply the new tools they learn, with the satisfaction of knowing that they are learning the most current tools; to instructors too, because we can help fix issues in real-time and see the excitement on the face of students when the piece of code works, or when the analysis does reveal new insights into the problem at hand.

Making a textbook out of these class notes, however, has presented unforeseen challenges. What issues will readers face when trying out the examples in this book? What questions will you have? Will the tools even be around by the time you open these pages? Will the concepts be moot?

There is no silver bullet for the first two questions and hopefully you will be using this textbook with the help of an instructor. However, I made painstaking efforts to think of questions you may have or issues you may run into and to proactively answer them – but this came at the cost of making the text less fluid, probably. The last questions, on technology, are actually the easiest to answer. First, R will probably disappear one day – like its predecessor, S, did; but very probably, similar functionalities and similar principles will carry over: R does borrow many of the features of S. Which brings me to the last question, my answer to which being: concepts never get moot. Once you've learned what *needs* to be done in data analysis, you know that whichever tool comes next will certainly offer you a way to do it. And even if the syntax is different, the principles will probably be similar. In other words, data analysis concepts (like math, like statistics) do not change much and I doubt they ever will. It's only how they will be expressed and carried out that will.

This does not mean, however, that we should reinvent the wheel and redo everything from scratch or from first principles. Yes, learning to calculate all the points on an efficient frontier or mastering the details of machine learning algorithms are useful skills, but they take entire textbooks to master. Some people like or want or need to know these skills, but the assumption made in this book is that you will want first and foremost to *apply* them, and to apply them quickly and reliably. The most recent software packages are therefore detailed, and even when one day they become obsolete, and even if your career brings you to using other software environments, you will know that there probably are tools out there that will accomplish the same analyses, and you will know, after reading this book, what they can do for you.

But enough about long-term plans. Let's get started!

Acknowledgments

My deep thanks to Riad Rehman for his detailed and insightful feedback, to Bruno Veras de Melo for his invaluable suggestions, and to Hammou El Barmi, Professor at Baruch College, and Hannah Roth for their continued support.

1

Your Working Environment

1.1 RStudio

The R language itself can be downloaded from the **R Project for Statistical Computing** at `https://www.r-project.org`. But we will focus on **RStudio** because it is a data science environment that simplifies many of the elements we are going to discuss. And (at least as of early 2022) it is free! So, even though any installation of R will be just fine to try out the material of this textbook (and you need to install R before installing RStudio), we recommend installing the RStudio desktop. The latest version can be downloaded from `https://www.rstudio.com/products/rstudio`.

A typical RStudio window is shown in Figure 1.1. The top left panel is probably the most important: it is where you will write your sequence of R commands in a file called a **script**. To do so, go to the top left drop-down menu and, under "File," select "New File" then finally "R script."

Note the different panels:

- As said, the top left is where your different scripts are written, each in different tabs.
- The bottom left, called **console**, is where the results of your R script will be displayed.
- The **environment** in the top right shows the list of all the variables and data table that you have defined, or that the system has defined for you, and their values.
- The bottom right panel displays different auxiliary information in different tabs, including the graphs you will be producing.

Saving and Executing a Script

Once you have saved a sequence of commands as an R script file, you can repeat its execution. Note that in the screenshot above, our extremely simple script still doesn't have a name: "Untitled1" is its default name, and the star and the red font are here to remind you that you haven't saved your work yet.

DOI: 10.1201/9781003320555-1

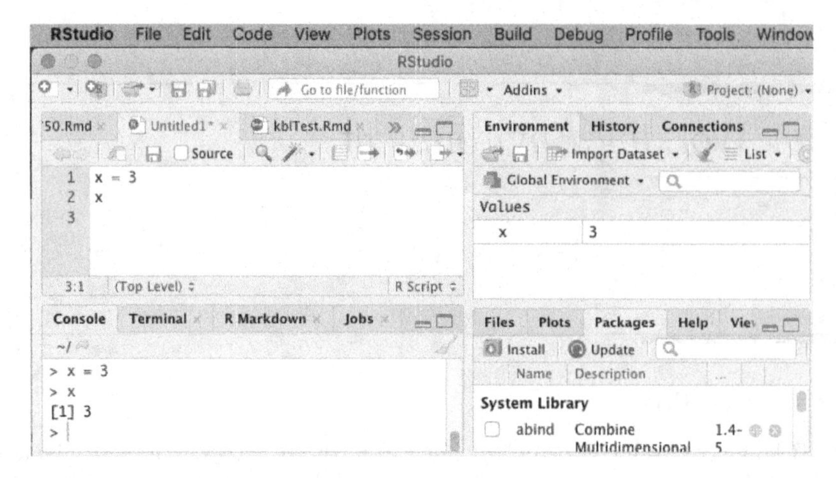

FIGURE 1.1 The RStudio environment

You can run all the commands contained in the script in many ways, including:

- From the panel containing your code, press Command + Option + R on a Mac (or an equivalent combination of keys on a Windows-based PC)
- From the console panel at the bottom left of your RStudio environment, by using the **`source()`** command on the full name (the name with the full path) of your R script, such as:

```
source("/Users/user/Documents/example.R")
```

1.2 R Notebooks

We saw earlier how to create a new R script, how to save it, and how to execute the commands it contains. Another way that is typically easier for incremental data analysis is to create a **notebook**. A notebook is just text interspersed with R commands, called **chunk**, with the output of the code, including any graph or figure, appearing right below the chunk. Regular text, code, and the output of executing that code are glued together by a piece of software called **Markdown**.

This will be more clear on an example. Go again to the top left drop-down menu and, under “File,” select “New File” then finally “R Notebook.” A default notebook will be created for you, as shown in Figure 1.2.

As you can immediately see, the text in the document appears in different colors, and a part appears against a gray background. Pay attention to these color cues as they tell you how Markdown understands them.

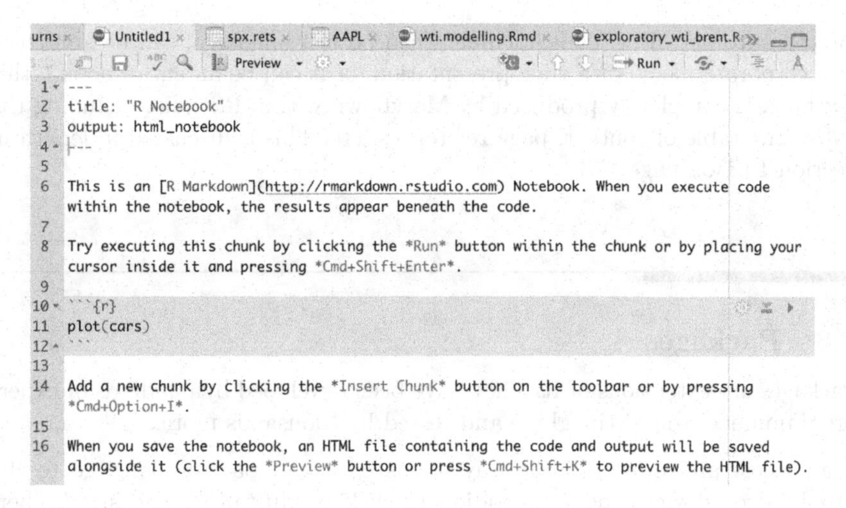

FIGURE 1.2 The RStudio Notebook

The document starts with a header, which begins and ends with three dashes seen on lines 1 and 4, respectively. This header tells Markdown what to do with the text; in this example, the header only gives a name to the document and tells Markdown that this is a notebook, i.e., that the output of the code should be placed in this same panel, under each code chunk.

In the same way that Markdown wouldn't know where the header is without the three dashes, Markdown wouldn't know what is text and what is R code if we didn't have markers for the start and end of code. As Figure 1.2 shows, line 10 marks the beginning of a chunk of R code using three successive back quotes followed by `r` within curly braces. (The "r" tells Markdown that the code is R code; Markdown is a generic tool that also works for, say, Python.) As Line 14 tells us, the markers for a new code chunk can be inserted by pressing the Command, Option, and i keys together – thus saving us the task of typing the three back quotes, the left curly brace, the letter r, the right curly brace, and the final three back-quotes. However we insert these markers, Markdown will understand what appears in-between as R code and confirm to us that it detected an R chunk by placing it against a gray background.

Let us now click somewhere in the code chunk and, as indicated on Line 8, let's press Command, Shift and Enter together. After a brief moment (the time for Markdown to call R, for R to execute the `plot()` command, and for Markdown to grab the output), the result of the command appears right below it. In this case, the result is of course a plot embedded in your text.

The next logical step after notebooks is to have an entire document assembling your thoughts and explanations on your code, your code itself, and the result of this code – all in a professional-looking document that Markdown generates for you automatically. That document can be just one page, can be an article or a corporate memo or a slide presentation, or it can be an entire book – this textbook is completely produced by Markdown within RStudio, including the index, the table of content, page references, etc. This is discussed in length in Section 19.1 on page 373.

1.3 Packages

Packages are extensions of R. They have been developed by hundreds of expert programmers around the globe and tested by thousands more.

The easiest and most reliable way to install a new package is to go to the "Tools" drop-down menu of RStudio and click on "Install Packages...". Then, type the name of the desired package in the middle field of the window that pops up. As of early 2022, that pop-up window also offers you to select which website you would like to download the package from. By default, the **CRAN** repository server is the place to go. (And unless you really know what you're doing, you probably don't want to download a package from anywhere else. See, however, below how to download packages from **github**.)

The other (and not recommended) option to install a package is to use the **`install.packages()`** command[1] from the **console** panel of RStudio:

```
install.packages("purrr")
```

Installing only means downloading the new package to your computer. *Once you've done it, you won't have to do it again* (until you change computers), so don't run the **`install.packages()`** function each time you run a piece of code. (I highly recommend installing packages from RStudio's menu anyway.)

Moreover, downloading a package doesn't mean it is made available to your R calculations. To do that, you have to explicitly tell R to include that package as an extension for the duration of your working session. This is done using the **`library()`** command:

```
library(tidyverse)
```

Until you do so, R will not recognize the functions or datasets in the package. In fact, throughout this textbook, I will assume you will *always* start your work by loading **`tidyverse`** as shown above.

[1] Note the plural in the command name, which is unusual in R.

Learning R in R: Swirl

The **swirl** package allows you to learn R and practice on a large number of small examples within R (or better, RStudio) itself. Just install the package as any other package, load it, and start it using the **swirl()** command:

```
library(swirl)
swirl()
```

This package is a great complement to this book because it emphasizes statistics and exploratory data analysis. To exit Swirl, type this:

```
bye()
```

Swirl tends to be sticky though, in that it doesn't want to stop. In my experience, you may need to use the **bye()** command in your console and also in the script you are working on.

1.4 Specialized Packages for Finance

We are going to see these different packages throughout this book, but a short overview of each of one is in order. First, **PerformanceAnalytics**, whose main page is:

https://cran.r-project.org/web/packages/PerformanceAnalytics/index.html

We will see that this package contains several interesting data sets, such as historical returns of hedge fund strategies (see page 14). Among the key functions offered by this package is **Return.calculate()**, which calculate returns out of successive asset prices or **chart.ECDF()** on page 86 that plots the **empirical CDF** of investment returns. More of these key functions are described on page 163 and after.

The **PortfolioAnalytics** package has a similar name, making it easy to confuse the two but has an entirely different purpose: it constructs sets of investments, called *portfolios*, that satisfy investment constraints and maximizes profit or minimizes risk. Its full documentation can be found at:

https://cran.r-project.org/web/packages/PortfolioAnalytics/vignettes/portfolio_vignette.pdf

The **RQuantLib** package is based on a popular finance library called **QuantLib**. Details on this package can be found at:

https://cran.r-project.org/web/packages/RQuantLib/index.html

The state of R libraries dedicated to **cryptocurrencies** is pretty sad. The main packages are `crypto2`, `Rbitcoin`, and `coinmarketcapr`. The first two are fragile, and the latter, which pulls data from https://coinmarketcap.com, does not provide meaningful functions or data unless you pay for a premium account on their website.

Finally, we should note `EdgarWebR` which provides an R interface to the SEC's (**Securities and Exchange Commission**) EDGAR database of corporation filings, including mutual funds. A description of this package can be found at:

https://cran.r-project.org/web/packages/edgarWebR/vignettes/edgarWebR.html

The EDGAR database provides mostly textual data coming from companies filings, so we will not use this package much in this book.

2

Reading Data in R

Most of the data we are going to use in this book (and in your career) are either read from files in the **comma-separated values** (**csv**) format, read from Excel files, or read directly off a website, typically through a specialized library. The simplest form consists of tables in csv files, so we will start with them.

2.1 Reading Input (Data) Files

Reading csv files is done using the `read.csv()` command:

```
data = read.csv("Documents/R/Datasets/myfile.csv")
```

Typically, your data sets would be stored in a dedicated directory, such as "R/Datasets" in the example above. Once you've decided on a folder, you can tell R this **working directory** is the default place where data files can be found, thus avoiding repeating the long full paths. You do that using the `setwd` ("set working directory") command. For instance:

```
setwd("Documents/R/Datasets")
```

You can then read files located in that directory by simply providing the file's name, as in:

```
data = read.csv("myfile.csv")
```

Depending on your application, however, it might be even easier to use `file.choose()`, which opens a new folder window to let you click through your documents and select the desired input file.

```
data = read.csv(file.choose())
```

As a concrete example, we are going to use the `MSCI AC World.csv` file.

```
acwi = read.csv("MSCI AC World.csv")
```

DOI: 10.1201/9781003320555-2

The **MSCI AC World** dataset lists the holdings, as of April 30, 2021, of the **MSCI All Country World Index**, usually referred to by its acronym, **ACWI**.

The newly created variable called **acwi** is a table with named columns, which R calls a **dataframe**. You can have a look at the first few rows of this table using the **head()** command. Here, we specify we need only the first two rows:

```
head(acwi, 2)
```

```
  Ticker                      Name Weight Market.Cap
1   AAPL                Apple Inc.   3.47    2193582
2   MSFT Microsoft Corporation   2.81    1899924
  Dividend.Yield Price.to.Cash.Flow Price.to.Book
1           0.62              30.73          33.4
2           0.85              31.17         14.13
  Price.to.Sales            GICS.Sector       Country
1           8.41 Information Technology United States
2          13.55 Information Technology United States
    ROE Return
1 73.69  78.98
2 40.14  40.72
```

It is a good habit to get a general idea of the range of values for each of the columns and how many data are missing (N/As). You get general statistics on the data using the **summary()** command applied to the entire dataframe, and a partial output is shown:

```
summary(acwi)
```

```
Dividend.Yield     Price.to.Cash.Flow    Price.to.Book
Length:2974        Length:2974           Length:2974
Class :character   Class :character      Class :character
Mode  :character   Mode  :character      Mode  :character
```

You will notice that some of the data in the dataframe are said to be of class "character," meaning they are considered as words instead of numbers. This problem happens frequently, and your data needs some pre-processing before using **summary()** – a pre-processing you'll have to do anyway. Said differently, you should always pay attention to the format of the data you loaded. Inspecting the data format can also be done using the **str()** command ("str" is short for "structure"), and a partial output is shown below:

```
str(acwi)
```

```
'data.frame':   2974 obs. of  12 variables:
 $ Weight             : num  3.47 2.81 2.29 1.21 1.11 ...
 $ Market.Cap         : num  2193582 1899924 1744112 ...
 $ Dividend.Yield     : Factor w/ 655 levels "--","0.01","0.02",.
```

```
$ Price.to.Cash.Flow : Factor w/ 2066 levels "--","0.24","0.54",
$ Price.to.Book      : Factor w/ 1107 levels "--","0.18","0.19",
```

We may also have noted that some numerical data were read as **factors**. What does that mean? Factor is the R name for categorical variables, i.e. discrete variables following an **ordinal scale** that classifies the data into distinct categories in which ranking is implied. Examples of factors include grades, such as A, B, C, D and F; or classifications of students as “freshman,” “sophomore,” “junior” and “senior.”

Numbers are usually understood as numbers by **read.csv()**, so why were they read as factors? Something threw the function off. What happened here is that some of the entries in the dataset consist of two dashes to indicate the value is missing. This is not a usual convention, and **read.csv()**, seeing these characters, assumed the entire column contained text. These different text entries, since they contain repetitions (from the above output, more than one stock has a **Dividend.Yield** of “0.01”), were transformed into categorical variables (i.e., factors).

We can convert these character strings to numerical values using the **as.numeric()** function:

```
acwi$Dividend.Yield = as.numeric(acwi$Dividend.Yield)
acwi$Price.to.Cash.Flow = as.numeric(acwi$Price.to.Cash.Flow)
acwi$Price.to.Book = as.numeric(acwi$Price.to.Book)
acwi$Price.to.Sales = as.numeric(acwi$Price.to.Sales)
acwi$ROE = as.numeric(acwi$ROE)
acwi$Return = as.numeric(acwi$Return)
```

Applying the command shows that the content now makes much more sense:

```
summary(acwi)
```

```
   Ticker                Name                 Weight
 Length:2974        Length:2974        Min.   :0.000
 Class :character   Class :character   1st Qu.:0.000
 Mode  :character   Mode  :character   Median :0.010
                                       Mean   :0.033
                                       3rd Qu.:0.030
                                       Max.   :3.470


   Market.Cap       Dividend.Yield Price.to.Cash.Flow
 Min.   :    535   Min.   : 0.0   Min.   :   0
 1st Qu.:   6338   1st Qu.: 0.9   1st Qu.:   8
 Median :  12395   Median : 1.9   Median :  14
 Mean   :  31708   Mean   : 2.4   Mean   :  31
 3rd Qu.:  27054   3rd Qu.: 3.4   3rd Qu.:  26
 Max.   :2193582   Max.   :20.7   Max.   :5324
                   NA's   :523    NA's   :127
```

```
 Price.to.Book     Price.to.Sales GICS.Sector
 Min.   :  0.18   Min.   :   0   Length:2974
 1st Qu.:  1.35   1st Qu.:   1   Class :character
 Median :  2.63   Median :   3   Mode  :character
 Mean   :  6.07   Mean   :   9
 3rd Qu.:  5.96   3rd Qu.:   6
 Max.   :280.36   Max.   :4102
 NA's   :64       NA's   :7
   Country               ROE              Return
 Length:2974        Min.   :-98.12   Min.   : -67.0
 Class :character   1st Qu.:  4.75   1st Qu.:  12.8
 Mode  :character   Median : 10.05   Median :  37.6
                    Mean   : 10.45   Mean   :  52.2
                    3rd Qu.: 17.22   3rd Qu.:  71.7
                    Max.   : 98.42   Max.   :1068.3
                    NA's   :77       NA's   :36
```

This reveals many key properties of, but also obstacles in, the data. First, most quantitative columns contain NA's; the two exceptions are the constituents' weights in the index and the market cap of the companies. It makes sense that all these entries were populated in this data file since it describes the index and since ACWI weights are proportional to the market caps of companies. In other words, the data provider had to know these data while the other fields (columns) were not required for the construction of the index.

The **weight** of a stock in a portfolio or an index is simply its share of the investment. If a portfolio worth $10,000 has $200 in Intel Corp., then Intel's weight in the portfolio is 200/10,000 = 2%. The sum of all the weights in a portfolio has to equal 100%, i.e., 1. The **Market.Cap** field reports the **market capitalization** of a company, which is the total value of all the stock in that company. If a company has sold a million shares and each share is valued at a thousand dollars, then the firm's market capitalization is a billion dollars. Market cap numbers are typically reported in millions of dollars, so the output of **summary()** above indicates outliers: at least one company as small as $535MM and at least one as large as 2.2 trillion dollars.

Companies can also offer regular cash payments, typically quarterly, to their shareholders. Such a payment is called **dividend**. Is a dividend of $2 a lot? It depends on the value of each share. If the value of each share is $100, then $2, paid four times a year, is high in proportion to the share value. The **dividend yield** is the ratio of the dollar dividend amount paid in a year divided by the dollar value of one share, and in this example, the dividend yield is:

$$\frac{4 \times \$2}{\$100} = 8\%,$$

which would be very high in practice as of Q1'22. Some companies pay no dividend, so dividend yields can go as low as 0 – but cannot be negative. The result of `summary()` reports outliers in terms of dividend yield, with a company sporting a 20.7% yield. This is suspicious enough to deserve investigation but can come from a massive drop in the stock's price, artificially (and probably, temporarily) inflating the dividend-over-stock-price ratio.

Further, **price-to-sales**, **price-to-cash-flow** and **price-to-book** relate the price of a share to, respectively: the total sales (also known as **revenue**) per share of that company; the **cash flow**, which is the cash left from the revenues after paying all expenses, taxes and interest on the debt (which would equal what is called **net income**), less dividends and maintenance (or budgeting for the replacement) of the existing equipment; and the accounting value of a company, called the **book value**, which equals the value of the firm's assets less its debt and liabilities[1]. The book value is also called **shareholder's equity**. The **return on equity**, shortened as **ROE**, equals the net income divided by shareholder's equity.

In the case of our **acwi** data set, `summary()` reports clearly suspicious outliers: The median and mean of price-to-cash-flow are 14 and 31, respectively, so a maximal value of 5,324 for that ratio would need to be investigated. It could be a data error or an extremely speculative company (think biotech with no product to sell yet). We can make the same observation for price-to-sales, whose maximum is 4,102 – anomalously higher than the median and mean of 3 and 9, respectively.

How anomalous are these values? Z-scores and the Mahalanobis distance, discussed on pages 35 and 42, respectively, will give us objective criteria to decide.

One suggestion, in such cases, is to check whether the same stocks are outliers on multiple metrics. If they are and the metrics are unrelated, then poor reporting on the company's part can be suspected. If the metrics are related, then maybe a common cause is responsible for the multiple outlying values; for instance, a sudden drop in stock price will cause dividend yield and price-to-sales to appear extremely high and extremely low, respectively. Whether or not these entries should be discarded depends on your application. If your goal is to investigate the relationship between dividend yield and price-to-sales, for example, then certainly, these outliers should be discarded.

As you may have noticed, shareholder's equity is an accounting term, whereas equity or stock typically refers to the shares in a company that trades on the stock market. In the US, a company whose stock trades on one of the exchanges of the stock market is said to be **publicly traded**, or **public** for short – but "public" does not mean the company is owned by the government, as in "public service."

---

[1]For an individual, the book value is called the **net worth**.

The **GICS** sectors build a taxonomy of the main public companies established by the **Global Industry Classification Standard** since 1999 by MSCI and Standard & Poor's. As of Q1'22, the GICS classification consists of 11 **sectors**, refined into 24 **industry group (GICS)**, themselves subdivided in 69 **industries** and 158 **sub-industries**.

If we now go back to the output of **summary()**, we observe that many columns exhibit outliers. Price-to-Sales, Price-to-Cash-Flow and Price-to-Book seem to have unreasonable values in some cases, which is not usual in ratios and in particular in financial ratios when the denominator is very small. Not infrequently, companies with almost no revenue or no accounting assets receive high valuation because investors expect fast growth or because some of their assets are not recognized by accounting standards. These cases however, might throw a wrench into any analysis we will try to perform on the constituents of the ACWI index, so we should stand ready to remove them.

Another observation we should make is that market capitalizations are extremely lopsided, and their distribution packs most of the data points on the left while exhibiting a very long right tail. This also means a large positive skew. We can verify this using the **hist()** function that produces the histogram of Figure 2.1:

```
hist(acwi$Market.Cap)
```

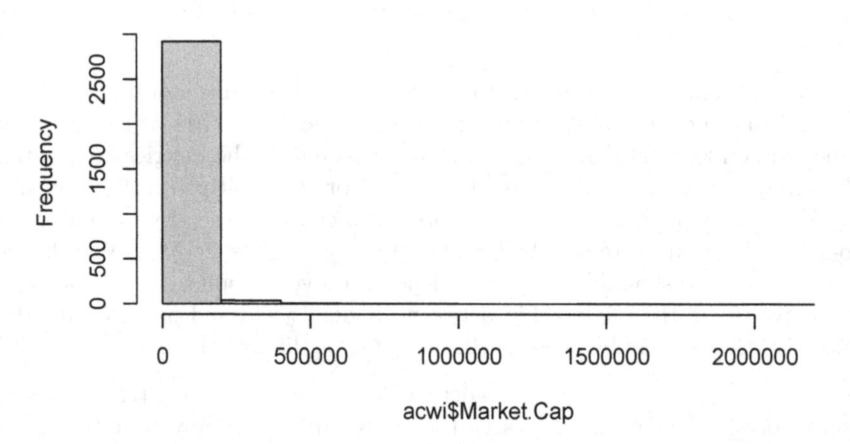

FIGURE 2.1 This histogram of market capitalizations exhibits a long right tail

and the skewness (or skew), more formally introduced on page 83, is calculated by the **skewness()** function of the **moments** package:

```
library(moments)
skewness(acwi$Market.Cap)
```

```
[1] 14.05
```

This very positive skew confirms our earlier observations.

One last note on **summary()**: you can apply it on an entire dataframe as we did earlier, but you can also get its statistics for a specific column:

```
summary(acwi$Market.Cap)
```

```
   Min. 1st Qu.  Median    Mean 3rd Qu.    Max. 
    535    6338   12395   31708   27054 2193582 
```

2.2 Reading Excel Files

Reading **Excel** files is as easy as any other file formats, thanks to the **openxlsx** library. As usual for packages a.k.a. libraries, make sure you have downloaded and installed that library once and for all. (Loading it into your RStudio session must be done every time, however.) Note that RStudio detects when a library is referenced and will display a banner suggesting to install it for you.

```
library(openxlsx)
```

```
data = read.xlsx("myfile.xlsx")
```

Sometimes, as in the `fredgraph.xlsx` file provided with this book, the data of interest appear after some comments. If you know the line at which the table of interest starts, you can specify that line using the **startRow** named parameter. The following command shows that the beginning of the created dataframe contains extraneous information that would require cleaning:

```
head(read.xlsx("fredgraph.xlsx"), 3)
```

```
               FRED.Graph.Observations   X2   X3   X4
1         Federal Reserve Economic Data <NA> <NA> <NA>
2     Link: https://fred.stlouisfed.org <NA> <NA> <NA>
3 Help: https://fredhelp.stlouisfed.org <NA> <NA> <NA>
```

In contrast, the following produces a ready-to-use data table:

```
head(read.xlsx("fredgraph.xlsx", startRow=14), 3)
```

```
  observation_date BAMLC0A1CAAAEY   DGS3 DGS3MO DGS10
1            42491          2.513 0.9719 0.2767 1.806
2            42522          2.435 0.8605 0.2727 1.644
3            42552          2.303 0.7945 0.3010 1.504
```

We will come back to these data from the Federal Reserve later in this book.

2.3 Reading Tables

read.table() is actually what **read.csv()** is based on. It can read files where data are separated by commas or any other symbol. It can read from URLs on the web or local files. It also automatically un-compresses files whose names end with the .gz suffix. For example, using the `stocks5.csv` also provided with this book:

```
Tstocks = read.table('stocks5.csv',
                     sep=',',
                     header=T)
```

This file will be used, among others, to illustrate Principal Component Analysis on page 281.

2.4 Packages Come With Datasets

Packages typically come with data sets, and you can check which ones using the **data()** command. The command however, lists all the data sets you have access to across all the packages you installed. To see the data sets provided by a specific package, specify the package's name as follows:

```
data(package = "dplyr")
```

The above only *lists* the data sets. Reading a specific data set, then, is done by specifying its name followed by the package that contains it. For example, the command below loads a data set named **edhec** from the **PerformanceAnalytics** package and places the data in a newly created dataframe that has the same name as the data set (i.e., **edhec** in this case).

```
data(edhec, package="PerformanceAnalytics")
```

That dataset, provided by the EDHEC Business School in France, provides the monthly returns of different hedge fund strategies from 1997 to 2019.

Another example is the **S&P 500** data from the **Ecdat** package. We load the series[2] of monthly returns of the stock index using the same **data()** function:

```
data(SP500, package="Ecdat")
```

[2]Technically, a **time series** as we will see on page 58.

Technical Detail You might notice that the data you just requested doesn't always appear immediately in the Environment top-right panel of RStudio and that all you see is a **<Promise>** message. That message says it all actually – R has a lazy (yes, that's the technical term) evaluation mechanism that will go fetch the data only when you actually use them. Try for example **head()** on the dataframe you see a **<Promise>** for.

Many of the datasets included in **Ecdat** are stored in the **RDA** format, which is R's data format. An example is the **Mishkin.rda** dataset of inflation rates that we will use on page 322. You can save your data in this format using the **save()** function, such as:

```
save(mydata, file="mybackup.rda")
```

Reciprocally, you load RDA data files using the **load()** function. For example:

```
load("mybackup.rda")
```

2.5 Reading XML Data

XML is a very flexible language to describe data and their structures, but it can be fiendishly difficult to parse for a human being. Since parsing XML file is done by software, we will only look at a simple (but artificial!) example:

```
<PersonList>
    <Person>
        <First>John</First>
        <Last>Doe</Last>
    </Person>
    <Person>
        <First>Mary</First>
        <Last>Foe</Last>
    </Person>
    <Person>
        <First>Shanice</First>
        <Last>Poe</Last>
    </Person>
</PersonList>
```

The easiest way to read such a simple structure is to use the **xmlToDataFrame()** function of the **XML** package:

```
library(XML)
(dataframe =
    xmlToDataFrame("persons.xml"))
```

```
    First Last
1    John  Doe
2    Mary  Foe
3 Shanice  Poe
```

We will not use XML data explicitly in the rest of this book.

2.6 JSON

One of the most promising way to read data is to connect directly to repositories in the **JSON** format. JSON JavaScript Object Notation is an open standard text-based format that uses human-readable text to store and transmit structured data.

The package that allows us to use the JSON format is **jsonlite**. You need to install and load it the usual way.

```
library(jsonlite)
```

You will find an example of a JSON dataset on page 28.

2.7 Chapter-End Summary

Finance, compared to other industries, is still a field where free repositories of up-to-date data are relatively scarce. Moreover, whatever the internet or the institution you may be working for offers comes in disparate formats. R and its different packages provide solutions in almost all cases, including functions to connect directly to a server and download data without intermediate files, for example using the JSON format.

In all cases, any dataset new to you should be inspected before processing so as to know (and, if necessary, convert) the data type of each field, to detect missing values, and to screen for outliers that may point to special cases you may need to handle separately in your data analysis.

3

Financial Data

3.1 Yahoo! Finance

QuantMod and tidyquant

The **QuantMod** package is part of **tidyquant**. They offer numerous functions to manipulate data, but for now, we will focus on the functions they offer to download data from repositories on the internet. Let's first load the package, after of course loading **tidyverse**, which is necessary throughout this book.

```
library(tidyverse)
library(tidyquant)
```

The **tidyquant** package allows us to pull historical prices from Yahoo Finance on most stocks and for a specific data range:

```
getSymbols("AAPL",
           from = "2021-04-01",
           to = "2021-04-09",
           get = "stock.prices")
```

```
[1] "AAPL"
```

No output is printed other than the name of the created dataframe, **AAPL**, but the data are there, stored in a dataframe. As we will discuss on page 58, the data take the form of a **time series** of type **xts** that contains dates and open/high/low/close prices. (The **open price** is simply the price at the open of a trading day, and the **close** is the price at which the last trade was recorded on that day.)

As said, **getSymbols()** creates a new object named "AAPL" – even though we did not use an assignment to that new variable. We can inspect the content of that variable by typing its name:

```
AAPL
```

```
           AAPL.Open AAPL.High AAPL.Low AAPL.Close
2021-04-01     123.7     124.2    122.5      123.0
2021-04-05     123.9     126.2    123.1      125.9
```

DOI: 10.1201/9781003320555-3

```
2021-04-06       126.5      127.1     125.7          126.2
2021-04-07       125.8      127.9     125.1          127.9
2021-04-08       128.9      130.4     128.5          130.4
           AAPL.Volume AAPL.Adjusted
2021-04-01    75089100         122.3
2021-04-05    88651200         125.2
2021-04-06    80171300         125.5
2021-04-07    83466700         127.1
2021-04-08    88844600         129.6
```

The last 3 arguments in the command above are optional: by default, **getSymbols** pulls stock prices, so you don't need the **get=** argument. Moreover, omitting the **from=** argument would indicate you are requesting data from the start of the history in Yahoo's database, and omitting the **to=** argument would download data up to the previous market close.

Technical Detail As of March 2021, **getSymbols** creates a new variable whose name is the stock's ticker. For example, the command above creates a new variable called **AAPL**. As discussed on page 69, this is called a **side-effect**. However, per the **quantmod** documentation, "this behavior is expect to change for getSymbols [in upcoming versions of the package] and all results will instead be explicitly returned to the caller unless **auto.assign** is set to **TRUE**.

By default, the function pulls data from **Yahoo! Finance**, and this can be made explicit (or changed) using the **src=** named argument; for example:

```
getSymbols("AAPL", src='yahoo')
```

However, **quantmod**'s other source of stock data, **Google**, is not available any more, so Yahoo currently seems to be our only option for stocks.

As you saw, **getSymbols** pulled the open/high/low/close prices (often shortened as **OHLC**), together with trading volumes and adjusted prices, which will be discussed on page 161. As is unfortunately too typical in finance, the unit in which volumes are expressed is unspecified. In the case of data from Yahoo Finance (and most data providers, in fact), volume refers to the number of shares of that stock that got changed hands and not the dollar amount.

Each of these columns can be extracted separately using, respectively, the **Op()**, **Hi()**, **Lo()**, **Cl()**, **Vo()** and **Ad()** functions. For example, extracting the time series (in the **xts** format discussed on page 58) of prices at market close can be done as follows:

```
Cl(AAPL)
```

```
           AAPL.Close
2021-04-01      123.0
```

```
2021-04-05        125.9
2021-04-06        126.2
2021-04-07        127.9
2021-04-08        130.4
```

We can also extract any of the OHLC prices of AAPL by using the **$** notation:

```
Apple.closing.prices = AAPL$AAPL.Close
```

Either way, the resulting time series can be plotted using the `plot()` function. The output is shown in Figure 3.1.

```
plot(Apple.closing.prices)
```

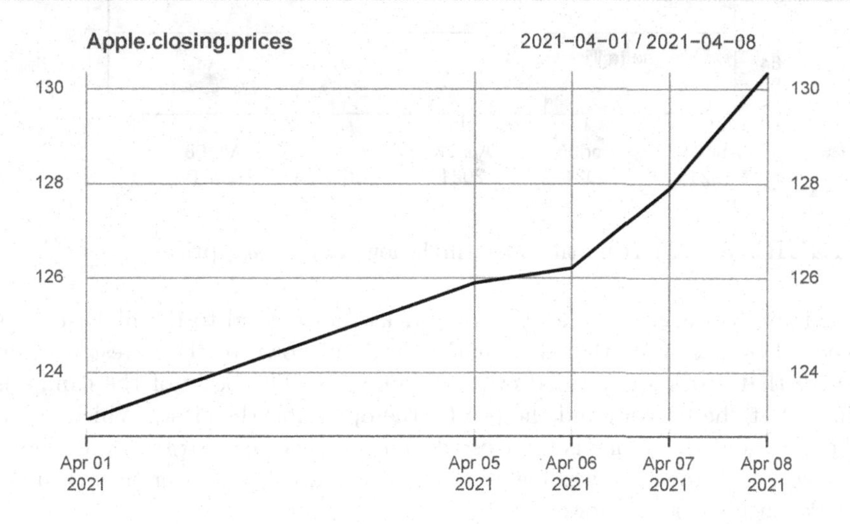

FIGURE 3.1 Simple example of plotting the closing prices of a stock

The `plot()` functions accept multiple parameters. The **line type** is specified using the `lty=` named parameter and can be specified using either text ("blank," "solid," "dashed," "dotted," "dotdash," "longdash," "twodash") or number (0, 1, 2, 3, 4, 5, 6). That way, writing:

```
plot(Apple.closing.prices, lty = "dashed")
```

is identical to writing:

```
plot(Apple.closing.prices, lty = 2)
```

Similarly, the line width is specified using the `lwd=` named parameter. We will see additional features of `plot()` later. See for example pages 74 and 86.

The OHLC prices are often depicted in the form of candlesticks, and this can be achieved using the `chartSeries()` function. The code below produces the graph in Figure 3.2.

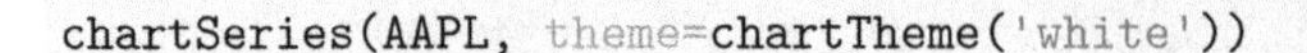

```
chartSeries(AAPL, theme=chartTheme('white'))
```

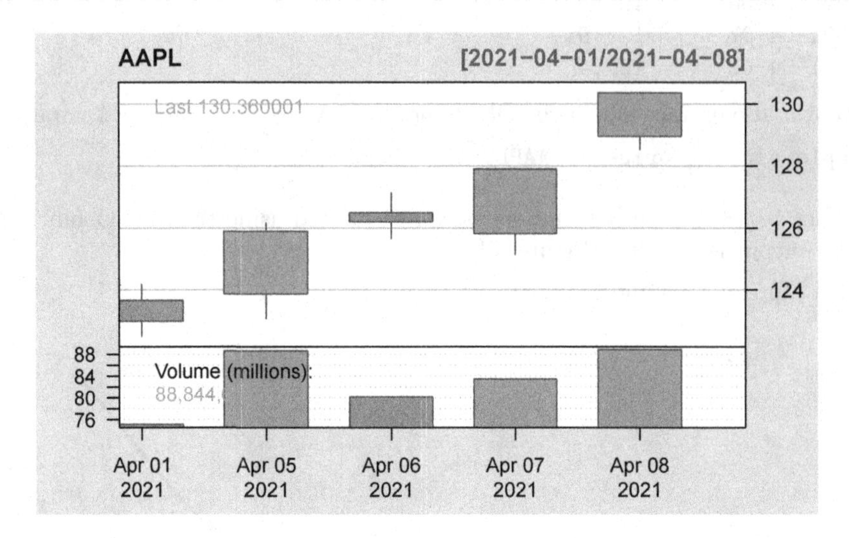

FIGURE 3.2 Charting the open, high, low and closing prices of a stock

In Figure 3.2, each "candle" has two wicks that extend to the highest price reached by the stock that day, at the top, and down to the lowest price at which that stock was traded on that same day. The body of the candle is limited at the bottom and the top by the open and the close. Which one is the open and which one is the close depends on the color: If the candle body is green, then the close was higher than the open; it is red or orange, then the stock finished the day lower.

Note also that you can pull data for more than one stock by passing to **getSymbols()** a vector of tickers as its first argument:

```
getSymbols(c("AAPL","IBM"),
           from = "2021-04-01",
           to ="2021-04-09")
```

```
[1] "AAPL" "IBM"
```

Similarly, we can get price histories for financial instruments other than stocks. For example, we can extract **bitcoin** data using its **BTC-USD** ticker:

```
getSymbols(
     Symbols = "BTC-USD",
     src = "yahoo",
     from = "2021-09-01",
     to = "2021-12-03")
```

```
[1] "BTC-USD"
```

This creates a time series object named **BTC-USD**.

Tidyquant

The **tq_get()** command of **tidyquant** also allows us to pull OHLC prices from **Yahoo Finance**:

```
library(tidyquant)
aapl = tq_get("AAPL",
              from = "2021-04-01",
              to = "2021-04-09",
              get = "stock.prices")
```

As for **getSymbols()**, **stock.prices** is the default value of the **get=** parameter, so you don't need it in this case. **tq_get()** reuses the code of **getSymbols()**, so not surprisingly the output is similar:

```
aapl
```

```
# A tibble: 5 x 8
  symbol date        open  high   low close   volume
  <chr>  <date>     <dbl> <dbl> <dbl> <dbl>    <dbl>
1 AAPL   2021-04-01  124.  124.  122.  123  75089100
2 AAPL   2021-04-05  124.  126.  123.  126. 88651200
3 AAPL   2021-04-06  126.  127.  126.  126. 80171300
4 AAPL   2021-04-07  126.  128.  125.  128. 83466700
5 AAPL   2021-04-08  129.  130.  129.  130. 88844600
# ... with 1 more variable: adjusted <dbl>
```

The output *looks* different, but we can check that the numerical values are actually the same: For example, the open price on the first day in the time series is:

```
aapl[1,3]
```

```
# A tibble: 1 x 1
   open
  <dbl>
1  124.
```

If it still looks different from what **AAPL** showed earlier, that's because this is a so-called **named vector**. The plain value of that named vector can be extracted as follows:

```
as.numeric(aapl[1,3])
```

```
[1] 123.7
```

which is indeed what we got on the first row of the **AAPL** dataframe on page 17.

The **stock.prices** option of the **get=** named parameter is only one of several available values, whose list can be obtained with **tq_get_options()** and its partial output reproduced below:

```
tq_get_options()
```

```
 [1] "stock.prices"        "stock.prices.japan"
 [3] "dividends"           "splits"
 [5] "economic.data"       "quandl"
 [7] "quandl.datatable"    "rblpapi"
```

dividends is one of the options, and we will study it next. Inelegantly, the other options refer not to other information you may want to pull on a given company but to other data providers:

- **quandl** and **quandl.datatable** pull data from **Quandl**, which we will discuss shortly;
- and last but not least, **rblpapi** downloads data from **Bloomberg**, assuming you have an account on the **Bloomberg Terminal**.

Dividends

Yahoo! also provides the historical dividends paid by a stock, and this time series can be downloaded using the **dividends** parameter:

```
tq_get("AAPL",
       get="dividends",
       from="2021-03-31",
       to="2022-02-22")
```

```
# A tibble: 4 x 3
  symbol date       value
  <chr>  <date>     <dbl>
1 AAPL   2021-05-07  0.22
2 AAPL   2021-08-06  0.22
3 AAPL   2021-11-05  0.22
4 AAPL   2022-02-04  0.22
```

These data can also be obtained using the **getDividends()** function:

```
(AAPL.div = getDividends("AAPL",
                         from="2021-03-31",
                         to="2022-02-22"))
```

```
           AAPL.div
2021-05-07     0.22
```

```
2021-08-06      0.22
2021-11-05      0.22
2022-02-04      0.22
```

Omitting the **to=** named variable, in either **tq_get()** or **getDividends()**, indicates you're requesting all dividends paid until the last market close.

Stock Splits

A **stock split** means that a share in a company becomes a certain number of "new shares" on the next trading day. For example, a "4-for-1" stock split means that one share becomes four new shares, each new share having one-fourth the value of the original share. Therefore, there is no magical creation of money here, and companies split their stocks for what are essentially cosmetic reasons.

In any case, we can find all the splits in Apple's stock using the following command – in this example, during the year 2020:

```
(aapl.splits = tq_get("AAPL", get="splits",
                      from = "2020-01-01", to="2022-02-22"))
```

```
# A tibble: 1 x 3
  symbol date       value
  <chr>  <date>     <dbl>
1 AAPL   2020-08-31  0.25
```

The value of 0.25 indicates that the new shares had 1/4 the value of the old share, so August 31 (a Monday) was the first day the new shares were traded[1]. Historical prices and trading volumes are adjusted for splits by most data providers, including Yahoo Finance.

Stock Market Indices and Their Constituents

The constituents of stock market indices can be pulled using the **tq_index()** function. Only a few indices are available through **tidyquant** though, and the current list can be obtained with **tq_index_options()**:

```
tq_index_options()
```

```
[1] "DOW"       "DOWGLOBAL" "SP400"     "SP500"
[5] "SP600"
```

For example, the constituents of the **S&P 500** index are:

```
(sp500 = tq_index("SP500"))
```

```
# A tibble: 505 x 8
```

[1]Apple's website indicates Friday, August 28, 2020, was the split date, which of course means the split was enforced after trading hours that Friday.

```
   symbol company       identifier sedol  weight sector 
   <chr>  <chr>         <chr>      <chr>   <dbl> <chr>  
 1 AAPL   Apple Inc.    03783310   20462~ 0.0690 Informa~
 2 MSFT   Microsoft ~   59491810   25881~ 0.0565 Informa~
 3 AMZN   Amazon.com~   02313510   20000~ 0.0358 Consume~
 4 TSLA   Tesla Inc     88160R10   B616C~ 0.0222 Consume~
 5 GOOGL  Alphabet I~   02079K30   BYVY8~ 0.0205 Communi~
 6 GOOG   Alphabet I~   02079K10   BYY88~ 0.0190 Communi~
 7 BRK.B  Berkshire ~   08467070   20733~ 0.0170 Financi~
 8 NVDA   NVIDIA Cor~   67066G10   23795~ 0.0143 Informa~
 9 UNH    UnitedHeal~   91324P10   29177~ 0.0136 Health ~
10 FB     Meta Platf~   30303M10   B7TL8~ 0.0131 Communi~
# ... with 495 more rows, and 2 more variables:
#   shares_held <dbl>, local_currency <chr>
```

The **identifier** column shows the unique **CUSIP** identifier for a company's stock. CUSIP stands for Committee on Uniform Securities Identification Procedures and is mostly used in the U.S. and Canada. You'll notice that Google issued two different types of shares, each with its own CUSIP. The **sedol** column provides the **SEDOL** identifier of the same stock. SEDOL stands for Stock Exchange Daily Official List and is mostly used in the UK. The international identifier is called **ISIN** (International Securities Identification Number), but that identifier is not reported in this dataset. Finally, **weight**s are defined on page 10 and their calculation is clarified on page 201.

3.2 Federal Reserve Economic Data (FRED)

The **Federal Reserve Economic Data** (**FRED**) repository maintained by the Federal Reserve Bank of St. Louis can be accessed using the **tq_get()** command of the **tidyquant** package seen earlier. The only differences are that you have to find the right code for the data you are looking for, and you have to specify what you want to **get** is **economic.data**, not a stock's price:

```
wti_price = tq_get("DCOILWTICO",
                   from = "2021-01-01",
                   to   = "2022-03-01",
                   get  = "economic.data")
```

To find the code for the data you're looking for, go to **https://fred.stlouisfed.org/categories**.

But oftentimes, it's easier to go to the FRED website, find the data you need (the Fed offers a plethora of time series with subtle variations, as is often the case when you deal with interest rates), and download them. As an exercise, we encourage you to go to one of FRED's default graphs and change the time series of this graph: First, go to `https://fred.stlouisfed.org/graph/?g=pleY` then edit the graph to find the following time series:

- ICE BofA AAA US Corporate Index Effective Yield, Percent, Monthly, Not Seasonally Adjusted
- 3-Month Treasury Constant Maturity Rate, Percent, Monthly, Not Seasonally Adjusted
- 3-Year Treasury Constant Maturity Rate, Percent, Monthly, Not Seasonally Adjusted
- 10-Year Treasury Constant Maturity Rate, Percent, Monthly, Not Seasonally Adjusted

Note that you'll have to change the frequency from daily (the default value) to monthly. Otherwise, you can use the file we downloaded for you, called `fredgraph.xlsx`. The file contains monthly data from May 1, 2016 to May 1, 2021.

3.3 Nasdaq

Daily prices of individual stocks can be downloaded for free from the Nasdaq website. This website offers a lot of data, often provided by external vendors, and some come for a fee. The best way to check if the data you need is available on Nasdaq's is `https://data.nasdaq.com/search`.

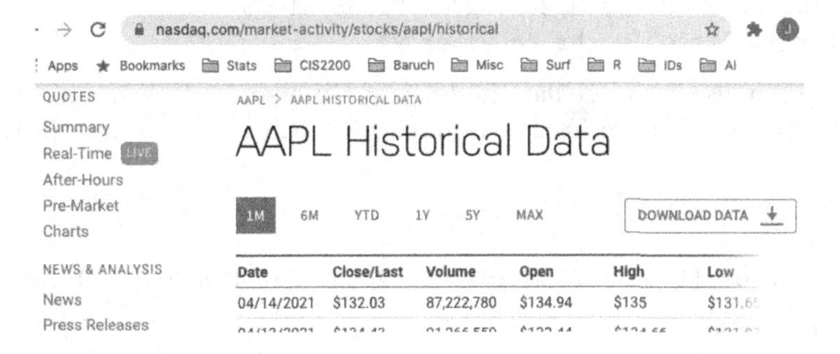

FIGURE 3.3 Nasdaq's webpage for historical stock prices and trading volumes

Just select the length of the history you desire (1 month is selected by default), then click on the "DOWNLOAD DATA" button on the right-hand side shown in Figure 3.3, and a csv file will appear in your download folder.

But Nasdaq also offers an entire package called **`Quandl`**. As usual, you need to install this package, then load it using the **`library()`** function:

```
library(Quandl)
```

The package is free, as of early 2022, and sending a few requests can be done without further consideration. However, after a few calls, you might get the following message:

```
Error: { "quandl_error": { "code": "QELx01", "message": "You have
  exceeded the anonymous user limit of 50 calls per day. To make
  more calls today, please register for a free Nasdaq Data Link
  account and then include your API key with your requests." } }
```

Once you have registered, you will get a key, i.e., a sequence of characters that becomes your "fingerprint" each time you send a query. You must "announce" yourself by presenting that key using the **Quandl.api_key()** function:

```
Quandl.api_key("X73B_4kF-5EyXROw11cf")
```

(Note that, for security reasons, the API key printed above is incorrect. You need to get your own key and insert it in the code above.)

You can then pull the time series you desire, assuming you know its identifier. For example, GDP data can be pulled from the **FRED** data based discussed earlier into a dataframe using the **raw** value for the **type=** named option of function **Quandl()** using the **FRED/GDP** identifier:

```
str(Quandl("FRED/GDP", type="raw"))
```

```
'data.frame':   300 obs. of  2 variables:
 $ Date : Date, format: "2021-10-01" ...
 $ Value: num  23992 23202 22741 22038 21478 ...
 - attr(*, "freq")= chr "quarterly"
```

But you can also get these data in different **time series** formats that we will discuss further on page 58. For example, you can get these data in the **ts** format:

```
str(Quandl("FRED/GDP", type="ts"))
```

```
 Time-Series [1:300] from 1947 to 2022: 243 246 250 260 266 ...
```

or in the **xts** format:

```
str(Quandl("FRED/GDP", type="xts"))
```

```
An 'xts' object on 1947 Q1/2021 Q4 containing:
  Data: num [1:300, 1] 243 246 250 260 266 ...
  Indexed by objects of class: [yearqtr] TZ: UTC
  xts Attributes:
 NULL
```

or in the **zoo** format:

```
str(Quandl("FRED/GDP", type="zoo"))
```

```
'zooreg' series from 1947 Q1 to 2021 Q4
  Data: num [1:300] 243 246 250 260 266 ...
  Index:  'yearqtr' num [1:300] 1947 Q1 1947 Q2 1947 Q3 1947 Q4 ...
  Frequency: 4
```

Corporate financial data are quite elaborate in Nasdaq's **MER/F1** database. Here is, for example, how to extract different financial metrics, over time, for **Deutsche Bank**. 2438 is the identifier for Deutsche Bank, and the name of each financial metric ("**indicator**"), the date at which the value was reported ("**reportdate**"), and the value itself ("**amount**") are the only data we are requesting here:

```
deutschebank =
  Quandl.datatable("MER/F1",
          compnumber=c("2438"),
          qopts.columns=c("indicator","reportdate","amount"))
```

The number of financial and accounting metrics is remarkable, as we can check using the **unique()** function. We display only the first few below:

```
head(unique(deutschebank$indicator))
```

```
[1] "Accrued Expenses Turnover"
[2] "Cash Flow Per Share"
[3] "Total Assets Per Share"
[4] "Revenue to Assets"
[5] "Return on Equity"
[6] "EPS - Net Income - Diluted"
```

The data can be stale, however. In the case of Deutsche Bank at least, the data offered by Nasdaq seems to end in 2015:

```
summary(deutschebank$reportdate)
```

```
        Min.      1st Qu.       Median         Mean
"2010-12-31" "2011-12-31" "2013-06-30" "2013-07-05"
     3rd Qu.         Max.
"2014-12-31" "2015-12-31"
```

This database is also used on page 107.

JSON API at Nasdaq

We introduced JSON on page 16. Few providers offer free JSON feed of financial data, but one exception is the history of oil prices offered by Nasdaq. However, it is not clear what other time series are offered by Nasdaq for free, and most of the good stuff seems to require a "premium" account.

As discussed earlier, the package that allows us to use the JSON format is **jsonlite**. You need to install and load it the usual way.

```
library(jsonlite)
```

Reading the data is then as simple as a **fromJSON()** request to the appropriate URL. For instance:

```
orb =
 fromJSON("https://data.nasdaq.com/api/v3/datasets/OPEC/ORB.json")
```

Notice that the data are pulled until the latest close as of the day you execute this command.

Unfortunately, the data arrive as character strings; the structure of the data is a bit intricate, too. Extracting dates and oil prices and converting them to the appropriate data type (more on this throughout this book) can be done as follows:

```
oil.prices = as.numeric(orb$dataset$data[,2])
oil.dates = as.Date(orb$dataset$data[,1])
```

Plotting historical oil prices is then easy, and the following command produces the graph in Figure 3.4.

```
plot(oil.dates, oil.prices)
```

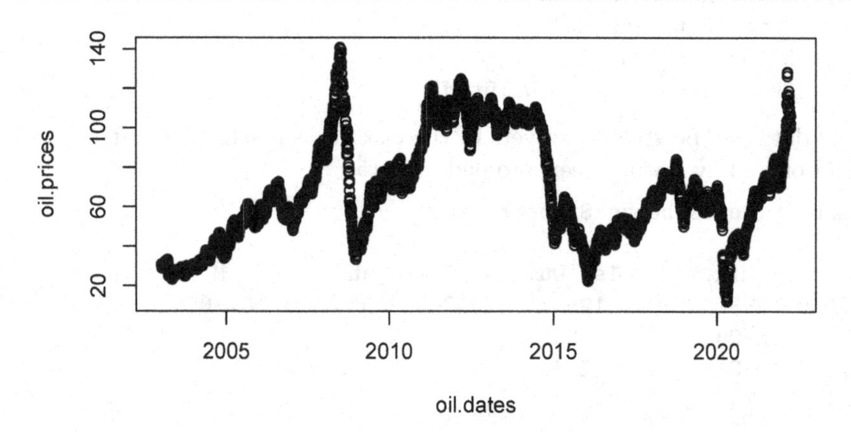

FIGURE 3.4 Historical oil prices pulled using JSON

3.4 Other Data Sources

Ecdat

The **Ecdat** package comes with historical returns on the S&P 500 index. You can pull them as follows:

```
data(SP500,package="Ecdat")
```

Note that this creates a dataframe with the same name as the data sets. This time series is old, however, and only contains daily observations from January 1981 to April 1991, so it probably has no other use than classroom support.

US Department of the Treasury

The **Department of the Treasury** publishes daily **yield curve** rates. This curve is based on the bid yields on actively traded Treasury securities at each close of the over-the-counter market:

```
https://www.treasury.gov/resource-center/data-chart-center/
   interest-rates/Pages/TextView.aspx?data=yield
```

Bloomberg

The **Bloomberg Terminal** is of course a fantastic source of data. If you have access to it, you may be familiar with Bloomberg's Excel add-in that lets you import their data directly into a spreadsheet. Similarly, the **Rblpapi** package lets you pull data directly into R data structures. For example:

```
library(Rblpapi)
blpConnect()
bdh("SPY US Equity", c("PX_LAST"), start.date = Sys.Date()-5)
```

pulls the end-of-session levels of the S&P 500 (ticker `SPX US Equity` on the Bloomberg Terminal) for the past 5 trading sessions[2]:

```
  date         PX_LAST
1 2021-04-26  417.61
2 2021-04-27  417.52
3 2021-04-28  417.40
4 2021-04-29  420.06
5 2021-04-30  417.30
```

[2]The very last value is the current intra-day value for the ticker.

Rbitcoin

The **Rbitcoin** package offers access to some data, but its functions are fragile. The following datasets may be of interest to you, however. The first dataset contains recent **bitcoin** trades expressed in US dollars. The market (i.e., the exchange) has to be specified, and **bitstamp** is one of the few available options that works for bitcoin prices expressed in dollars.

```
library(Rbitcoin)
trades = market.api.process(market = 'bitstamp',
                            currency_pair = c('BTC', 'USD'),
                            "trades")
```

You can then plot the prices over the provided period, resulting in Figure 3.5.

```
Rbitcoin.plot(trades)
```

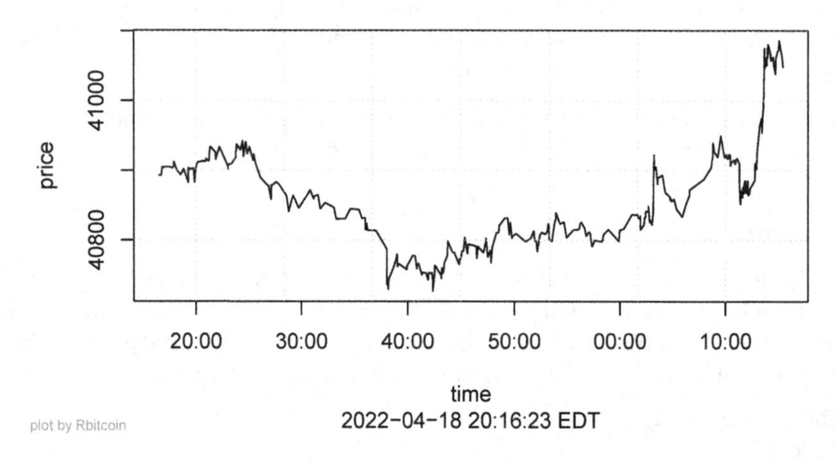

FIGURE 3.5 Bitcoin prices provided by Rbitcoin

The other available exchange is **kraken** for bitcoin prices in euros. Other exchanges mentioned in the package documentation are now closed.

4

Introduction to R

4.1 Expressions

You're now familiar with RStudio, or an equivalent environment. We can then go to the window where you type your commands, called the **console**. There, you can enter numerical calculations, what we would call expressions, such as:

```
1 / 20.33 * 3000
```

```
[1] 147.6
```

```
sin(pi / 2)
```

```
[1] 1
```

Note that the accounting convention of separating multiples of 1,000 is not accepted by R, so a million has to be written `1000000` or `1e6` (i.e., 1×10^6), but not `1,000,000`. Also, percentages have to be written in decimal form, so 12.3% must be written `0.123`.

4.2 Creating New Variables

Variables are, in their simplest forms, like mathematical variables: they contain values. You can assign values using either `=` or `<-`.

```
x = 3*4
y <- 2* 10
z = 2^4
```

Many R users like the `<-` symbol, but it might look artificially geeky, and this book will prefer `=`. One benefit of `<-` however is that it is consistent with its symmetrical form: indeed, in R, it is legal to write `->` into a variable, as in:

```
2*7 -> an.even.number
2*8 -> next.even.number
```

DOI: 10.1201/9781003320555-4

Compared to mathematical variables however, variables in R are much more general and can be used to store any object, such as a table of data, an expression, an executable mathematical function, or even a graph!

Variable names must start with a letter. The rest of their names can only contain letters, numbers, and the underscore (_) and period (.) characters.

If you're new to programming, keep in mind that it is in your interest that names be descriptive. To do that, you will want to give your variables names made of two or three words – but unfortunately in R (and in all programming languages I can think of), these words can't just be separated by spaces as we do in English. So either you glue them together into something like "mynewvariable," but that would be hard to read, or you use a non-letter character to separate the words. For example, you can use . or _ to separate words to create names like **`simple.variable`** or **`simple_variable`**. You can also change capitalization, as in **`simpleVariable`**. So all three commands below are legal statements in R:

```
product_of_two_numbers = 3*pi
another.style = 2* 10
yetAnotherStyle = 12
```

All these names are legal names, and each programmer has their stylistic preference. As a personal opinion, separating words by only uppercase letters make some variable names difficult to read. Using the underscore is fine but requires, on most keyboards, two fingers to get the symbol. In contrast, the period is easy to reach with one finger – a minuscule saving of energy that over time adds up to a lot. In any case, pick the one pattern you prefer and try to use it consistently.

Note also that the value of a variable is not printed unless you request it, which you do by typing the variable's name:

```
product_of_two_numbers
```

```
[1] 9.425
```

You can also display the value or the content of a variable using the **`print()`** command, which also accepts optional arguments to specify, for example, the number of significant digits:

```
print(product_of_two_numbers, digits=3)
```

```
[1] 9.42
```

Note that, in the last few examples, we've had to define variables before typing their names to get their value. But R offers a (confusing, IMO) syntax to do both at the same time by wrapping the entire command by **`()`**, i.e. a pair of parenthesis. In this example, we define **`piSquared`** *and* print its value:

```
(piSquared = pi**2)
```

```
[1] 9.87
```

4.3 Data Types and Type Conversion

If you have some experience with other programming languages, you may have noticed that we haven't had to **declare** variables before using them, and in particular, we haven't had to specify their type in advance: R figures this out on its own.

R represents data differently according to their types, though. These types are:

- `num` stands for numerical. That's the default for all numbers, even if they "look" like integers to us
- `int` stands for integers – but we'll have to specify that type using the `L` specifier if we want to contrast it from a generic numeric; please see below.
- `complex` stands for complex numbers
- `dbl` stands for doubles or real numbers.
- `chr` stands for character vectors or strings.
- `dttm` stands for date-times (a date + a time).
- `lgl` stands for logical or **boolean** values that are either TRUE or FALSE.
- `fctr` stands for factors, which R uses to represent categorical variables with fixed possible values, as discussed on page 9.
- And `date` is self-explanatory.

There is also a special symbol to denote a missing value: `NA` (all uppercase. R is case sensitive, by the way). `NA`s can pop up because of errors in calculations, but more typically because some values were missing in a data set. We can test whether a value is `NA` using the `is.na()` function. `NA`s will have to be removed using different features we will see as we make progress: one is the `na.omit()` function, another is the `na.rm=` option of many data manipulation functions, and yet another is the `na.locf()` function that fills `NA`s with the previous non-`NA` values.

Note that a `NA` is different from `NULL`, which means that the value in question simply doesn't exist (for example, an empty data structure), rather than being existent but unknown. An example of a situation where a `NULL` occurs is the creation of an empty vector using the `c()` function explained on the next page:

```
c()
```

```
NULL
```

Dates

Dates are cumbersome animals. They can be represented in many ways: as an integer, counting from a reference day; that is Excel's representation. Or as a string, with months noted by their English name or by a number from 1 to 12; the order can be year, month, day, or month, day and year as is customary in the U.S. And that string may or may not contain the time of the day as well. This is illustrated by the two standard ways to get today's date in R:

```
date()
```

```
[1] "Mon Apr 18 20:16:23 2022"
```

and:

```
Sys.Date()
```

```
[1] "2022-04-18"
```

But notice the different data types:

```
str(date())
```

```
 chr "Mon Apr 18 20:16:23 2022"
```

```
str(Sys.Date())
```

```
 Date[1:1], format: "2022-04-18"
```

Complex Numbers

Note first that `3+1i` will be understood as you expect, but `3+i` will not. In the latter case, R assumes you want to add the value of a variable called `i`, which may or may not have been defined (a big trap indeed). Otherwise, complex arithmetic works naturally in R. For example, you can multiply complex numbers:

```
(3+1i)*(2+2i)
```

```
[1] 4+8i
```

4.4 Vectors

Vectors are what they sound like: a series of data of any one type – typically, numbers. A vector typically comes out of reading data, but it can also be constructed using the `c()` function. For example:

```
simple.vector = c(2, -3, 8)
```

The usual arithmetic operations extend to vectors of numbers:

```
v = c(11, 12, 13)
w = c(1, 2, 3)
v + w
```

```
[1] 12 14 16
```

```
v / w
```

```
[1] 11.000  6.000  4.333
```

```
w^v
```

```
[1]       1    4096 1594323
```

If one of the operands in the arithmetic operation is not a vector but a single value (a so-called **`scalar`**), then that single value is extended to the calculations on all the elements of the vector. For example, adding 1 to the vector v we just created means that this value of 1 should be added to all the entries of v:

```
(y = 1+v)
```

```
[1] 12 13 14
```

Mean, Standard Deviation and Z-score

We will have large sections of this textbook devoted to statistics and a chapter dedicated to functions, but you should already get familiar with the most frequent stats functions. Considering the data below, **`mean()`** (arithmetic mean, a.k.a. **average**), **`median()`**, **standard deviation `sd()`** and **variance `var()`** can be calculated in quite a natural way:

```
x = c(0, 1, 1, 2, 3, 5, 8, 13, 21, 34)
mean(x)
```

```
[1] 8.8
```

```
median(x)
```

```
[1] 4
```

```
sd(x)
```

```
[1] 11.03
```

```
var(x)
```

```
[1] 121.7
```

Z-scores are critical in data analysis because they quickly identify outliers by measuring how far a value is from its mean, expressed in standard deviations. More formally, the z-score z or a value x with mean μ and standard deviation σ is:

$$z = \frac{x - \mu}{\sigma} \tag{4.1}$$

This definition has two key benefits: first, if the initial variable follows a normal distribution (see page 78), then the z-score follows the standard normal distribution of mean 0 and standard deviation 1. Second, z-scores are unitless: if you calculate the z-score of a variable expressed in dollars, then subtracting its mean will give dollars and then dividing by the standard deviation, also expressed in dollars, will make the result unit-less.

How do you calculate a z-score? You can simply apply its equation:

```
x = c(2, -3, 1, -1, 0, 2, 36, 3, 1, -2, 0, 1)
(x-mean(x))/sd(x)
```

```
 [1] -0.12783 -0.60721 -0.22371 -0.41546 -0.31959
 [6] -0.12783  3.13195 -0.03196 -0.22371 -0.51134
[11] -0.31959 -0.22371
```

Or you can use the **scale()** function – here, only the first results are shown:

```
head(scale(x), 3)
```

```
        [,1]
[1,] -0.1278
[2,] -0.6072
[3,] -0.2237
```

The second key question is: How do we interpret z-scores? Most statisticians consider that an **outlier** must be 3 standard deviations away from its mean, i.e. have a z-score of 3. In the example above, 36 would be an outlier.

We now have a more formal tool to decide which of the price-to-book values were outliers. We calculate them as follows:

```
pb.zscores =
  (acwi$Price.to.Book - mean(acwi$Price.to.Book, na.rm=TRUE)) /
    sd(acwi$Price.to.Book, na.rm=TRUE)
```

Then, using the **which()** function, we can find which stocks are outliers – but there are too many of them to print here. Moreover, six stocks have price-to-book ratios more than 9 standard deviations away from the index average! These stocks are:

```
acwi$Name[which(pb.zscores > 9)]
```

```
[1] "United Parcel Service, Inc. Class B"
```

```
[2] "S&P Global, Inc."
[3] "O'Reilly Automotive, Inc."
[4] "DocuSign, Inc."
[5] "Citrix Systems, Inc."
[6] "Adani Green Energy Limited"
```

Map, Apply and Reduce

We often need to **map** or **apply** the same function to all the elements of a vector, either to have as many ending values (e.g., calculate the square root of each of the elements of the vectors) or to calculate a single *aggregate* value. The former can be done using the `sapply()` function of base R.

```
x = c(2, 8, 25, 18)
sapply(x, sqrt)
```

```
[1] 1.414 2.828 5.000 4.243
```

(Note that, in this simple example, `sqrt(x)` would have worked fine to get the square root of each element of `x`.)

To perform the latter, the `reduce()` function combines all the elements of a vector using the operator of your choosing.

```
reduce(x, `+`)
```

```
[1] 53
```

Note that the `+` sign is bracketed by left quotes, not right/straight quotes.

Indexing

You can refer to elements of a vector using square brackets, either specifying a single entry:

```
x[4]
```

```
[1] 18
```

or a subset:

```
x[2:4]
```

```
[1]  8 25 18
```

(The `:` operator, which produces a vector, is discussed soon on page 46.)

You can also specify specific entries, such as to the 1st, 3rd and 2nd in the example below:

```
x[ c(1, 3, 2) ]
```

```
[1]  2 25  8
```

Very conveniently, you can refer to all the elements of a vector except the first using the negative sign `-`. So the "`-1` notation" keeps all the elements of a vector but the first one (or all the rows or columns of a matrix except the first one, as we will see).

```
x[-1]
```

```
[1]  8 25 18
```

You can also extract all the elements except the last one, again using the negative sign. The position of the last element is not always known, however, so you often need the **`length()`** function to find its index:

```
x[ -length(x) ]
```

```
[1]  2  8 25
```

4.5 Matrices

Matrices can be created using the **`matrix()`** function. For example, the following line of code places the vector of the first 15 integers, listed using the colon : operator discussed on page 46, in a matrix of 5 columns. Elements of the matrix are filled one row at a time, so the first 5 integers fill out the top row, the next 5 integers fill the second row, etc.

```
(myMatrix = matrix(1:15, ncol = 5))
```

```
     [,1] [,2] [,3] [,4] [,5]
[1,]    1    4    7   10   13
[2,]    2    5    8   11   14
[3,]    3    6    9   12   15
```

Matrix elements can be retrieved using the square-bracket notation to specify the row number and/or the column number. Rows come first and columns come second after a comma:

```
myMatrix[2,3]
```

```
[1] 8
```

If a row or column number is not specified, then the entire row or column is provided:

```
myMatrix[2,]
```

```
[1]  2  5  8 11 14
```

```
myMatrix[,3]
```

```
[1] 7 8 9
```

The "-1 notation" seen on page 38 extends to matrices (and to dataframes), so all the columns but the first one is denoted as:

```
myMatrix[, -1]
```

```
     [,1] [,2] [,3] [,4]
[1,]    4    7   10   13
[2,]    5    8   11   14
[3,]    6    9   12   15
```

Since the notation extends to dataframe, removing the leftmost column of a dataframe (which often contains non-numeric data we may need to remove for calculations) can also be done using the `-1` notation.

Operations on matrices include addition and subtraction, and the usual symbols `+` and `-` are extended to matrices. Matrix multiplication is trickier as you might remember from algebra classes and is denoted by `%*%`. Finally, matrix transposition is performed using the `t()` function, such as:

```
t(myMatrix)
```

```
     [,1] [,2] [,3]
[1,]    1    2    3
[2,]    4    5    6
[3,]    7    8    9
[4,]   10   11   12
[5,]   13   14   15
```

To calculate sums or averages across the columns or rows of a matrix (or a dataframe), use one of the **`colSums()`**, **`rowSums()`**, **`colMeans()`** or **`rowMeans()`** functions. For instance, the two statements below apply the sum and mean function, respectively along either of the two dimensions:

```
colSums(myMatrix)
```

```
[1]  6 15 24 33 42
```

```
rowMeans(myMatrix)
```

```
[1] 7 8 9
```

This idea can be generalized thanks to the **`apply()`** function. The arguments to this function are: the name of the matrix, the direction along which the operation should be applied, and the name of the operation, such as **sum()** or **mean()**. For instance, the means of each column of **myMatrix** are:

```
apply(myMatrix, 1, mean)
```

```
[1] 7 8 9
```

We get of course the same result as **rowMeans(myMatrix)**.

Note also that matrices are different from dataframes in that their columns have no names: there are matrices in the mathematical sense. But sometimes, in particular when applying algebraic operations on matrices, you will need to convert a dataframe using **as.matrix()**. For example, the sample file `stocks5.csv` we first read on page 14 contains 8 stocks and 31 monthly returns.

```
stocks = read.csv(file = 'stocks5.csv')
```

You can verify that **as.matrix()** works as intended thanks to the **class()** function – compare the output of the two statements below:

```
class(stocks)
```

```
[1] "data.frame"
```

```
class(as.matrix(stocks))
```

```
[1] "matrix" "array"
```

A relevant function is **outer()**, which creates a matrix out of two vectors by performing an operation (of your choice) on all possible combinations of two vectors. The function is called after the outer product in algebra, which would be expressed in R as:

```
x = c(10, 100, 100)
y = c(1, 2, 3)
outer(x, y, FUN="*")
```

```
     [,1] [,2] [,3]
[1,]   10   20   30
[2,]  100  200  300
[3,]  100  200  300
```

Note that the **FUN=** parameter specifies the function to be applied to each combination of the two input vectors.

This outer product could have also been performed using the **%*% matrix multiplication** operator with the help of the **t() transpose** function:

```
x %*% t(y)
```

```
     [,1] [,2] [,3]
```

```
[1,]   10   20   30
[2,]  100  200  300
[3,]  100  200  300
```

As an exercise, we can put most of these concepts to work to calculate a **covariance matrix**: We first calculate the means of each column, which correspond to the average returns of each stock:

```
(column.means = apply(stocks, 2, mean))
```

```
     GM       F     JPM    BAML    GOOG    AMZN
0.01097 0.00129 0.01323 0.01419 0.01613 0.03323
   AAPL     CRM
0.02484 0.02839
```

We then construct a matrix of $n = 31$ rows (one per observation) and 8 columns (one per stock) by multiplying a "vertical" vector of n ones with a "horizontal" vector of 8 means. Note that we use the matrix-multiplication operator **%*%**:

```
n = nrow(stocks)
mean.matrix = matrix(data=1, nrow=n) %*% column.means
```

(The **mean.matrix** could also have been built using the **outer()** function discussed earlier.)

We "center" each return by subtracting their respective means:

```
D = as.matrix(stocks) - mean.matrix
```

And finally we calculate the covariance matrix. Note that we need the transpose function **t()** seen on page 39.

```
cov.matrix = (n-1)^-1 * (t(D) %*% D)
```

Of course, R offers a built-in function, **cov()**, to calculate a **covariance matrix** directly from the dataframe of stock returns and without having to explicitly create matrices:

```
print(cov(stocks), digits=2)
```

```
         GM       F     JPM   BAML    GOOG    AMZN
GM   0.0060 3.9e-03 0.00235 0.0032 1.3e-03 0.00138
F    0.0039 5.7e-03 0.00236 0.0028 3.5e-05 0.00099
JPM  0.0024 2.4e-03 0.00365 0.0038 1.2e-03 0.00218
BAML 0.0032 2.8e-03 0.00384 0.0048 1.8e-03 0.00301
GOOG 0.0013 3.5e-05 0.00123 0.0018 3.2e-03 0.00302
AMZN 0.0014 9.9e-04 0.00218 0.0030 3.0e-03 0.00696
AAPL 0.0020 1.4e-03 0.00122 0.0019 1.7e-03 0.00302
CRM  0.0011 7.0e-04 0.00096 0.0018 1.9e-03 0.00380
       AAPL     CRM
```

```
GM   0.0020 0.00109
F    0.0014 0.00070
JPM  0.0012 0.00096
BAML 0.0019 0.00176
GOOG 0.0017 0.00191
AMZN 0.0030 0.00380
AAPL 0.0071 0.00223
CRM  0.0022 0.00359
```

A related concept, the **correlation matrix**, can be calculated using the `cor()` function.

```
print(cor(stocks), digits=2)
```

```
       GM      F  JPM BAML   GOOG AMZN AAPL  CRM
GM   1.00 0.6624 0.50 0.61 0.2852 0.21 0.30 0.24
F    0.66 1.0000 0.52 0.54 0.0082 0.16 0.22 0.15
JPM  0.50 0.5151 1.00 0.92 0.3594 0.43 0.24 0.26
BAML 0.61 0.5388 0.92 1.00 0.4683 0.52 0.32 0.43
GOOG 0.29 0.0082 0.36 0.47 1.0000 0.64 0.36 0.56
AMZN 0.21 0.1566 0.43 0.52 0.6369 1.00 0.43 0.76
AAPL 0.30 0.2176 0.24 0.32 0.3590 0.43 1.00 0.44
CRM  0.24 0.1540 0.26 0.43 0.5611 0.76 0.44 1.00
```

We can elegantly plot the correlation matrix using the `corrplot()` function offered by the package of the same name. The output can be seen in Figure 4.1.

```
library(corrplot)
corrplot(cor(stocks), method = 'number')
```

Please see on page 250 for other ways to use `corrplot()`.

Extension of z-scores: Mahalanobis distance

We saw on page 35 that z-scores tell us how far, in standard deviations, a value is from the mean of its distribution. This is the one-dimensional case, but we may be interested in detecting which points in a multi-dimensional space are outliers. Indeed, we can imagine that a 3-D point is not necessarily an outlier along the first dimension, or the second, or the third, but an outlier in aggregate over all three. The **Mahalanobis distance** is a multi-dimensional generalization of the z-score. Like the z-score, it centers data around their means, but in a multi-variate case, it takes not only the variance of each vector into account (like the z-score does) but also their covariance.

Let's put that in practice on price-to-cash-flow, price-to-book and price-to-sales of the `acwi` data set. These three price ratios are in columns 6 through 8:

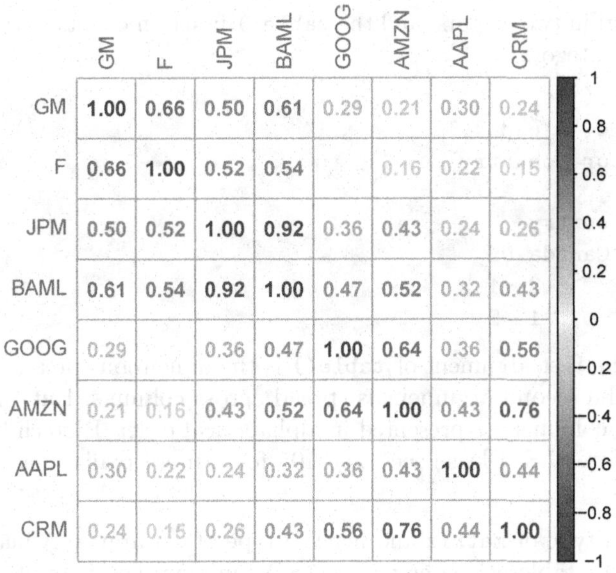

	GM	F	JPM	BAML	GOOG	AMZN	AAPL	CRM
GM	1.00	0.66	0.50	0.61	0.29	0.21	0.30	0.24
F	0.66	1.00	0.52	0.54		0.16	0.22	0.15
JPM	0.50	0.52	1.00	0.92	0.36	0.43	0.24	0.26
BAML	0.61	0.54	0.92	1.00	0.47	0.52	0.32	0.43
GOOG	0.29		0.36	0.47	1.00	0.64	0.36	0.56
AMZN	0.21	0.16	0.43	0.52	0.64	1.00	0.43	0.76
AAPL	0.30	0.22	0.24	0.32	0.36	0.43	1.00	0.44
CRM	0.24	0.15	0.26	0.43	0.56	0.76	0.44	1.00

1
0.8
0.6
0.4
0.2
0
−0.2
−0.4
−0.6
−0.8
−1

FIGURE 4.1 Correlation matrix visualized using corrplot

```
price.ratios = na.omit(acwi[6:8])
```

We call the **mahalanobis()** function with the three arguments it needs:

```
mah = mahalanobis(price.ratios,
                  colMeans(price.ratios),
                  cov(price.ratios))
```

Interpreting the absolute values of the Mahalanobis distance is harder than it is for z-scores, but we can look for the stocks whose distance is in the top tenth of 1%:

```
acwi$Name[which(mah > quantile(mah, .999))]
```

```
[1] "DuPont de Nemours, Inc."
[2] "Hisamitsu Pharmaceutical Co., Inc."
[3] "Commercial Bank (Q.S.C.)"
```

Tables

A **contingency table** counts the number of elements in each category defined in each column and row. As an example, we can create a contingency table reporting on the number of tech and biotech stocks in the US and Canada based on a sample of six companies. The characteristics of these six companies are contained in two vectors, and the **table()** function counts how many firms fall in each category:

```
sector  = c("Tech","Biotech","Tech","Tech","Biotech","Biotech")
country = c("Canada","Canada","US","US","US","Canada")
table(sector, country)
```

```
         country
sector    Canada US
  Biotech      2  1
  Tech         1  2
```

Note that the first argument of **table()** is the dimension spread down across rows, and the second argument is spread across columns. But the values of rows and of columns are presented in alphabetical order: Biotech before Tech, Canada before U.S. Please see page 105 for a more realistic application of **table()**.

A **table** is a type of **matrix** and hence a type of dataframe. A matrix can be converted to a table using the **as.table()** function. For example:

```
(simpleMatrix = matrix(c(1,2,3,4),
                       ncol=2,
                       byrow = TRUE))
```

```
     [,1] [,2]
[1,]    1    2
[2,]    3    4
```

```
simpleTable = as.table(simpleMatrix)
```

Notice the differences in underlying structures between the matrix and the table:

```
str(simpleMatrix)
```

```
 num [1:2, 1:2] 1 3 2 4
```

```
str(simpleTable)
```

```
 'table' num [1:2, 1:2] 1 3 2 4
 - attr(*, "dimnames")=List of 2
  ..$ : chr [1:2] "A" "B"
  ..$ : chr [1:2] "A" "B"
```

In particular, default names were given to columns and rows. This can be seen also by displaying the table's content:

```
simpleTable
```

```
  A B
```

```
A 1 2
B 3 4
```

We can change column and row names using the **colnames()** and **rownames()** functions. These functions are generic and apply also to dataframes.

```
colnames(simpleTable) = c("Left","Right")
rownames(simpleTable) = c("Top","Bottom")
simpleTable
```

```
       Left Right
Top       1     2
Bottom    3     4
```

The **outer()** function can be applied to tables and is very versatile because the **FUN=** parameter is not limited to multiplication or to arithmetic operations for that matter. In particular, all the parameters that follow the function are passed by **outer()** to that function. For example:

```
tb = c("Top","Bottom")
lr = c("Left","Right")
outer(tb, lr, paste, sep=" ")
```

```
     [,1]          [,2]
[1,] "Top Left"    "Top Right"
[2,] "Bottom Left" "Bottom Right"
```

Note that, as for **sapply()** for example, the **sep=** parameter needed by **paste()** is provided as an additional argument to **outer()**.

4.6 Lists

Lists are more general than vectors. First, each element can be of any type: unlike vectors, R doesn't coerce all the elements of a list into one same type:

```
myList = list(3.14, "Moe", c(1, 1, 2, 3), 42)
myList
```

```
[[1]]
[1] 3.14

[[2]]
[1] "Moe"

[[3]]
[1] 1 1 2 3
```

```
[[4]]
[1] 42
```

You may have noticed that the vector `(1,1,2,3)` became a sublist of the main list. In other words, lists can be nested, with lists within lists.

Nested lists are not easy to examine in R, and their structure is best visualized using the `str()` command. For example, to compare the two lists, one way is as follows:

```
str(myList)
```

```
List of 4
 $ : num 3.14
 $ : chr "Moe"
 $ : num [1:4] 1 1 2 3
 $ : num 42
```

Also, list are often easiest to work on when they are flattened. As shown earlier, the `unlist()` function is useful in that regard:

```
unlist(myList)
```

```
[1] "3.14" "Moe"  "1"    "1"    "2"    "3"    "42"
```

`unlist()` is also useful to convert one-column or one-row dataframes into vectors.

Sequences of Numbers

Creating a sequence of numbers spread at regular intervals happens quite often in data analysis, possibly because we need to index a vector (in which case the numbers will be integers) or because we need to evaluate a function at repeating intervals. This can be done in different ways. The first method is to use the: operator to produce a sequence where each number is the prior number plus one

```
1:10
```

```
 [1]  1  2  3  4  5  6  7  8  9 10
```

In the above example, the increment was positive one because the first number was less than the second. However, in the opposite case:

```
5:1
```

```
[1] 5 4 3 2 1
```

the increment is negative one.

Numbers around the colon operator do not need to be integers. For example

```
0.3:2.4
```

```
[1] 0.3 1.3 2.3
```

Note that R requires decimal numbers starting with a zero to have this zero. What I mean is that 0.3 should be written as **0.3**, not **.3**.

Another way to generate a sequence is to use the **seq()** function:

```
seq(1,10)
```

```
 [1]  1  2  3  4  5  6  7  8  9 10
```

This function takes several optional arguments. The third and optional argument is the increment; in the example below, this increment is set to 0.2:

```
seq(0.3, 1.14, 0.2)
```

```
[1] 0.3 0.5 0.7 0.9 1.1
```

Another optional argument is **length.out=**, which specifies how many elements you want between the lower and upper bound. R will calculate the increment that allows to produce exactly that number of values:

```
seq(1, 10, length.out = 5)
```

```
[1]  1.00  3.25  5.50  7.75 10.00
```

A related function is **rep()** (short for repeat), which allows us to conveniently put the same constant into long vectors. The call form is rep(x,times), which creates a vector of times multiplied by length(x) elements – that is, times copies of x:

```
rep(2, times=3)
```

```
[1] 2 2 2
```

Comparisons

All the usual math operators are of course available in R. The only trap, that is very common across programming languages, is that *testing* for equality is done using *two* equal signs to form the **==** operator. In contrast, a single **=** is reserved to *force* equality, which is also known as an **assignment**:

```
x = 2
x == 2
```

```
[1] TRUE
```

We can also test whether x is less than 2, less than or equal to 2, or different from 2 using the **<**, **<=** and **!=** operators, respectively:

```
x < 2
```

```
[1] FALSE
```

```
x <= 2
```

```
[1] TRUE
```

```
x != 2
```

```
[1] FALSE
```

These comparison operators extend, element-wise, to vectors. For example, assuming we have the two vectors z and s:

```
z = c(0, 1, 2, 3, 5, 8, 13, 34)
(s = 0:7)
```

```
[1] 0 1 2 3 4 5 6 7
```

Comparing each pair of elements can be done using the same operators:

```
z < s
```

```
[1] FALSE FALSE FALSE FALSE FALSE FALSE FALSE FALSE
```

```
z == s
```

```
[1]  TRUE  TRUE  TRUE  TRUE FALSE FALSE FALSE FALSE
```

Any and All

In other circumstances, we don't need to know which elements of a vector satisfy a certain condition but simply to know if **any()** or **all()** of the elements work out. Continuing with the same value for vector z:

```
any(z > 8)
```

```
[1] TRUE
```

```
any(z > 88)
```

```
[1] FALSE
```

Likewise, for the **all()** function:

```
all(z > 0)
```

```
[1] FALSE
```

```
all(z > -1)
```

```
[1] TRUE
```

Subset of a vector

Specifying a subset of a vector can be done in many ways. The most general specifies whether or not each element should be kept using a vector of true/false values:

```
z = c(0, 12, 10, 8)
z[ c(TRUE, FALSE, FALSE, TRUE) ]
```

```
[1] 0 8
```

This convention of using true/false vectors as selectors of elements of another vector means that any calculation that returns a vector of boolean values of the right length can select elements of another vector. For example, to find all the elements of z such that the value is less than 9, we build a true/false vector using **z<9** and that vector, not displayed below, is used to select the appropriate elements of **z**:

```
z[ z < 9 ]
```

```
[1] 0 8
```

The **ifelse()** function provides the classic if-then-else construct of all programming languages. The first argument is the condition, the second the "then" clause, and the third the "else" case. For example, if we wanted to replace any negative number by 0, we could say "if (a given element in) v is less than 0, then use 0, otherwise use that same element of v":

```
v = c(1, -2, 5, 18)
(v = ifelse(v<0, 0, v))
```

```
[1]  1  0  5 18
```

We can subset a vector, i.e. select some of its elements, using the square-bracket notation. For example, if we need to select all elements greater than the median:

```
v = c(3, 6, 1, 9, 11, 16, 0, 3, 1, 18, 2, 9, 6, -4)
v[ v > median(v)]
```

```
[1]  6  9 11 16 18  9  6
```

Note that the expressions in the brackets can be arbitrarily complex. For instance, to select all elements in the lower and upper 5%, we can write:

```
v[ (v < quantile(v, 0.05)) | (v > quantile(v, 0.95)) ]
```

```
[1] 18 -4
```

This example used the **|** operator, which means "or" when indexing. If you wanted "and," you would use the **&** operator.

Technical Pause: & versus &&, | vs ||

Both `&` and `&&` mean "and," `|` and `||` mean "or." But `&&` is used as a conjunction between two Boolean scalars (not vectors) while `&` is a vector operator. So you would write:

```
TRUE && FALSE
```

```
[1] FALSE
```

but you would write:

```
c(TRUE, FALSE) & c(TRUE, TRUE)
```

```
[1]  TRUE FALSE
```

`&&` is typically used in an `if()` statement:

```
c1 = 1 < 10
c2 = 2 > 4
if (c1 && c2) { print("Both conditions are true") }
if (c1 || c2) { print("At least one condition is true") }
```

```
[1] "At least one condition is true"
```

Subsetting

Subsetting means selecting only those elements of a vector that satisfy a certain property. For example, we could select all elements that exceed + or - 1 standard deviations from the mean. We can do this subsetting inside the bracket notation `[]` used to index vectors:

```
v[ abs(v - mean(v)) > sd(v)]
```

```
[1] 16 18 -4
```

Similarly, we could select all elements that are neither `NA` nor NULL:

```
v = c(1, 2, 3, NA, 5)
v[!is.na(v) & !is.null(v)]
```

```
[1] 1 2 3 5
```

Keep also in mind that the `subset()` function exists explicitly. For example:

```
v = c(1, 2, 3, NA, 5)
subset(v, v>2)
```

```
[1] 3 5
```

Which

which() gives the positions, not the values, of vector elements satisfying a condition:

```
b = c(7, 2, 4, 3, -1, -2, 3, 3, 6, 8, 12, 7, 3)
which(b<7)
```

```
[1]  2  3  4  5  6  7  8  9 13
```

Compare it with the "such that" notation, which returns the *values*, not the *indices*, of the elements satisfying the condition:

```
b[b<7]
```

```
[1]  2  4  3 -1 -2  3  3  6  3
```

The expression inside **which()** can be arbitrarily complex as long as it is **boolean**:

```
which((10:18)%%2 == 0)
```

```
[1] 1 3 5 7 9
```

Two variants are worth noting: **which.min()** and **which.max()**, which provide the indices of the smallest and largest elements in a vector:

```
which.min(b)
```

```
[1] 6
```

```
which.max(b)
```

```
[1] 11
```

which.min() and **which.max()** can also be useful for some crude optimization, as we will see on page 75.

For a logical vector x with both FALSE and TRUE values, **which.min(x)** and **which.max(x)** return the index of the first FALSE or TRUE, respectively.

Concatenating Strings and Values into a Single String

We will soon graph lines of best fit and will want to add the R-squared to the caption. But how do you build text that contains a (variable!) result? The answer is to piece together (which is called **concatenation** in computer science) the text you want with the variable whose value needs to be inserted as text. This is done in R using the **paste()** function:

```
Rsq = 0.88
paste("R-squared =", Rsq)
```

```
[1] "R-squared = 0.88"
```

Exercise 4.10.1 on page 65 has you adjust the command above to improve the formatting and produce "R-squared = 88 %".

Note that, by default, the **sep**arator when concatenating text is the space character, but we can specify any text:

```
paste(1,'two',3,'four',5,'six')
```

```
[1] "1 two 3 four 5 six"
```

```
paste(1,'two',3,'four',5,'six',sep = "_")
```

```
[1] "1_two_3_four_5_six"
```

Cutting

Going back to numerical vectors, many data science problems involve bucketing a vector's values into fixed ranges or intervals delimited by **breaks**. For example, one may need to bucket numbers into four groups:

- negative values, so the breaks are minus infinity and 0;
- from 0 excluded to 3 included;
- from 3 excluded to 9 included;
- values greater than 9 (9 is the "left" break; up to infinity, the "right" break)

Such a bucketing is performed using the `cut()` function and its `breaks=` optional argument:

```
b = c(7, 2, 4, 3, -1, -2, 3, 3, 6, 8, 12, 7, 3)
cut(b, breaks=c(-Inf, 0, 3, 9, Inf))
```

```
 [1] (3,9]    (0,3]    (3,9]    (0,3]    (-Inf,0]
 [6] (-Inf,0] (0,3]    (0,3]    (3,9]    (3,9]
[11] (9, Inf] (3,9]    (0,3]
Levels: (-Inf,0] (0,3] (3,9] (9, Inf]
```

Note that the buckets each number falls into are enumerated in order and that the possible intervals are listed on the last line.

4.7 Data Frames

Data frames (also spelled dataframes) are essentially tables, but more precisely, a collection of columns of values, each column with their own name. This type of data comes with a plethora of functions crafted to facilitate their slicing and dicing that we are going to discover throughout this textbook. (You may also encounter the word **tibble**, which is just a more recent implementation of dataframes.)

Almost all the dataframes you'll use will come from reading data files, but in a few cases, creating a data frame "from scratch" is useful. A simple way to create a dataframe is to make vectors the columns of the new dataframe using the **data.frame()** function:

```
v1 = c(1, 2, 3)
v2 = c(3, 4, 5)
(df = data.frame(v1, v2))
```

```
  v1 v2
1  1  3
2  2  4
3  3  5
```

You'll notice that the names of the vectors become the names of the columns. These vectors can also be constructed inside the call to **data.frame()**:

```
(simple.dataframe = data.frame(v1 = 1:4,
                               v2 = letters[1:4],
                               months = month.name[3:6]))
```

```
  v1 v2 months
1  1  a  March
2  2  b  April
3  3  c    May
4  4  d   June
```

The **str()** command is extremely useful to get details on the **str**ucture of a data set:

```
str(df)
```

```
'data.frame':	3 obs. of  2 variables:
 $ v1: num  1 2 3
 $ v2: num  3 4 5
```

A similar function, **glimpse()**, is provided by the **dplyr** library (itself included in **tidyverse**, which I assume you always load):

```
glimpse(df)
```

```
Rows: 3
Columns: 2
$ v1 <dbl> 1, 2, 3
$ v2 <dbl> 3, 4, 5
```

A data frame's dimensions and its column names can be pulled using **dim()** and **names()**, respectively:

```
dim(df)
```

```
[1] 3 2
```

```
names(df)
```

```
[1] "v1" "v2"
```

Selecting One or More Columns

To select a column simply add the name (header) of the column, preceded by the dollar sign **$**:

```
simple.dataframe$v2
```

```
[1] "a" "b" "c" "d"
```

This dollar-sign operator is probably the most frequently used operator in R. So much so that having to repeat the name of the dataframe name to the left of the **$** sign quickly becomes tedious. To reference the same dateframe repeatedly, use the **attach()** function to make the columns of a dataframe available simply using their names. For example, after attaching the dataframe:

```
attach(simple.dataframe)
```

we can simply write:

```
v2
```

```
[1] 3 4 5
```

You can select one or several columns by their names using the "square bracket notation." For example:

```
simple.dataframe[ , "months"]
```

```
[1] "March" "April" "May"   "June"
```

Note that there is nothing between [and the comma: it means we are requesting all **rows**. Columns are the second dimension in a data frame, so column names come **after** the coma. We can select multiple column names by putting them in a vector:

```
simple.dataframe[ , c("months", "v2")]
```

```
  months v2
1  March  a
2  April  b
3    May  c
4   June  d
```

Note that the result is displayed, which is the default behavior in R when you type a variable's name.

You can also select columns by their positions

```
simple.dataframe[ , c(2, 1)]
```

```
  v2 v1
1  a  1
2  b  2
3  c  3
4  d  4
```

Combining Data Frames

Combining data frames can be done using **rbind()** and **cbind()**. In both cases, the order of the arguments matter. Let's start with **rbind()**: if the existing data frame comes first followed by the new row, then the new row is attached at the "bottom" of the data frame:

```
new.row = c(6, 7)
rbind(df, new.row)
```

```
  v1 v2
1  1  3
2  2  4
3  3  5
4  6  7
```

If the existing data frame comes second, then the new row is bound at the top of the data frame:

```
new.top = c(6, 7)
rbind(new.top, df)
```

```
  v1 v2
1  6  7
2  1  3
3  2  4
4  3  5
```

Note that, in either case, the data frame **df** does not change unless there is an assignment. For example, to modify the content of **df**, we could write:

```
df = rbind(new.top, df)
```

Likewise, **cbind()** attaches a new column to the data frame. If the existing data frame appears first, then the additional column is attached to the right of the existing column; if it appears second, then the additional column becomes the leftmost columns.

```
some.months = month.name[1:3]
cbind(df, some.months)
```

```
  v1 v2 some.months
1  1  3     January
2  2  4    February
3  3  5       March
```

Note that the variable's name is used as the header of the new column.

A new column can also be added using the **$** operator:

```
df$months = some.months
df
```

```
  v1 v2   months
1  1  3  January
2  2  4 February
3  3  5    March
```

Note also that the new column must be of the same length as the existing columns of **df**.

A More Realistic Example of a Dataframe

As alluded to in the beginning of this section, almost all the data frames used in practice come from data sets your code reads. In this short section, we will use the `MSCI AC World.csv` data set that we introduced on page 7. As before, we read the file into a variable called **acwi** which was, as a matter of fact, a data frame.

Selecting a column, for example the first column, can be done with the **select()**:

```
select(acwi, 1)
```

We are not printing the result of course since the first column of our data set is very long. But we should look at the structure of what is created. We can do that by placing the result in a variable and using **str()**:

```
first.column = select(acwi, 1)
str(first.column)
```

```
'data.frame':	2974 obs. of  1 variable:
 $ Ticker: chr  "AAPL" "MSFT" "AMZN" "FB" ...
```

We see that **select()** created a data frame, not a vector of values.

Column selection can be also done by specifying the column name:

```
select(acwi, Ticker)
```

Note that, for our convenience, the name of the column in **select()** does not need to be placed between quotes even though it's a character string and not an R object.

There are many other ways to select a column. One would use the base syntax of R:

```
acwi['Ticker']
```

Note that the column name is here placed between quotes (simple or double – both are accepted). Not doing so would produce an error:

```
# Incorrect
acwi[Ticker]
```

```
Error in `[.data.frame`(acwi, Ticker) : object 'Ticker' not found
```

We can also select multiple columns:

```
select(acwi, Ticker, Weight, Dividend.Yield)
```

or on the contrary, we can deselect a column by name using a negative- sign:

```
select(acwi, -Price.to.Sales)
```

We can deselect a range of columns by specifying a range (using the : operator separating the leftmost and rightmost columns) and placing the negative sign in front of the range. Pay attention however that the negative sign applies to the entire range, so the range has to be placed between parentheses:

```
select(acwi, -(Names:Price.to.Sales))
```

Like we could select multiple column specifications separated by commas, we can also deselect them:

```
select(acwi, -Name, -Weight, -(Dividend.Yield:Price.to.Sales))
```

This last subset of our **acwi** dataframe will be useful in the rest of this textbook, so we save it in a shorter dataframe we call **acwi.short**:

```
acwi.short = select(acwi, -Name, -(Dividend.Yield:Price.to.Sales))
head(acwi.short, 3)
```

```
  Ticker Weight Market.Cap            GICS.Sector
1   AAPL   3.47    2193582 Information Technology
2   MSFT   2.81    1899924 Information Technology
3   AMZN   2.29    1744112 Consumer Discretionary
        Country   ROE Return
```

```
1 United States 73.69  78.98
2 United States 40.14  40.72
3 United States 27.44  40.15
```

The **acwi.short** dataset will be used again in particular on page 95.

4.8 Time Series

Time series extend data frames with an explicit timestamp for each row (each observation, presumably). They appear in R under different names, but don't let that confuse you. Just know that the original **ts()** function and data structure of the **stats** package has more recently been extended. In particular, the **xts** and **zoo** data structures and packages are generally preferable. But all you really need to know is that time series will pop up in your work as **ts**, **xts** or **zoo** objects, and we will highlight their differences only when necessary. Actually, we already saw on page 26 how to use **Quand** to pull time series in the **xts** format. For example:

```
gdp.xts = Quandl("FRED/GDP", type="xts")
```

To extract the data for a certain date range, specify that range as a string and pass it using the square-bracket **[]** operators. The start and end of the range are separated by a slash, the **/** operator. As a first example, if you want to pull data up to a date (here, a year), use **/** in that string and leave the first part empty:

```
gdp.xts["/1948"]
```

```
        [,1]
1947 Q1 243.2
1947 Q2 246.0
1947 Q3 249.6
1947 Q4 259.7
1948 Q1 265.7
1948 Q2 272.6
1948 Q3 279.2
1948 Q4 280.4
```

To extract the data for a given date range:

```
gdp.subset = gdp.xts["2016/2017"]
```

You can extract the first calendar year of data using the **first()** function:

```
first(gdp.xts, '1 year')
```

```
          [,1]
1947 Q1 243.2
1947 Q2 246.0
1947 Q3 249.6
1947 Q4 259.7
```

We can group data by calendar convention, such as years, using the **split()** function:

```
gdp.yearly = split(gdp.subset, f="years")
lapply(gdp.yearly,FUN=mean)
```

```
[[1]]
[1] 18695

[[2]]
[1] 19480
```

Time series offers numerous more functionalities, which we are going to explore on different examples, that of two stocks obtained using **getSymbols()**:

```
getSymbols("AAPL", from = "2020-12-10", to   = "2021-03-31")
```

```
[1] "AAPL"
```

We are going to focus on the adjusted price:

```
apple = AAPL$AAPL.Adjusted
```

We can also convert the time series to monthly data, either using **to.period()** or, even more simply, **to.monthly()**

```
to.period(apple, period="months")
```

```
           apple.Open apple.High apple.Low apple.Close
2020-12-31      122.3      135.7     120.9       131.7
2021-01-29      128.5      142.1     125.7       131.0
2021-02-26      133.1      136.4     120.3       120.5
2021-03-30      127.0      127.0     115.7       119.2
```

(Note that **xts** also offers **to.weekly()**, **to.yearly()**, etc.)

But sometimes, time series have misaligned dates. This is frequent in equity markets when stock exchanges have different holiday calendars. Let's for example pull data on Apple and on the Industrial and Commercial Bank of China (ICBC):

```
getSymbols(
     Symbols = c("AAPL","1398.HK"),
     src = "yahoo",
```

```
       from = "2021-11-24",
       to   = "2021-11-28")
```

```
[1] "AAPL"    "1398.HK"
```

Let's look at their content. Apple's is easy:

```
AAPL
```

```
           AAPL.Open AAPL.High AAPL.Low AAPL.Close
2021-11-24     160.8     162.1    159.6      161.9
2021-11-26     159.6     160.4    156.4      156.8
           AAPL.Volume AAPL.Adjusted
2021-11-24    69463600         161.7
2021-11-26    76959800         156.6
```

The time stamps of a time series can be read using the **time()** function:

```
time(AAPL)
```

```
[1] "2021-11-24" "2021-11-26"
```

And the **start()** and **end()** date of the time series can be extracted using these respective functions:

```
start(AAPL)
```

```
[1] "2021-11-24"
```

```
end(AAPL)
```

```
[1] "2021-11-26"
```

Getting ICBC's data is a bit tricky because its ticker is 1398.HK, so the symbol that was created starts with a number, which is not a valid variable name in R: Starting an expression with a number makes R treat it as a number. We thus have to tell R to treat these numbers as characters, which we do using the back-quote ' symbol:

```
`1398.HK`
```

```
           1398.HK.Open 1398.HK.High 1398.HK.Low
2021-11-24         4.28         4.31        4.24
2021-11-25         4.26         4.28        4.23
2021-11-26         4.24         4.25        4.18
           1398.HK.Close 1398.HK.Volume
2021-11-24          4.26      111086413
2021-11-25          4.25       79213566
2021-11-26          4.19      110361676
           1398.HK.Adjusted
2021-11-24             4.26
```

```
2021-11-25                4.25
2021-11-26                4.19
```

We notice that only one of the two time series has data on Thanksgiving. That's of course because US markets were closed that day. But let's assume that we need to calculate the correlation of the closing prices of these two time series:

```
icbc  = `1398.HK`$`1398.HK.Close`
apple = AAPL$AAPL.Close
```

We need first to align the time series so that the values in each fall on the same days. We can use the **merge()** function:

```
(icbc.and.apple = merge(icbc,apple))
```

```
           X1398.HK.Close AAPL.Close
2021-11-24           4.26      161.9
2021-11-25           4.25         NA
2021-11-26           4.19      156.8
```

We see that some data are missing. In some circumstances, filling in the missing value by the previous valid number is acceptable, and we can do this using the **na.locf()** function:

```
na.locf(icbc.and.apple)
```

```
           X1398.HK.Close AAPL.Close
2021-11-24           4.26      161.9
2021-11-25           4.25      161.9
2021-11-26           4.19      156.8
```

In other circumstances, we prefer to fill in an approximated (interpolated) value using **na.approx()**:

```
na.approx(icbc.and.apple)
```

```
           X1398.HK.Close AAPL.Close
2021-11-24           4.26      161.9
2021-11-25           4.25      159.4
2021-11-26           4.19      156.8
```

Or, we could simply remove all the data for that date:

```
merge(icbc,apple, all=FALSE)
```

```
           X1398.HK.Close AAPL.Close
2021-11-24           4.26      161.9
2021-11-26           4.19      156.8
```

Converting a Time Series to a Dataframe

In some cases, you will want to convert a time series to a dataframe. For example, you can convert the time series of Apple's dividends as follows:

```
aapl.div.df =
  data.frame(date=index(AAPL.div), dividend = coredata(AAPL.div))
```

Key Functions for Time Series

To recapitulate, here are some of the key functions to manipulate time series and the pages where you can find detailed explanations and examples:

What You Need	Function Name	Page
Vector of time stamps	`time()`	60
Start and end of time series	`start()`, `end()`	60
Extracting a time frame from a series	bracket notation. E.g., ["2019/2022"]	58
Convert to weekly, monthly, etc.	`to.period()`, `to.weekly()`, etc.	59
Combine two time series	`merge()`	61, 172
Splitting series by time period	`split()`	59

4.9 Data Wrangling

Many providers of financial analytics offer to upload or download portfolios of stocks and bonds from or to spreadsheets or CSV files. However, they all have their preferred format, so you might end up having to perform the low-value-added but necessary task of converting from one format to the other – a tedious but necessary process called **wrangling**, to which many books are dedicated.

For example, a provider like **Factset** would, by default, export holdings in the following format:

Stock Name	2021-02-08	2021-02-09	2021-02-10
Apple	10	12	9
Amazon	30	31	32
Google	100	101	102

In contrast, the **PORT** utility of the **Bloomberg Terminal** requires holding files to follow the following format:

Stock Name	Date	Quantity
Apple	2021-02-08	10
Amazon	2021-02-08	30
Google	2021-02-08	100
Apple	2021-02-09	12

Exercise 4.10.2 shows how to do this wrangling using basic R functions seen earlier, but the simplest way to do the format conversion is to use the **tidyr** package:

```
library(tidyr)
```

The next step is to read the data from the file provided with this textbook. Note that we will be using the **openxls** library, so you first need to install the package if you haven't already.

```
library(openxlsx)
```

```
stock.data = read.xlsx("stocks.xlsx")
```

```
head(stock.data)
```

```
   Stock 44235 44236 44237 44238
1  Apple    10    12     9    13
2 Amazon    30    31    32    33
3 Google   100   101   102   103
```

Using the **head()** command, we see that column names are numbers instead of dates. That's because, per Microsoft's website, "Excel stores dates as sequential serial numbers so that they can be used in calculations. By default, January 1, 1900 is serial number 1, and January 1, 2008 is serial number 39448 because it is 39,447 days after January 1, 1900." We will fix this in a moment.

Let's focus for now on what columns we want to reorganize. Those are all the date columns indeed, i.e., all the columns *but* the one called **Stock**. Said differently, we want the new table to contain one new row for each of the possible date columns of the old table, with each column name becoming a new entry in a newly created column that we will call "Date." The values stored across columns in the old table will be stored down one column in the new table, and we will call this new column "Quantity." Note that the content of the **Stock** column is repeated as needed on each row. We can do all this in one shot using the **pivot_longer()** function of **tidyr**:

```
wrangled.data =
   stock.data %>%
     pivot_longer(!Stock,
                  names_to = "Date",
                  values_to="Quantity")
head(wrangled.data, 3)
```

```
# A tibble: 3 x 3
  Stock Date  Quantity
  <chr> <chr>    <dbl>
1 Apple 44235       10
2 Apple 44236       12
3 Apple 44237        9
```

Note that the first argument of **pivot_longer()** is the names of the columns we want to reshuffle. Out of convenience, these names do not need to be put between quotes because R knows which columns we are referring to since the dataframe, **stock.data**, is provided as context. Moreover, since we want to reshuffle all the columns *except* the one called **Stock**, we wrote the first argument to **pivot_longer()** as "NOT Stock," with the exclamation point meaning NOT.

However, the table still looks unfinished since the dates are in character format. The last step, then, is to convert the dates from strings to integers, and then the integers to dates using **as.Date()**:

```
wrangled.data$Date = as.Date(strtoi(wrangled.data$Date))
head(wrangled.data, 3)
```

```
# A tibble: 3 x 3
  Stock Date       Quantity
  <chr> <date>        <dbl>
1 Apple 2091-02-10       10
2 Apple 2091-02-11       12
3 Apple 2091-02-12        9
```

Mission accomplished!

The **tidyr** package offers multiple other functions to help wrangle data, in particular **pivot_wider()** that performs the operation opposite to **pivot_longer()**. See Exercise 4.10.3.

4.10 Exercises

4.10.1 Formatting

We saw how to create a text with variable content:

```
Rsq = 0.88
paste("R-squared =", Rsq)
```

```
[1] "R-squared = 0.88"
```

Adjust the command above to display a percentage, i.e. "R-squared = 88 %", instead of the decimal form.

4.10.2 Format Conversion

Coming back to the FactSet-to-Bloomberg example, write a piece of R code that reads a csv file in the FactSet format, convert the data format to the Bloomberg PORT one, and write a csv file in the latter format – all *without* using **tidyr**.

■ SOLUTION

We will illustrate many of the functions seen before by converting the FactSet portfolio format to the Bloomberg one.

The table is "horizontal," with dates across instead of the more usual arrangement of having dates and data down the table. We first extract the dates, which are stored as the names of the columns, using **colnames()**:

```
columnNames = colnames(stock.data)
(dates = columnNames[-1])
```

```
[1] "44235" "44236" "44237" "44238"
```

We see that the dates come up as strings.... To convert back to integer (which is Excel's way), we use the **strtoi()** (string to integer) function:

```
(datesAsIntegers = strtoi(dates))
```

```
[1] 44235 44236 44237 44238
```

Note that dates as integers are sufficient for Excel. But R has a **Date** datatype, so we can convert these integers to the Date datatype:

```
(datesOk = as.Date(datesAsIntegers, origin="1899-12-30"))
```

```
[1] "2021-02-08" "2021-02-09" "2021-02-10" "2021-02-11"
```

Now the task is to convert all these data into a table where each row contains exactly one stock name, one date and the quantity for that stock on that day. We follow that order in what follows. We first extract the stock names:

```
(stockNames = stock.data[,1])
```

```
[1] "Apple"  "Amazon" "Google"
```

Since the PORT format repeats the names, in the original order, as many times as they are dates, we repeat the list of stock names using the `rep()` function:

```
(namesColumn = rep(stockNames, each = length(datesAsIntegers)))
```

```
 [1] "Apple"  "Apple"  "Apple"  "Apple"  "Amazon"
 [6] "Amazon" "Amazon" "Amazon" "Google" "Google"
[11] "Google" "Google"
```

```
(datesColumn = rep(datesAsIntegers, length(stockNames)))
```

```
 [1] 44235 44236 44237 44238 44235 44236 44237 44238
 [9] 44235 44236 44237 44238
```

We now read the quantities. Note that the FactSet format is "horizontal," so we remove the first column (which contains the firms' names) using the `-1` notation, extract the quantities, then transpose them using the `t()` function seen on page 39.

```
(quantities = stock.data[,-1])
```

```
  44235 44236 44237 44238
1    10    12     9    13
2    30    31    32    33
3   100   101   102   103
```

```
(quantColumn = as.vector(t(quantities)))
```

```
 [1]  10  12   9  13  30  31  32  33 100 101 102 103
```

We can now bind the dates column, then the quantity column:

```
finalDataframe = cbind(namesColumn, datesColumn)
(finalDataframe = cbind(finalDataframe, quantColumn))
```

```
      namesColumn datesColumn quantColumn
 [1,] "Apple"     "44235"     "10"
 [2,] "Apple"     "44236"     "12"
 [3,] "Apple"     "44237"     "9"
 [4,] "Apple"     "44238"     "13"
 [5,] "Amazon"    "44235"     "30"
 [6,] "Amazon"    "44236"     "31"
 [7,] "Amazon"    "44237"     "32"
 [8,] "Amazon"    "44238"     "33"
 [9,] "Google"    "44235"     "100"
```

```
[10,] "Google"     "44236"     "101"
[11,] "Google"     "44237"     "102"
[12,] "Google"     "44238"     "103"
```

We can now produce the final file:

```
write.csv(finalDataframe, "stockOutput.csv")
```

■ END OF SOLUTION

4.10.3 Wrangling Using pivot_longer

Coming back to the FactSet-to-Bloomberg example, use **pivot_wider()** of the **tidyr** package to convert a dataframe in the Bloomberg format back into the FactSet format.

4.10.4 Computing Daily Returns From Daily Prices

If the price of a stock at the close of 4 consecutive (daily) trading sessions is $53.12, $56.73, $55.23 and $59.06, calculate the 3 daily returns. The return on an investment is the profit or loss you made (end value ev less starting value sv) divided by the starting value sv. So the return in the period is:

$$\text{return} = \frac{ev - sv}{sv} = \frac{ev}{sv} - 1$$

■ SOLUTION

Define a vector with the four indicated values, then divide the 2nd entry by the 1st entry, the 3rd by the 2nd and the 4th by the 3rd... But do that in one shot: divide all the entries **except the first one** by all the entries **except the last one**, and finally subtract 1:

```
dailyStockPrice = c(53.12, 56.73, 55.23, 59.06)
dailyStockPrice[-1] /
  dailyStockPrice[-length(dailyStockPrice)] -1
```

```
[1]  0.06796 -0.02644  0.06935
```

Note, however, that accurate calculation of a stock's return requires taking into account dividends, corporate actions, etc. This is done through **adjusted prices** that will be discussed on page 161. Calculation of returns can be done using pre-canned functions, such as **periodReturn()** from the **quantmod** package that we will investigate on page 163, or the **Return.calculate()** function of the **PerformanceAnalytics**, discussed on page 163.

■ END OF SOLUTION

4.10.5 Histogram of Apple's Daily Returns

After downloading from Nasdaq the csv file of daily prices, at close, of Apple's stock, calculate daily returns and plot its histogram. When trying this out, make sure, when downloading, that the values are saved as numbers so that the dollar sign for stock prices does not pollute your data.

■ Solution

```
aapl = read.csv("AAPL_nasdaq.csv")
```

```
aaplPrices = aapl$Close.Last
dailyReturns = aaplPrices[-1] /
  aaplPrices[-length(aaplPrices)] - 1
hist(dailyReturns, breaks=30)
```

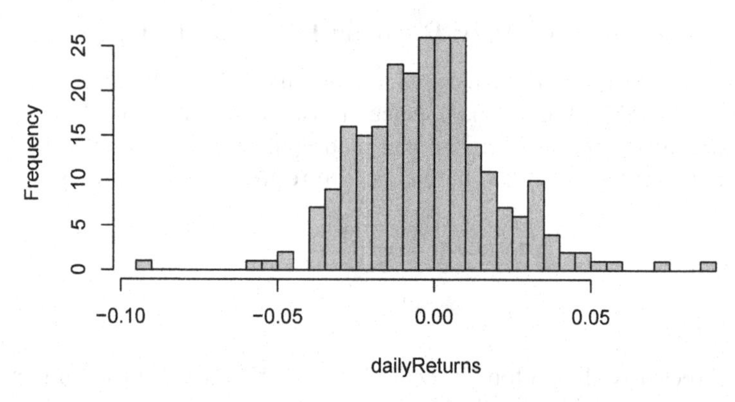

■ End of solution

5

Functions

We have already used the **sin()**, **log()**, **sqrt()** and **mean()** functions without discussing them much. That's because their use is very natural: essentially, using them is done using a notation that is close to that of math. But their uses go way beyond math, and their versatility requires quite a bit of care. Moreover, a user can define their own functions to, among other uses, avoid repeating code for repetitive tasks. Therefore a dedicated chapter is not too much.

Before we proceed, keep two things in mind. First, we use a bold courier font for function names and add a pair of parentheses to highlight the fact that these are functions, not other elements of R code. And second, when running the examples of this textbook, don't forget to systematically load the **tidyverse** package.

5.1 Calling Existing Functions

The notation is **function.name** `(` arguments separated by commas `)`. The values passed to the function are called **arguments** or **parameters**.

In contrast to math however, functions can change things: they can change the value of variables and data frames. Functions can display results and they can input data, from the user or from a data set. In computer science, we say they have **side-effects**. Note that typically R functions do not change any values of their arguments, so changing a variable's value requires making an assignment to it. For example, to remove `NA`s from a data frame, you need to put the result of **na.omit()** back into that data frame:

```
my_dataframe = na.omit(my_dataframe)
```

An example of a function with side-effect is **getSymbols()**, which is discussed on page 17, and which by default creates a new variable and assigns content to it.

Another difference is that the number of parameters passed to the function changes, with some arguments being optional. As in math, the order in which

DOI: 10.1201/9781003320555-5

values are passed to a function tells us which value is for which argument. But when there is an ambiguity, R allows for **named arguments**.

Actually, we will argue that using an argument's name makes your code much easier to read (even for yourself, a few weeks after writing it), even in cases when naming is not required.

Let's take the example of the **format()** function, which helps present data in a way that is pleasant to the human eye. Typing:

```
?format
```

we get something that contains the following:

```
format(x, trim = FALSE, digits = NULL, nsmall = 0L,
       justify = c("left", "right", "centre", "none"),
       width = NULL, na.encode = TRUE, scientific = NA,
       big.mark   = "",   big.interval = 3L,
```

We learn a few important things

- **format()** accepts arguments such as **digits**, **big.mark** and **scientific**; even without understanding what they mean yet (that's explained farther down on the help page, and part of it in this textbook as well), we get an intuitive sense of what is available.
- This plethora of options makes naming arguments critical to legibility. A command such as:

```
format(x, trim=TRUE, digits=4)
```

is much more clear than the equivalent version below, even though it meant typing a few more characters that were not absolutely necessary.

```
format(x, TRUE, 4)
```

Making sure *how* the arguments must be passed is also critical. To see why, let's go back to the **mean()** function.

What do you think is the value of **mean(9,10,11)**? No, it's not 10. It's 9. The **mean()** function computes the mean of the *first* argument. The second and third arguments are reserved for other purposes.

To pass multiple items into a single argument, we put them in a vector with the **c()** operator. Then, **mean(c(9,10,11))** will return 10, as you might expect.

To make things more confusing, some functions, such as **mean()**, take one argument but other functions, such as **max()** and **min()**, take multiple arguments and apply themselves across all arguments. Just look out for that trap.

When in doubt, it is always a good idea to refer to the help page on a function. You do that by typing a question mark followed by the function's name:

```
?mean
```

That help page tells us we can remove **NA**s thanks to the 3rd function argument. Great! So one might think that we can calculate the mean of a vector containing **NA**s as follows:

```
vector = c(1, 3, 5, NA)
```

```
mean(vector, TRUE)
```

Try it – We're getting an error...
As typed above, **FALSE** is the second argument

There are two ways to fix it: The first solution is to give a value to the second argument, so **FALSE** gets in 3rd position:

```
mean(vector, 0, TRUE)
```

```
[1] 3
```

The other solution is to use R's **named argument** feature:

```
mean(vector, na.rm=TRUE)
```

```
[1] 3
```

Solution #2 is not necessarily shorter to write but much easier to read and understand! Not only do we clearly see what is true (it is the **na.rm=** argument), but Solution #1 had a zero in the middle that had no meaning at all and was only used as a spacer.

5.2 Creating New Functions

Creating Our First Function

Creating a function is easy: It consists in defining a new "object," stating that this object is a function, listing the different arguments that are needed, calculating whatever stuff you want to calculate, and returning the fruit of all this hard work. The value returned by a function is the last expression in the function's body. We can also make that return value explicit using the **return()** function.

```
mysquare = function(x) {
    my_calc = x*x
    return(my_calc)
  }
```

The above can be simplified by the way. For example, this function definition would work as well:

```
mysquare = function(x) {
    return(x*x)
  }
```

Finally, once the function is defined, using it (also known as calling it) is as simple as for any other function:

```
mysquare(5)
```

```
[1] 25
```

But functions are not limited to mathematical calculations and can serve in many ways, even prosaic ones such as enforcing a specific numerical format on your results – let's say using a comma to separate thousands and no more than 2 decimals:

```
numformat = function(x) {
    format(x, digits = 2, big.mark = ",")
  }
numformat(3452345)
```

```
[1] "3,452,345"
```

```
numformat(.12358124331)
```

```
[1] "0.12"
```

5.3 Function Composition (a.k.a Piping)

Assume you need to calculate $log(\sqrt{v})$ for different components of a vector v. Here is a first way to do it:

```
v = c(1, 3, 5, 9)
sv = sqrt(v)
log(sv)
```

```
[1] 0.0000 0.5493 0.8047 1.0986
```

Alternatively, we could write:

```
log(sqrt(v))
```

```
[1] 0.0000 0.5493 0.8047 1.0986
```

The first method makes it easier to see what happens at each step. For example, it allows you to print each intermediate variable. But it has a few shortcomings:

- You need to create intermediate variables that can take up memory space, and you won't need them later.
- Their names tend to be a bit meaningless.
- It is easy to refer to the wrong intermediate variable; for example, in the first method, we could easily type log(v) instead of log(sv), and the code would run fine – we would never know there was a mistake until we peruse the code.

The second method avoids some of these problems, but composing the third function on top of log and sqrt would begin to be hard to read.

What we want is function composition, like in math: When we apply function g then f on x, meaning $f(g(x))$, we write $(f \circ g)(x)$. So in math, $f \circ g$ is a function, a new function made from applying g then f. In our example, we would like to be able to write something like $log \circ sqrt$. The **magrittr** package lets us do just that[1]. The syntax however doesn't use $\circ$ – if only because there is no such key on computer keyboards. Instead, R created the **pipe %>%** operator. It is similar to $\circ$ but, unlike in math, functions are applied in the left-to-right order in which they appear:

```
sqrt(v) %>% log
```

```
[1] 0.0000 0.5493 0.8047 1.0986
```

Because typing a variable's name means requesting its content, the above can even be written as:

```
v %>% sqrt %>% log
```

```
[1] 0.0000 0.5493 0.8047 1.0986
```

We can now easily compose a third function – even one we created ourselves:

```
sqrt(v) %>% log %>% numformat
```

```
[1] "0.00" "0.55" "0.80" "1.10"
```

A note on code formatting: It is customary to place each piped command on a different line, with each line (but the last) ending with %>%. So a better style for the previous command is:

```
v %>%
  sqrt %>%
```

[1] In theory, you would need to explicitly load **magrittr** using the **library(magrittr)** command. But we highly recommend you systematically load **tidyverse**, which will make sure that the functions of **magrittr** are available, among others.

```
log %>%
numformat
```

```
[1] "0.00" "0.55" "0.80" "1.10"
```

5.4 Optimization

In finance and economics, functions often need to be optimized – meaning that we want to find for which input value the function reaches its maximum or minimum. Not surprisingly, this function is called **optimize()** and, here again, the British spelling is also accepted: **optimise()** does exactly the same.

Here's how you find the minimum of an (almost) arbitrary function:

```
f = function(x) { (x-0.3)^2 }
```

Let's quickly plot **f()** to see what it looks like:

```
x = seq(-4, 4, length=100)
y = f(x)
plot(x,y, type="l")
```

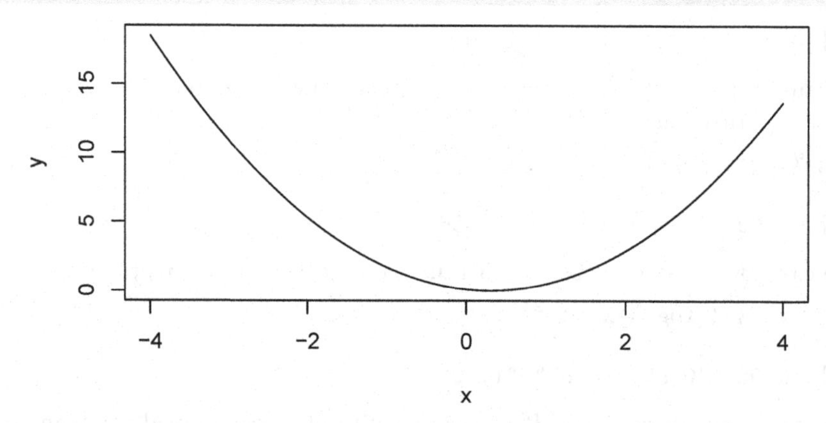

(Note the **type=** optional parameter of the **plot()** function introduced on page 19.)

Clearly (even without graphing it), the function's minimum is reached at $x = 0.3$, and the value of f at that point is 0. Let's see how **optimize()** does it: It takes the function to optimize as its first argument, and the interval on which to optimize as its second.

```
optimize(f, c(-4, 4))
```

```
$minimum
```

```
[1] 0.3

$objective
[1] 2.773e-32
```

You probably noticed that, by default, **optimize()** finds the minimum. Finding the maximum only requires to set the corresponding parameter to TRUE:

```
optimize(f, c(-4, 4), maximum=TRUE)
```

```
$maximum
[1] -4

$objective
[1] 18.49
```

Note that **optimize()** does a better job than just using **which.min()**, which we saw on page 51. However, the latter can still give a good approximation, thanks to the **x** and **y** vectors we just calculated. The minimum value is reached for:

```
x[which.min(y)]
```

```
[1] 0.2828
```

and the value of $f(x)$ is:

```
y[which.min(y)]
```

```
[1] 0.0002949
```

Among other financial applications, **optimize()** will allow us on page 300 to calculate the implied volatility of an option.

5.5 Manipulating Character Strings

Many tasks in the world of finance involve processing textual information. That may be less glamorous than detecting patterns in stock prices and crafting trading strategies, or at least more prosaic, but it is definitely a type of task that data scientists in finance are also expected to master.

To illustrate what R let us do, consider again our **ACWI** data set that we first read on page 8. We know that the first few company names are:

```
head(acwi$Name)
```

```
[1] "Apple Inc."                "Microsoft Corporation"
[3] "Amazon.com, Inc."          "Facebook, Inc. Class A"
```

```
[5] "Alphabet Inc. Class C"  "Alphabet Inc. Class A"
```

Imagine you are asked to separate stocks meeting a specific pattern. It can be, for instance, to find all stocks that include "Inc." in their formal name, or all "Class A" constituents of the index. One way to perform this operation is to use the **str_extract()** function of **tidyverse**:

```
head(str_extract(acwi$Name, "Inc"))
```

```
[1] "Inc" NA    "Inc" "Inc" "Inc" "Inc"
```

As you can see, that function provides a confirmation that the string was found (and we will see later how that can be useful), but often you'll want a true/false vector – typically, to select only those rows in the data frame where `Name` matches your criterion. You can get a boolean value using **str_detect()**:

```
head(str_detect(acwi$Name, "Inc"))
```

```
[1]  TRUE FALSE  TRUE  TRUE  TRUE  TRUE
```

Now, you might be interested in finding **either** Inc or Corporation in the index constituents. To look for either of two substrings in a vector (or data frame column) of strings can be done using the **| alternation** operator:

```
head(str_detect(acwi$Name, "Inc|Corporation"))
```

```
[1] TRUE TRUE TRUE TRUE TRUE TRUE
```

But what about finding, for example, Class A or Class C shares? The space between "Class" and the letter actually doesn't bother these functions:

```
head(str_extract(acwi$Name, "Class A|Class C"))
```

```
[1] NA         NA         NA         "Class A" "Class C"
[6] "Class A"
```

The alternation operator constructed our first **regular expression**. Regular expressions are ways to express patterns, or general structures, in text. Examples of regular expressions include: Lines whose first character is an A; Words that consist of a vowel followed by a number; or lines that contain either one word or another, as we just saw.

To illustrate this, assume you want to find records in `acwi` for companies whose name starts with an A. What you want, in your regular expression, is a marker for the concept of "beginning of a line." That marker is the caret ^ operator:

```
head(str_extract(acwi$Name, "^A"))
```

```
[1] "A" NA  "A" NA  "A" "A"
```

Omitting the caret would return all the names that contain the letter A anywhere in the text – in which case, "Facebook, Inc. Class A" would be a match since the full name contains the letter A.

Notice that regular expressions are case-sensitive, hence:

```
stocks = c("Apple", "apple")
str_detect(stocks, "Apple")
```

```
[1]  TRUE FALSE
```

To make the search case-insensitive, the easiest way is to make explicit, using the function **regex()**, that the text **"Apple"** is a regular expression and to use that function's optional **ignore_case** argument:

```
str_detect(stocks, regex("Apple", ignore_case = TRUE))
```

```
[1] TRUE TRUE
```

Continuing our tour of regular expressions, you may have noticed earlier that we searched for **"Inc"**, not for **"Inc."** even though the formal names of the top companies in our data set include Inc. with a period. That's because the period marker is a special character in regular expressions: The . marker stands for "any character except new line." Likewise, the \\d marker stands for "any digit." To illustrate, assume we are given a list of bonds with the following **Moody's** ratings:

```
Moody.ratings = c("Aa1", "Aaa", "A1", "Baa2", "Aa3")
```

Assume then that we are looking for bonds with ratings Aa1, Aa2 or Aa3 but neither Aaa nor lesser in quality, using the A.\\d regular expression:

```
str_detect(Moody.ratings, "A.\\d")
```

```
[1]  TRUE FALSE FALSE FALSE  TRUE
```

That regular expression should be parsed as the letter A, lower- or upper-case, followed by any character (the period) and followed by one digit.

If the period denotes "any character," you may wonder, then how can we search for patterns that include a period? (This is not an unrealistic question, since for example the MSCI ACWI index includes Berkshire Hathaway Inc., and more specifically its Share Class B whose ticker is BRK.B.)

The solution is the double-backslash \\ **escape** character. We saw it before the letter d to denote not "d" but a digit. Likewise, \\. denotes the period itself. For example, consider this vector of codes of some sort:

```
searched = c("A.b", "Ab", "Ab.", "X.3", "2.b", ".ab", "abd.124")
```

If we wanted to find all such codes that include a character followed by a period, itself followed by any character, we would write:

```
str_detect(searched, ".\\..")
```

```
[1]  TRUE FALSE FALSE  TRUE  TRUE FALSE  TRUE
```

If we wanted to search for codes including any character followed by a period, then followed by a digit, the command would instead be:

```
str_detect(searched, ".\\.\\d")
```

```
[1] FALSE FALSE FALSE  TRUE FALSE FALSE  TRUE
```

5.6 Key Statistics Functions

As we repeat throughout this textbook, data analysis would be just fancy hand-waiving without statistics. This textbook does not intend to provide an introduction to statistics, and the reader is strongly encouraged to pick up one of the many excellent books on the topic and/or to google up concepts as they appear, if need be.

The first key concept we need to review quickly is the **normal distribution**, also known as **Gaussian distribution**. That distribution is by far the most widely used, for several reasons. First, it is **symmetrical**, meaning that values have the same probabilities to be more or less than their average, or mean. Second, it needs only two parameters to be described: the mean we just alluded to, and a measure of the dispersion of values; the most commonly measure of that dispersion is the **standard deviation**. Third, the normal distribution tends to occur frequently in nature, in parts because of a mind-boggling result called the **Central Limit Theorem**: if you take lots of values coming from *any* distribution, their sum will have a normal distribution. And fourth and maybe foremost: the normal distribution is widely used in finance because it is the only one most practitioners are familiar with. This is unfortunately a self-fulling prophecy.

The second key concept we need to review is the **probability density function**, or **PDF**. The PDF of a distribution tells which values a variable, picked at random from that distribution, most probably takes. An example of a PDF is graphed in Figure 5.1. The PDF in the figure is that of the **standard normal distribution** of mean 0 and standard deviation 1. Its x-axis is constructed as a sequence of 100 evenly-spaced values ranging from -4 to 4, thanks to the `seq()` function seen earlier. The PDF is provided by the `dnorm()` function (we'll come back soon on where this name comes from) applied to the `x`-values with a `mean=` of 0 and a standard deviation set to `sd=` 1:

```
x = seq(-4, 4, length=100)
```

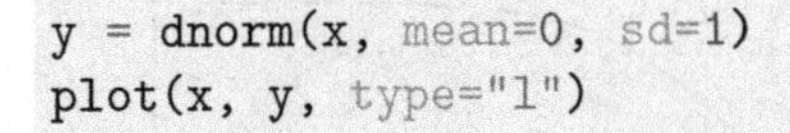

```
y = dnorm(x, mean=0, sd=1)
plot(x, y, type="l")
```

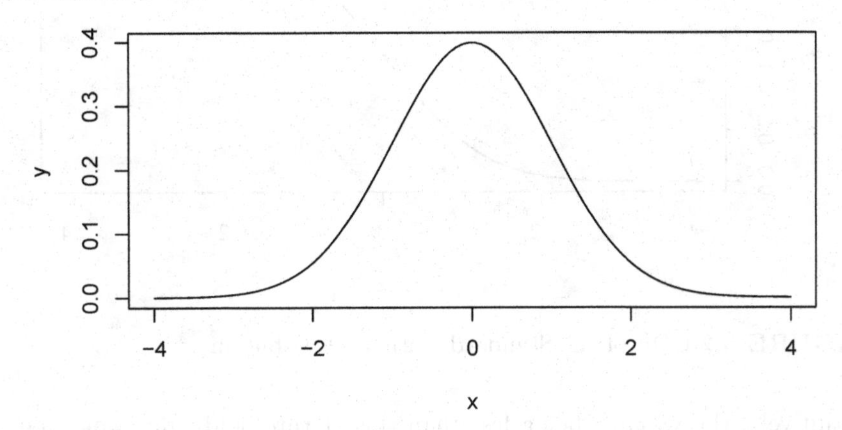

FIGURE 5.1 Probability density of the standard normal distribution plotted with plot()

But that doesn't immediately tell us the probability that a random value falls within a certain interval. For that, we need the **cumulative distribution function (CDF)**, which is called **pnorm()** in R. So for example, **pnorm(-1, mean=μ, sd=σ)** is $P_{N(\mu,\sigma)}(X < -1)$, the probability that a variable following a normal distribution of mean μ and standard deviation σ is less than the "threshold" value of -1. Because the CDF is cumulative, the probabilities (the result of the CDF) increase as the "threshold" value increases, which can be seen in Figure 5.2 produced by the code below:

```
x = seq(-4, 4, length=100)
y = pnorm(x, 0, 1)
plot(x, y, type="l")
```

A classic result from statistics is that, in a normal distribution, 68.27% of data points fall inside of 1 standard deviation from the mean, and so about 31.6% fall outside – approximately 15.8% on either side. Written differently, for a normal distribution of mean μ and standard deviation σ, the *cumulative* probability of all the points less than $\mu - \sigma$ (i.e., less than 1 standard deviation from the mean) is 15.8%: $P_{N(\mu,\sigma)}(X < \mu - \sigma)$ is about equal to 15.8%, whatever the values of the mean and standard deviation. We can indeed verify that $P_{N(0,1)}(X < -1)$ is about 15.8%:

```
pnorm(-1, mean=0, sd=1)
```

```
[1] 0.1587
```

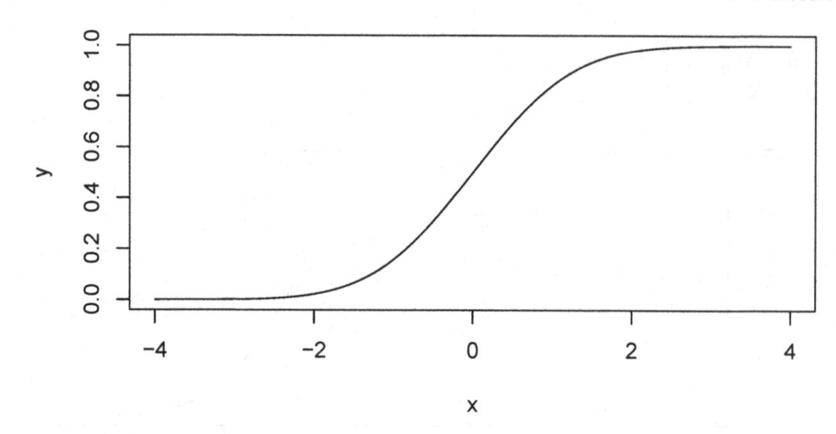

FIGURE 5.2 CDF of the Standard Normal Distribution

Intuitively, the variable being less than this -1 threshold value would be a much rarer event if the variable mean was 2. We can verify the value of $P_{N(2,1)}(X < -1)$ as follows:

```
pnorm(-1, mean=2, sd=1)
```

```
[1] 0.00135
```

If would, however, be a more frequent event if the dispersion of the variable was larger. For example, with a standard deviation of 12.3, $P_{N(0,12.3)}(X < -1)$ becomes:

```
pnorm(-1, mean=0, sd=12.3)
```

```
[1] 0.4676
```

Moreover, we can verify that the probability $P_{N(\mu,\sigma)}(\mu - \sigma < X < \mu + \sigma)$ is 68.27% by noting that this expression can be rewritten as $P_{N(\mu,\sigma)}(X < \mu + \sigma) - P_{N(\mu,\sigma)}(X < \mu - \sigma)$. Then we can calculate its value as follows, for example on $N(5,3)$:

```
pnorm(8, mean=5, sd=3) - pnorm(2, mean=5, sd=3)
```

```
[1] 0.6827
```

By default, **pnorm()** calculates the probability of the **lower tail**, which means the probability that a value picked at random is lower than the threshold value. The optional parameter **lower.tail** allows us to change that. It is TRUE by default but can be set to FALSE to calculate the **upper tail**:

```
pnorm(-1, mean=0, sd=1, lower.tail=FALSE)
```

```
[1] 0.8413
```

We can verify that the $P(X < -1) + P(X > -1) = 1$, or equivalently that the entire area under the distribution curve is 1:

```
pnorm(-1, mean=0, sd=1) +
  pnorm(-1, mean=0, sd=1, lower.tail=FALSE)
```

```
[1] 1
```

Finally, the inverse of the CDF, also known as the **quantile** function, is named **qnorm()** in R in the case of a normal distribution. We saw that the area under the curve of the standardized normal distribution to the left of -1 standard deviations is 15.8% and can be calculated as **pnorm(-1, mean=0, sd=1)** = 0.1587. Then, by definition, the mapping back from 0.158 to -1 is provided by **qnorm()**:

```
qnorm(0.1586553, 0, 1)
```

```
[1] -1
```

Naming Convention

Now is a good time to explain the function names seen so far in this section. R adopted the following convention: The first letter gives the function type and is one of the below; the rest is the shortened name of the distribution:

- d for "**density**," the density function (PDF)
- p for "**probability**," the **cumulative distribution function** (**CDF**)
- q for "**quantile**," the **inverse CDF**
- r for "**random**," to generate a random variable having the specified distribution

For example, the function **pnorm()** we saw earlier is named as "p + norm," which tells us it is the CDF of the normal distribution. This naming convention applies to all distributions. For example, since **lnorm** is the short name for the **log-normal distribution**, **rlnorm()** generates numbers randomly picked from a log-normal distribution. As a refresher, a variable has a log-normal distribution if the logarithm of that variable is normally distributed. This implies in particular that the variable cannot have negative value: the log of zero is negative infinity, but the log of negative numbers is undefined. Likewise, to pick random numbers from a **t distribution**, the function we need is **rt()**.

The arguments of **rlnorm()** are the number **n** of data points we want to be generated (this is the first argument of all the "r" functions seen above), the mean and the standard deviation **sd**. (The mean and standard deviation are optional, and their default values are 0 and 1, respectively.) The three log-normal distributions defined below have standard deviations 1, 0.2 and 0.1, respectively:

```
x1 = rlnorm(n=400, sd=1)
x2 = rlnorm(n=400, sd=1/5)
x3 = rlnorm(n=400, sd=1/10)
```

Similarly, **dlnorm()** is the PDF of the log-normal distribution, and it can be visualized as follows (with a standard deviation of 1), resulting in Figure 5.3.

```
x = seq(0, 4, length=100)
y = dlnorm(x, sd=1)
plot(x, y, type="l")
```

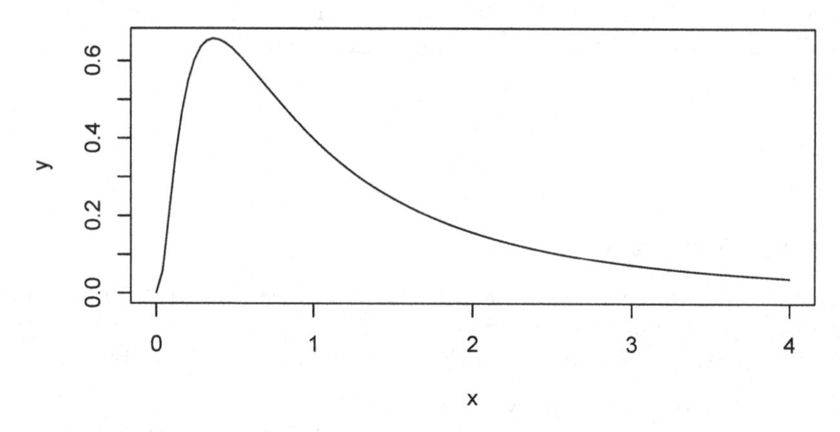

FIGURE 5.3 PDF of a log-normal distribution

Let's now plot histograms for each of these distributions so we can compare their shapes. To place these three plots side-by-side, we use the **par()** function whose **mfrow=** argument takes a vector specifying the number of rows (a single one in our case) and the number of plots on each row (3).

```
par(mfrow=c(1,3))
hist(x1, breaks=20)
hist(x2, breaks=20)
hist(x3, breaks=20)
```

Figure 5.4 shows clearly that the left plot (the distribution of **x1**) has most of its values on the left side of the value range but a long right tail. In contrast, the rightmost histogram (of **x3**) has left and right sides that are relatively similar or symmetrical. We can also verify on the figure that no value on the x-axis is negative, per the key property of log-normal distributions discussed earlier.

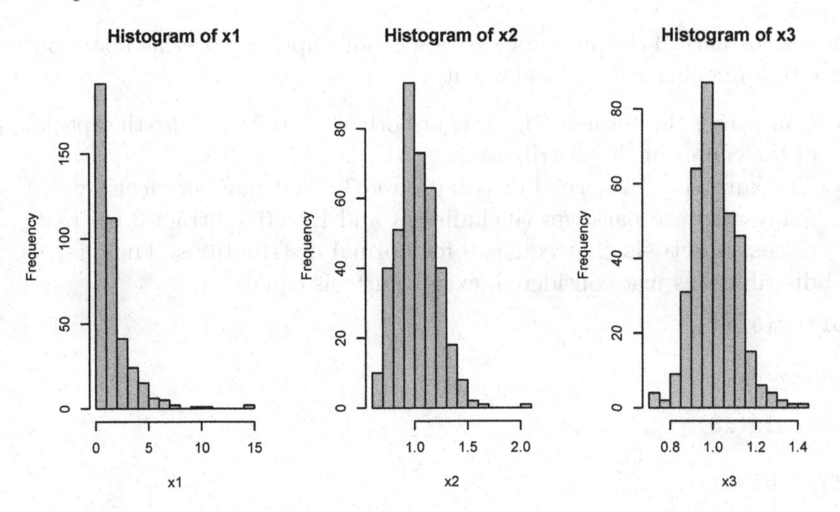

FIGURE 5.4 Log-normal distributions with three different standard deviations

Skew and Kurtosis

What we just described is called **skew** or **skewness**. The distribution of **x1** has a right or positive skew, while the **x3** distribution is almost symmetrical and thus has a skew close to 0:

```
library(moments)
skewness(x1)
```

```
[1] 3.336
```

```
skewness(x3)
```

```
[1] 0.4102
```

(The **skewness()** function is provided by the **moments** package.)

A variable sampled from a normal distribution has of course a skewness close to 0:

```
x.normal = rnorm(150)
skewness(x.normal)
```

```
[1] -0.1023
```

Another trait of distributions is the thickness of their tails – i.e., the relative frequency of extreme events, where "extreme" is defined as values that are very low or very high relative to most of the rest of the data. And that trait is called **kurtosis**.

The mathematical definition of kurtosis is not important to this book, but we have to remember two facts about it:

- It measures the "fatness" of either or both of the tails but also the "peakness" of the center of the distribution.
- The kurtosis of a normal distribution is 3. That number "feels" weird, so many software packages (including R and Excel) subtract 3 to create an **excess kurtosis** that equals 0 for normal distributions. Thus, for the 3 distributions just considered, excess kurtosis equals:

```
kurtosis(x1)
```

```
[1] 17.24
```

```
kurtosis(x3)
```

```
[1] 1.049
```

```
kurtosis(x.normal)
```

```
[1] 0.4551
```

Before we end on this topic, keep something in mind. Do skew and kurtosis matter in finance? The answer is that they are critical. Remember that if you lose 50%, you have to make 100% to just break even. So positive skews of investment returns are essential, and more so when kurtosis is high. But more on this in another chapter.

5.7 Empirical Distributions

The frequent question is to compare the distribution of a variable (in our case, returns most frequently) to the normal distribution. Let's look at the daily prices of Google's stock, at market close, during a short period:

```
google.data = read.csv("HistoricalQuotesLong.csv")
google.close.prices = google.data[, 2]
```

For reasons explained on page 228, financial quantitative models typically use **log returns**. If the price of a security changed from P_0 to P_1, then the log return is defined as $\log\left(\frac{P_1}{P_0}\right)$, which equals $\log(P_1) - \log(P_0)$. In R, we can thus calculate the log returns of the vector of close prices using the following expression:

```
google.logreturns = diff(log(google.close.prices))
```

The distribution of these returns can easily be plotted using the `hist()` function. The graph appears in Figure 5.5.

```
hist(google.logreturns, breaks=50, freq=FALSE)
```

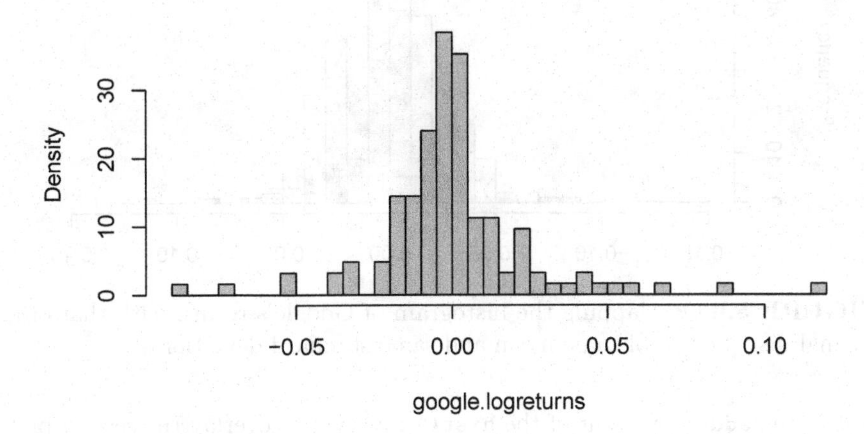

FIGURE 5.5 Histogram of Google's returns

The mean and standard deviation of these returns are:

```
(google.avg  = mean(google.logreturns))
```

```
[1] 0.0004883
```

```
(google.sd   = sd(google.logreturns))
```

```
[1] 0.02515
```

Now the question is: how does that **empirical distribution** compare to the normal distribution with same mean and same standard deviation (more precisely: to 300 values sampled from the normal distribution of same mean and same deviation)? We can plot the two histograms, overlaid, as follows. The result of this code appears in Figure 5.6.

```
normal.samples = rnorm(300, google.avg, google.sd)
breakpoints = seq(-0.15, 0.15, by=0.01)
hist(normal.samples,
     breaks=breakpoints,
     xlab="",
     main="Log-returns of Google (red) vs Normal (gray)")
hist(google.logreturns,
     breaks=breakpoints,
     add=TRUE,
     col="red",
     main="")
```

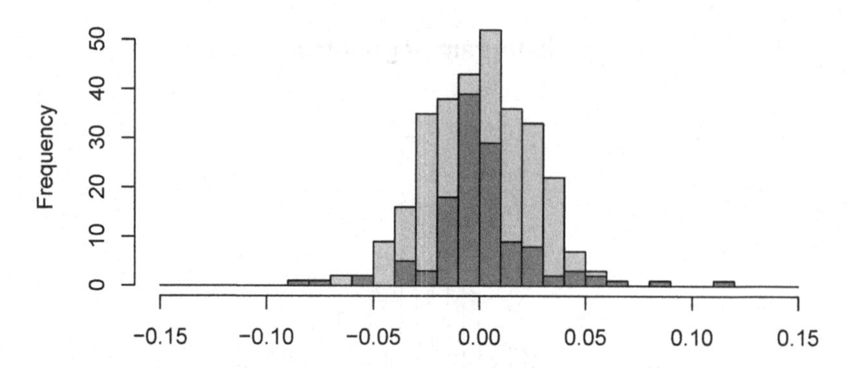

FIGURE 5.6 Overlapping the histogram of Google's return with that of a normal distribution of same mean and same standard deviation

Notice the **add=** argument of the **hist()** function to overlay the second plot on top of the first.

But the usefulness of a histogram depends much on the bin size: make it too small, and you have few data points in each bucket – up to the extreme case of at most one data point per bin. Make it too wide, and you lose most features of the distribution. So a clearer tool to appreciate the difference between two distributions, or between empirical data and the normal distribution, is to use the **empirical cumulative distribution function**, often abbreviated as **empirical CDF** or **ECDF**, using function **ecdf()** of the **stats** package. The graph in Figure 5.7 is constructed in two steps: first plotting the CDF of the normal distribution using data points we already defined, then plotting the ECDF of the sample being considered using the **plot()** function introduced on page 19.

```
plot(ecdf(normal.samples), col="red", main="", xlim=c(-0.1,0.1))
par(new=TRUE)
plot(ecdf(google.logreturns), xlim=c(-0.1,0.1))
```

Note that we had to specify one same range for the x-axes of the two plots, using the **xlim=** optional argument of the **plot()** function, to ensure that the two ECDFs would be comparable.

A simpler way to do all this in one command is to use the **chart.ECDF()** function of the **PerformanceAnalytics** package. The output is shown in Figure 5.8.

```
chart.ECDF(google.logreturns)
```

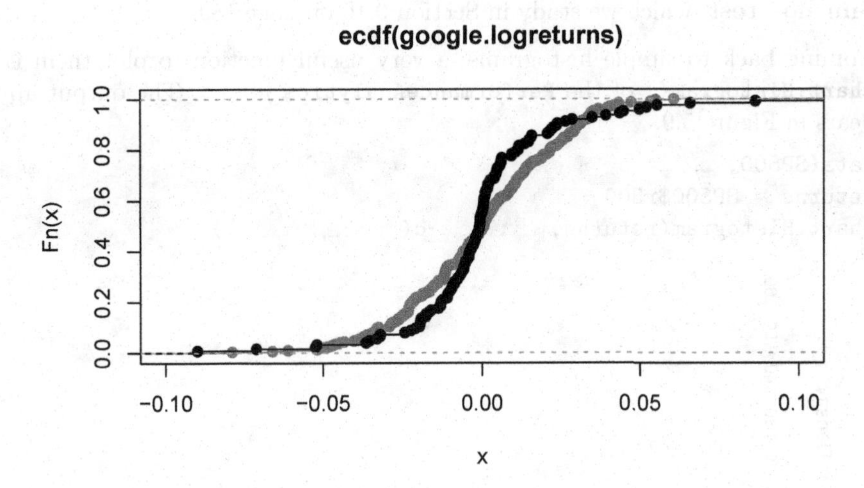

FIGURE 5.7 ECDF of Google's return compared to the CDF of the normal distribution, in red

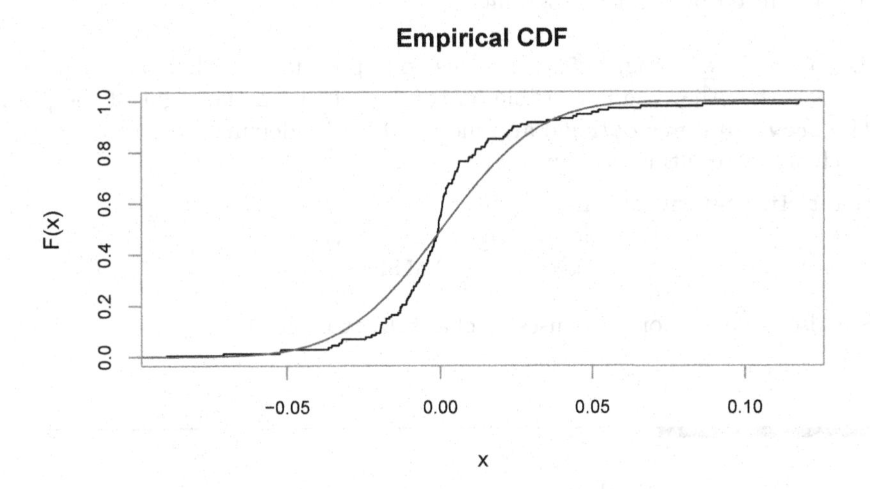

FIGURE 5.8 Using chart.ECDF() to produce the ECDF of Google's returns

The gap between the two CDFs is one way to quantify the discrepancy between the two distributions and is, in fact, the intuition behind the **Kolmogorov-Smirnov test**, which we study in Section 9.10 on page 189.

Coming back to simple histograms, a very useful function to plot them is **chart.Histogram()** of the **PerformanceAnalytics** library. The output appears in Figure 5.9.

```
data(SP500, package="Ecdat")
returns = SP500$r500
chart.Histogram(returns, methods=c("add.normal"))
```

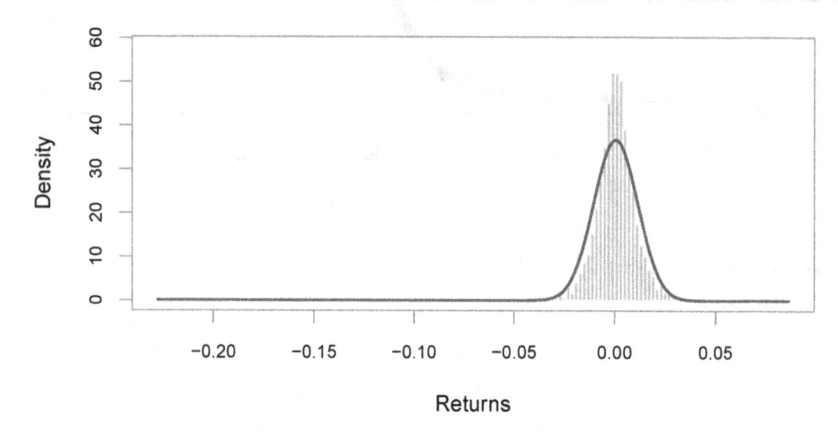

FIGURE 5.9 Histogram of SP500 returns overlaid with the normal distribution produced by chart.Histogram()

As you can see in Figure 5.9 that this code produced, outliers are very few – so few that they are not visible on the far left or the far right of the plot. The **show.outliers** optional argument, which by default is set to true, can be changed, resulting in Figure 5.10.

```
chart.Histogram(returns,
                methods=c("add.normal"),
                show.outliers = FALSE)
```

See also page 314 for other uses of **chart.Histogram()**.

5.8 Chapter-End Summary

R and its extension packages offer a very large menu of functions for text and number manipulation, for statistics and for graphing, among others. Many more will be discussed in the rest of this book, but the main concepts and functions seen in this chapter were PDFs, CDFs and ECDFs, and how to plot

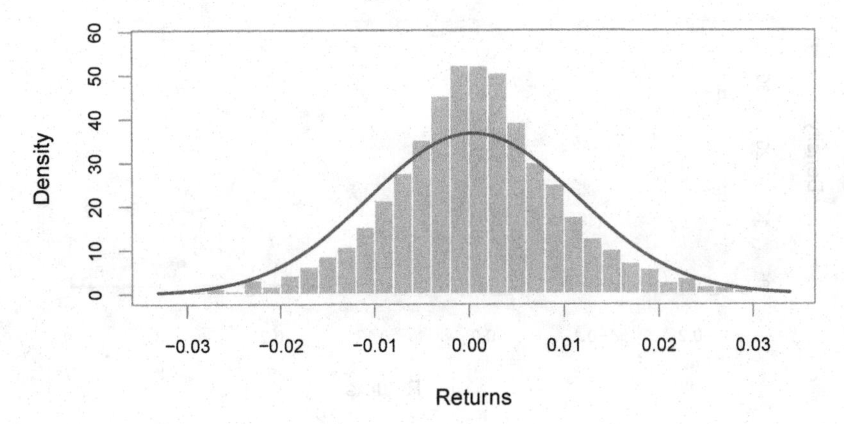

FIGURE 5.10 Same histogram of SP500 returns, still overlaid with the normal distribution, but with outliers removed

them for different distributions. If the existing functions are not enough, we also saw how to create our own.

5.9 Exercises

5.9.1 Histogram of Oil Returns

On page 28, we pulled daily historical data on oil price as follows:

```
orb =
  fromJSON("https://data.nasdaq.com/api/v3/datasets/OPEC/ORB.json")
oil.prices = as.numeric(orb$dataset$data[,2])
```

Compute their daily percentage changes (returns). Plot the histogram of these returns and compare it to a normal distribution.

■ Solution

```
oil.returns =
  oil.prices[-1]/oil.prices[-length(oil.prices)]-1
chart.Histogram(oil.returns,
                methods=c("add.normal"))
```

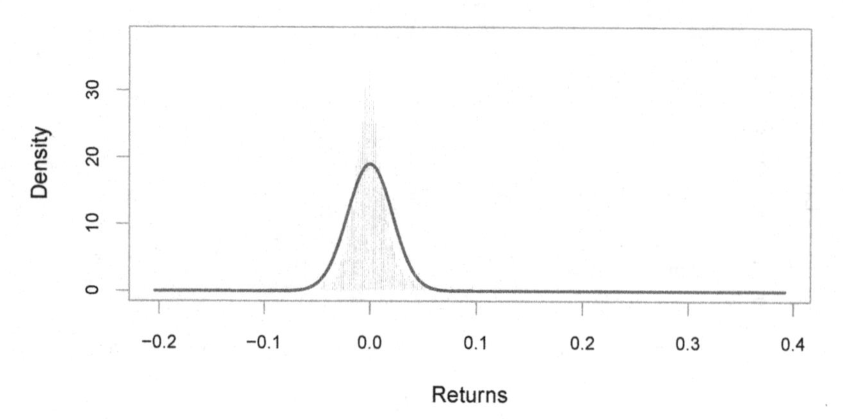

■ END OF SOLUTION

5.9.2 ECDF of Oil Prices

Continuing with oil, plot the ECDF of oil prices and that of their daily returns.

■ SOLUTION

```
chart.ECDF(oil.prices)
```

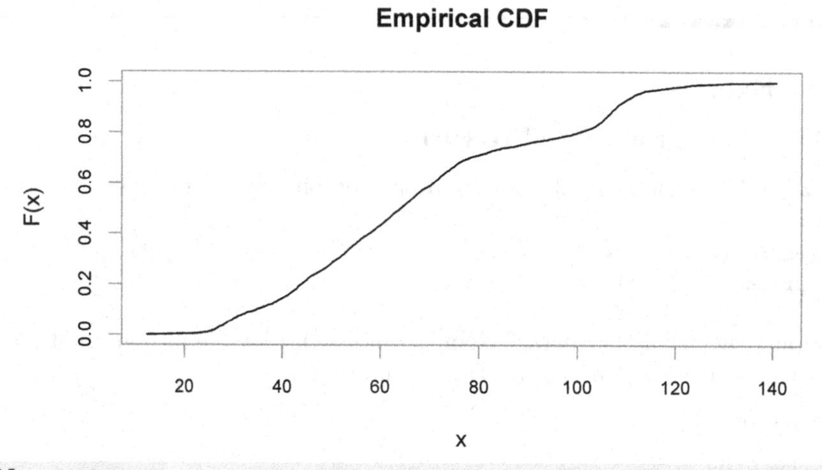


```
oil.returns =
  oil.prices[-1]/oil.prices[-length(oil.prices)]-1
chart.ECDF(oil.returns)
```

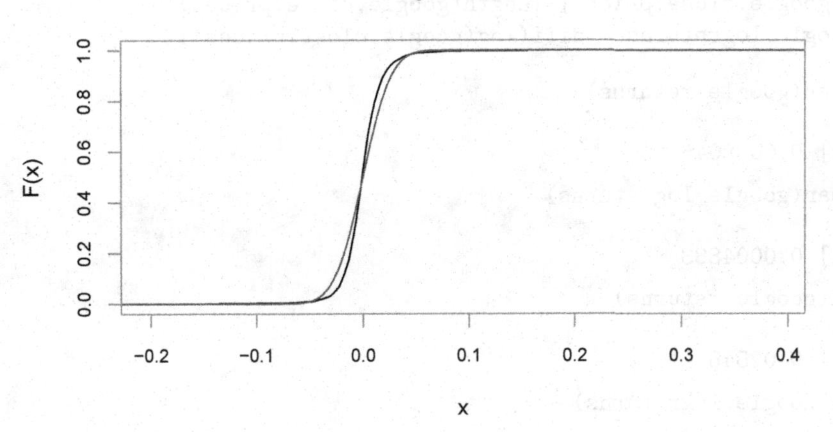

■ END OF SOLUTION

5.9.3 Peak of Oil Prices

Based again on the Nasdaq data on oil prices, find the date at which oil reached its peak price.

■ SOLUTION

```
oil.dates[which.max(oil.prices)]
```

```
[1] "2008-07-03"
```

■ END OF SOLUTION

5.9.4 Qnorm

What is probably wrong with the command below? Why does it execute instead of causing an error?

```
qnorm(0,1586553, 0, 1)
```

```
[1] -Inf
```

5.9.5 Returns vs Log Returns

Compare the mean and standard deviations of the returns and log-returns of Google.

■ SOLUTION

```
google.close.prices = google.data[, 2]
google.returns =
  google.close.prices[-1] /
```

```
  google.close.prices[-length(google.close.prices)] -1
google.logreturns = diff(log(google.close.prices))
```

```
mean(google.returns)
```

```
[1] 0.0008046
```

```
mean(google.logreturns)
```

```
[1] 0.0004883
```

```
sd(google.returns)
```

```
[1] 0.02546
```

```
sd(google.logreturns)
```

```
[1] 0.02515
```

■ END OF SOLUTION

5.9.6 Skew and Kurtosis

Write R code to calculate the skewness and kurtosis of the three log-normal distributions seen earlier with standard deviations 1, 0.2 and 0.1.

5.9.7 Function to Calculate Returns

Create a function that, given a vector of prices, returns the vector of returns. Verify using `all()` that you find the same results as in the previous exercise.

■ SOLUTION

```
calc.returns = function(price.vector) {
  r = price.vector[-1]/price.vector[-length(price.vector)] - 1
  return(r)
}
oil.returns.2 = calc.returns(oil.prices)
all(oil.returns == oil.returns.2)
```

```
[1] TRUE
```

■ END OF SOLUTION

5.9.8 Risk Limit

Assume you're the risk manager at a bank whose trading desk makes on average \$89,000 per day. The P&L follows a normal distribution with a standard deviation of \$135,000. Calculate what loss can be expected 10% of the time.

■ SOLUTION

```
qnorm(0.10, mean=89000, sd=135000)
```

```
[1] -84009
```

A loss of -84009 dollars or worse can be expected on 10% of trading days.

■ END OF SOLUTION

5.9.9 Probability of Reaching a Profit Target

With the same assumptions as the previous exercise, calculate the frequency at which you can expect profits of $200,000 or more.

■ SOLUTION

```
pnorm(200000, mean=89000, sd=135000, lower.tail = FALSE)
```

```
[1] 0.2055
```

■ END OF SOLUTION

5.9.10 Finding Most Significant Outlier

We saw on page 36 how to calculate the z-scores of price-to-book in the ACWI data set. Use **which.max()** to find the most outlying stock on that metric.

■ SOLUTION

```
acwi$Name[which.max(pb.zscores)]
```

```
[1] "O'Reilly Automotive, Inc."
```

■ END OF SOLUTION

6

Data Transformation

As you have probably guessed, dataframes are R's workhorse. This is not surprising since they are, in essence, tables that contain your data. They come with a multitude of functions to find and extract specific data, to filter data sets into smaller ones according to your needs, or on the contrary merge data sets into larger ones. We will also review additional packages that enrich R's capabilities even further – the main one being the **tidyverse** package that you should always load before trying out the examples of this book.

6.1 Selecting Rows: Slicing

We start with the first basic operation, slicing, and illustrate it on the **acwi.short** dataframe produced on page 57.

The **slice()** function allows us to select a row, or set of rows, specified by their row numbers. Using the : notation, we can for example specify the **range** of integer numbers from 2 to 4, and extract the 3 corresponding rows from a dataframe:

```
slice(acwi.short, 2:4)
```

```
  Ticker Weight Market.Cap            GICS.Sector
1   MSFT   2.81    1899924 Information Technology
2   AMZN   2.29    1744112 Consumer Discretionary
3     FB   1.21     926153 Communication Services
        Country   ROE Return
1 United States 40.14  40.72
2 United States 27.44  40.15
3 United States 25.42  58.80
```

When the dataframe is produced from previous calculations, you can pipe it into **slice()**. Remember that the **pipe** operator **%>%** seen on page 73 passes the result from its left to the function on its right. So the command below is equivalent:

DOI: 10.1201/9781003320555-6

```
acwi.short %>% slice(2:4)
```

6.2 Group_by

As we saw earlier, the companies in this data set reside in various countries, operate in different sectors of the economy, etc. Depending on your application, you may need to group stocks according to these categories. The function you need then is **group_by()**:

```
group_by(acwi.short, Country)
```

```
# A tibble: 2,974 x 7
# Groups:   Country [50]
   Ticker Weight Market.Cap GICS.Sector    Country   ROE
   <chr>   <dbl>      <dbl> <chr>          <chr>   <dbl>
 1 AAPL     3.47   2193582. Information ~ United~ 73.7
 2 MSFT     2.81   1899924. Information ~ United~ 40.1
 3 AMZN     2.29   1744112. Consumer Dis~ United~ 27.4
 4 FB       1.21    926153. Communicatio~ United~ 25.4
 5 GOOG     1.11   1617417. Communicatio~ United~ 19
 6 GOOGL    1.1    1579420. Communicatio~ United~ 19
 7 TSLA     0.83    683191. Consumer Dis~ United~  4.78
 8 688910   0.82    556984. Information ~ Taiwan  29.8
 9 JPM      0.73    465598. Financials     United~ 11.3
10 BMMV2K   0.72    763015. Communicatio~ China   27.2
# ... with 2,964 more rows, and 1 more variable:
#   Return <dbl>
```

Note two things: First, the term **Country** was written without double quotes, yet R understood you were not referring to a variable named **Country** but to a column in the dataset. This small details make life more comfortable for heavy users of R, saving them the need to added double quotes – but unfortunately that feature is not available for all functions.

The second point to notice is that, mercifully, R did not display the entire dataset and stopped after 10 rows. All 2,974 entries are still present, however: **group_by()** only copy of the **acwi.short** dataframe with a slightly different structure to maintain the grouping. The only sign of that new structure is the **Groups: Country [50]** line inserted at the top of the output.

6.3 Filter

We can also select observations (rows) by their value thanks to **filter()**. For example:

```
filter(acwi.short,  Market.Cap > 1700000 )
```

```
  Ticker Weight Market.Cap            GICS.Sector
1   AAPL   3.47    2193582 Information Technology
2   MSFT   2.81    1899924 Information Technology
3   AMZN   2.29    1744112 Consumer Discretionary
4 BJTM27   0.03    1886985                 Energy
        Country   ROE Return
1 United States 73.69  78.98
2 United States 40.14  40.72
3 United States 27.44  40.15
4  Saudi Arabia 18.26  11.99
```

Note that the Saudi company Aramco, ticker BJTM27, is one of the largest in the world in terms of market capitalization but that its weight in the stock index is very low. Index constructors like MSCI often take other factors into consideration, such as how much of the stock is floating freely on markets (as opposed to held by the main owner, such as a State).

Multiple conditions can be specified, separated by a comma. This is equivalent to saying that *both* **GICS.Sector == "Information Technology"** *and* **Market.Cap > 1000000** must be satisfied.

```
filter(acwi.short,
       GICS.Sector == "Information Technology",
       Market.Cap > 1000000)
```

```
  Ticker Weight Market.Cap            GICS.Sector
1   AAPL   3.47    2193582 Information Technology
2   MSFT   2.81    1899924 Information Technology
        Country   ROE Return
1 United States 73.69  78.98
2 United States 40.14  40.72
```

Note that we requested to *test for equality* and hence used the **==** operator: We did not *assign* the value 1 to a variable called **GICS.Sector**.

This could also be written using the logical **&** operator:

```
filter(acwi.short,
      GICS.Sector == "Information Technology" & Market.Cap > 1000000)
```

```
  Ticker Weight Market.Cap              GICS.Sector
1   AAPL   3.47    2193582 Information Technology
2   MSFT   2.81    1899924 Information Technology
        Country   ROE Return
1 United States 73.69  78.98
2 United States 40.14  40.72
```

Similarly, we can ask for either one of two (or more!) conditions be satisfied using the | operator:

```
filter(acwi.short, Country == "Kuwait" | Country == "Portugal")
```

```
   Ticker Weight Market.Cap              GICS.Sector
1  688952   0.03      19647               Financials
2  410359   0.02      21956                Utilities
3  650313   0.02      20085               Financials
4  B1FW75   0.01       9593                   Energy
5  B1Y1SQ   0.01      11491         Consumer Staples
6  660008   0.01       8660 Communication Services
7  689030   0.01       6021              Industrials
8  B15DYL   0.00       6912               Financials
9  BOOPQY   0.00       2718              Real Estate
10 688951   0.00       2159               Financials
    Country    ROE Return
1    Kuwait   6.54  19.72
2  Portugal   8.68  35.00
3    Kuwait   7.43  36.98
4  Portugal -14.54   0.45
5  Portugal  15.67   8.30
6    Kuwait  14.16  16.79
7    Kuwait   3.71  74.05
8    Kuwait   5.35  41.20
9    Kuwait   4.53  26.77
10   Kuwait   4.42   3.13
```

which can be made easier to read, and to extend to more conditions on countries, using the **%in%** operator and specifying a set using the **c()** function. The command below would thus produce the exact same output:

```
filter(acwi.short, Country %in% c("Kuwait","Portugal"))
```

6.4 Arrange

arrange() works similarly to **filter()** except that instead of selecting rows, it changes their order. It takes a dataframe and a set of column names (or more complicated expressions) to order by. If you provide more than one column name, each additional column will be used to break ties in the values of preceding columns. For instance, the three lowest ROEs are found as follows, with ties broken by returns:

```
arrange(acwi.short, ROE, Return) %>%
  head(3)
```

```
  Ticker Weight Market.Cap            GICS.Sector
1   EXPE   0.04      25306 Consumer Discretionary
2    TXG   0.02      21458            Health Care
3 BZCNB4   0.03      31682 Consumer Discretionary
        Country    ROE Return
1 United States -98.12 148.28
2 United States -93.64 147.65
3       Germany -92.52  88.48
```

Use **desc()** to re-order in descending order:

```
arrange(acwi.short, desc(ROE)) %>%
  head(3)
```

```
  Ticker Weight Market.Cap            GICS.Sector
1     MO   0.14      88422       Consumer Staples
2   QCOM   0.24     156705 Information Technology
3 B5B1TX   0.03      54238              Materials
        Country   ROE Return
1 United States 98.42  21.66
2 United States 94.63  76.43
3        Russia 91.75  23.77
```

6.5 Rename

This **rename()** function allows to rename one (or more) columns:

```
copy.of.acwi.table = rename(acwi.short, Return.on.Equity = ROE)
```

Note that everything else in the data frame is copied in the output as well; in this example, the new data frame is stored in a new one. Note also that you could have renamed multiple column in one shot, for example using the following command:

```
copy.of.acwi.table = rename(acwi.short,
```

```
                    Return.on.Equity = ROE,
                    Sector = GICS.Sector)
```

Of course, the above could have been written using the pipe notation:

```
another.copy.of.acwi.table = acwi.short %>%
  rename(Return.on.Equity = ROE)
```

A related function is **colnames()**, which accesses the column names so you can read or overwrite them (we saw **colnames()** on page 45):

```
(df = data.frame(x1=c(1,2), x2=c(3,4)))
```

```
  x1 x2
1  1  3
2  2  4
```

```
colnames(df) = c("First Column", "Second Column")
df
```

```
  First Column Second Column
1            1             3
2            2             4
```

6.6 Mutate

Oftentimes, it is useful to create new columns storing new values you need to compute. One way to do that is to use the **mutate()** function. For example:

```
acwi.short %>%
  mutate(mega.cap = Market.Cap > 1000000 ) %>%
  head(3)
```

```
  Ticker Weight Market.Cap            GICS.Sector
1   AAPL   3.47    2193582 Information Technology
2   MSFT   2.81    1899924 Information Technology
3   AMZN   2.29    1744112 Consumer Discretionary
        Country   ROE Return mega.cap
1 United States 73.69  78.98     TRUE
2 United States 40.14  40.72     TRUE
3 United States 27.44  40.15     TRUE
```

But the alert reader might observe that adding a new column to a dataframe can simply be done using the dollar-sign notation, such as:

```
acwi.short$another.mega.cap.column =
  acwi.short$Market.Cap > 1000000
head(acwi.short, 2)
```

```
  Ticker Weight Market.Cap            GICS.Sector
1   AAPL   3.47    2193582 Information Technology
2   MSFT   2.81    1899924 Information Technology
        Country   ROE Return another.mega.cap.column
1 United States 73.69  78.98                    TRUE
2 United States 40.14  40.72                    TRUE
```

So what's the benefit of **mutate()**? The main one is that the function can be piped. For one, the first version of the code did not have to repeat the name of the dataframe, whereas the second one requires it before each dollar sign, making the code verbose. But the main benefit comes from the ability to create new columns in the middle of a series of piped commands. For example, assume we want to check if the MSCI ACWI index is market cap-weighted and find outliers. We can calculate the weight-over-market cap ratio on the fly (multiplied by 10,000 to make results more readable), store it in a column, and use the name of that newly created column to sort the data in **desc()**ending order using the **arrange()** function:

```
acwi.short %>%
  mutate(weight.over.marketcap =  Weight/Market.Cap*10000) %>%
  arrange(desc(weight.over.marketcap)) %>%
  head(3)
```

```
  Ticker Weight Market.Cap            GICS.Sector
1   TCOM   0.03       2929 Consumer Discretionary
2   ATHM   0.01       2777 Communication Services
3    IFF   0.05      15203              Materials
        Country   ROE Return another.mega.cap.column
1         China -3.12  51.71                   FALSE
2         China 20.58  12.88                   FALSE
3 United States  5.81   8.50                   FALSE
  weight.over.marketcap
1               0.10242
2               0.03600
3               0.03289
```

As we can see, one stock has, as of the time this data set was pulled, a weight in the index that was not commensurate with its market capitalization. This can happen when a company sees its market value drop or increase significantly between two reconstitutions of the index.

6.7 Summarize

To collapse many values down to a single summary metric, use **summarize()** (the British spelling, **summarise()**, is also accepted.)

You specify what metric(s) you want a summary for. Here, the mean. You can specify the name of the new column being created. For example, we can calculate the average weight of index constituents:

```
summarize(acwi.short, average.weight = mean(Weight))
```

```
  average.weight
1        0.03327
```

Note that you don't need to put quotes around **Weight** even though it's just a name defined in **acwi.short**, not an R object. This is just a convenience offered to the user. Note also that we don't even have to say where **Weight** is defined: **summarize()** understands that it has to refer to one of the columns of the dataframe passed as the first argument.

We should also highlight that **summarize()** does not change **acwi.short**; in particular, it does not add a data field containing the average weight. Note also that you can ask **summarize()** to remove any NA's before performing calculations:

```
summarize(acwi.short, average.weight = mean(Weight, na.rm=TRUE))
```

```
  average.weight
1        0.03327
```

Also, **summarize()** is fully compatible with the pipe operator %>%, so you can write:

```
acwi.short %>%
  summarize(average.weight = mean(Weight))
```

```
  average.weight
1        0.03327
```

Let's now calculate the average return for the whole index:

```
acwi.short %>%
  summarize(average.weight = mean(Return, na.rm=TRUE))
```

```
  average.weight
1          52.22
```

Note that we need to remove the NA's that are present in the **Return** column of the data. We did that by setting the optional **na.rm=** parameter of **mean()**

to TRUE. But that parameter is also available in **summarize()**, so we could have equivalently written this:

```
acwi.short %>%
  summarize(average.weight = mean(Return), na.rm=TRUE)
```

```
  average.weight na.rm
1             NA  TRUE
```

Importantly, let's highlight the fact that multiple summaries can be performed at the same time. (Otherwise, we don't absolutely need **summarize()** and could calculate the mean of returns by other means, no put intended.) For example:

```
acwi.short %>%
  summarize(index.return = mean(Return, na.rm = TRUE),
            median.ROE   = median(ROE, na.rm = TRUE))
```

```
  index.return median.ROE
1        52.22      10.05
```

Another interesting statistic that we may want to aggregate is how many representatives do each country have in the ACWI index? To count the number of distinct (unique) tickers, use **n_distinct()** seen on page 104:

```
acwi.short %>%
  group_by(Country) %>%
  summarize(count.by.country = n_distinct(Ticker))
```

```
# A tibble: 50 x 2
   Country        count.by.country
   <chr>                     <int>
 1 Argentina                     3
 2 Australia                    64
 3 Austria                       5
 4 Belgium                      13
 5 Brazil                       53
 6 Canada                       89
 7 Chile                        14
 8 China                       709
 9 Colombia                      5
10 Czech Republic                3
# ... with 40 more rows
```

(Another relevant function is of course **unique()** seen on page 27.)

Which countries have the most companies represented in the ACWI index? We need to sort the counts we just calculated. To sort these counts, pipe the previous result into **arrange()**, by **desc()**ending order:

```
acwi.short %>%
  group_by(Country) %>%
  summarize(count.by.country = n_distinct(Ticker)) %>%
  arrange(desc(count.by.country))
```

```
# A tibble: 50 x 2
   Country        count.by.country
   <chr>                     <int>
 1 China                       709
 2 United States               620
 3 Japan                       301
 4 Korea                       106
 5 India                        96
 6 Canada                       89
 7 United Kingdom               88
 8 Taiwan                       87
 9 France                       73
10 Australia                    64
# ... with 40 more rows
```

You'll notice that R was smart enough not to list all the countries. Note also that many results come out in R in the form of a dataframe, not a vector. For example, the list of the top 5 countries, by number of companies, in the ACWI index could be done as follows. Note that we use the **group_by()** and **filter()** functions seen earlier. The **n_distinct()** functions counts the number of unique (distinct) values for the specified variable (here, **Ticker**).

```
(top.countries = acwi %>%
  group_by(Country) %>%
  summarize(count.by.country = n_distinct(Ticker)) %>%
  arrange(desc(count.by.country)) %>%
  head(5) %>%
  select(Country))
```

```
# A tibble: 5 x 1
  Country
  <chr>
1 China
2 United States
3 Japan
4 Korea
5 India
```

The result is a dataframe (or tibble). So testing if a country belongs to that list will not work as planned:

```
"China" %in% top.countries
```

```
[1] FALSE
```

A solution is to convert the one-column dataframe into a vector using the **unlist()** function:

```
"China" %in% unlist(top.countries)
```

```
[1] TRUE
```

To simplify the examples that follow, let's quickly create a small subset of our data:

```
acwi.subset = select(acwi.short, Market.Cap, GICS.Sector, Country)
acwi.subset =
  filter(acwi.subset, Country %in% c("China", "Japan"))
```

We can then use the **table()** function seen on page 43

```
table(acwi.subset$Country)
```

```
China Japan
  709   301
```

We of course get the same counts as we did earlier. We can calculate the percentage represented by each country (in our subset, however) using **prop.table()**:

```
prop.table(table(acwi.subset$Country))
```

```
China Japan
0.702 0.298
```

The last examples were instances of **summary tables**, but there are even simpler ways to build them as we will see on page 106.

6.8 Contingency Tables

As discussed on page 43, a **contingency table** reports statistics on two categorical variables. Such tables can easily be built in R using the same **table()** function we just discussed:

```
table(acwi.subset$GICS.Sector, acwi.subset$Country)
```

```
                         China Japan
  Communication Services    30    16
```

```
Consumer Discretionary      79    44
Consumer Staples            62    32
Energy                      18     3
Financials                 103    27
Health Care                 87    24
Industrials                100    69
Information Technology      98    36
Materials                   65    23
Real Estate                 45    18
Utilities                   22     9
```

As we saw earlier, these data can be reported as percentages using **prop.table()**.

Tabulation Packages

The simpler way to produce clean-looking tables is, as often, to use a specialized package. For this, the **gtsummary** package is a good start. For example, let's load the package a create a subset of our ACWI example dataset:

```
library(gtsummary)
```

We can then produce summary statistics, in Figure 6.1, on sectors and returns using the **tbl_summary()** function:

```
acwi.subset %>% tbl_summary()
```

We can create more sophisticated and more elegant tables with more information using the same function. We can exclude reporting on missing data using the optional **missing=** parameter. We can also modify the table header and put labels in bold using **modify_header()** and **bold_labels()**, respectively. The output appears in Figure 6.2.

```
  tbl_summary(
    acwi.subset,
    by = Country,
    missing = "no"
  ) %>%
  modify_header(label = "ACWI") %>%
  bold_labels()
```

Characteristic	N = 1,010[1]
Market.Cap	8,231 (4,875, 17,648)
GICS.Sector	
Communication Services	46 (4.6%)
Consumer Discretionary	123 (12%)
Consumer Staples	94 (9.3%)
Energy	21 (2.1%)
Financials	130 (13%)
Health Care	111 (11%)
Industrials	169 (17%)
Information Technology	134 (13%)
Materials	88 (8.7%)
Real Estate	63 (6.2%)
Utilities	31 (3.1%)
Country	
China	709 (70%)
Japan	301 (30%)

[1] Median (IQR); n (%)

FIGURE 6.1 Using tbl-summary()

6.9 Aggregate

The **aggregate()** function allows us to perform in one call what would typicall require using **group_by()** followed by **summerize()**. To see that function in action, let's pull a larger dataframe directly from the **Nasdaq MER/F1** database seen on page 27.

```
mer.data =
  Quandl.datatable("MER/F1",
        paginate = TRUE,
        qopts.columns=c("ticker", "reportdate", "reporttype",
                        "indicator","amount"))
```

Note that, because we are requesting a large amount of data, the **paginate=** optional argument had to be set to true; it means that the **Quandl.datatable()** function reads the data in multiple "pages" instead of one large bulk request.

ACWI	China, N = 709[1]	Japan, N = 301[1]
Market.Cap	8,113 (4,676, 17,488)	8,648 (5,642, 18,578)
GICS.Sector		
Communication Services	30 (4.2%)	16 (5.3%)
Consumer Discretionary	79 (11%)	44 (15%)
Consumer Staples	62 (8.7%)	32 (11%)
Energy	18 (2.5%)	3 (1.0%)
Financials	103 (15%)	27 (9.0%)
Health Care	87 (12%)	24 (8.0%)
Industrials	100 (14%)	69 (23%)
Information Technology	98 (14%)	36 (12%)
Materials	65 (9.2%)	23 (7.6%)
Real Estate	45 (6.3%)	18 (6.0%)
Utilities	22 (3.1%)	9 (3.0%)

[1] Median (IQR); n (%)

FIGURE 6.2 Second example using tbl-summary()

From the data we received, we keep only those pertaining to total revenues reported in annual shareholder reports:

```
mer.data = mer.data %>%
  filter(indicator == "Total Revenue", reporttype=="A")
```

Note that it would of course be more efficient to request only the total revenues and only the annual report, instead of extracting all the data then discarding most of them. Unfortunately, **`Quandl.datatable()`** does not let us use any column name as a filter, so we can't for example request **`Quandl.datatable("MER/F1", indicator = "Total Revenue",..)`**.

Now that we have the necessary data, we can be interested in the average revenue of all companies that the Nasdaq repository keeps tabs on. The **`aggregate()`** function must know what we will be summarizing (here, the **`amount`** column), how we want it summarize (here, the **`FUN`**ction is **`mean()`**), and **`by=`** which variable the data should be bucketed (below, the data is grouped by **`reportdate`**). Finally, the **`na.rm=`** optional argument indicates, as usual, that entries containing `NA`'s should be discarded.

```
revenues = aggregate(mer.data$amount,
                     FUN=mean,
                     by=list(mer.data$reportdate),
                     na.rm=TRUE)
```

Visualizing this time series quickly reveals what is at best a seasonal pattern or at worst an issue in our data. The plot appears in Figure 6.3.

```
plot(revenues, type="l")
```

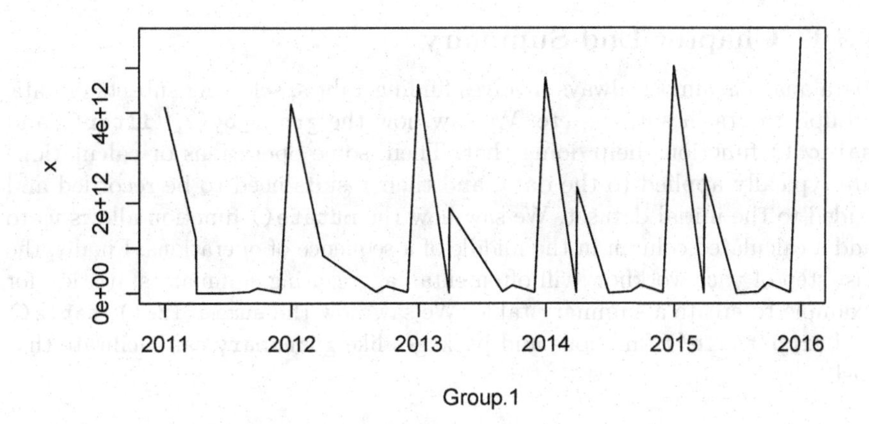

FIGURE 6.3 Total revenues across companies in the MER/F1 database

This often happens when one large entry distorts the average and that entry does not appear with the same periodicity, or when the number of records fluctuate dramatically. To understand what is happening, let's use **aggregate()** again to count, using **n_distinct()**, the number of distinct companies that reported at each date:

```
aggregate(mer.data$ticker,
          by=list(mer.data$reportdate),
          FUN=n_distinct,
          na.rm=TRUE) %>%
  head()
```

```
     Group.1  x
1 2010-12-31 23
2 2011-03-31  1
3 2011-04-02  1
4 2011-06-30  1
5 2011-08-31  1
6 2011-09-24  1
```

We clearly see the problem: in Figure 6.3, all but the peaks can be discarded as they all correspond to a single firm.

6.10 Chapter-End Summary

Data analysis almost always involves finding values, selecting subsets of data, grouping data in subsets, etc. We saw how the **group_by()**, **filter()** and **select()** functions help doing that. Then, some operations or calculations are typically applied to the data, and their results need to be recorded and added to the initial dataset. We saw how the **mutate()** function allows us to add a calculated column in the middle of a sequence of operations. Finally, the last step of your workflow will often entail aggregating summary statistics, for example to create a summary table. We saw how the **summarize()**, **table()** and **prop.table()** functions and packages like **gtsummary** can facilitate that task.

6.11 Exercises

6.11.1 Filtering on Either of Two Conditions

Find the stocks in the **acwi.short** dataframe created earlier whose market cap is greater than $1.7bn *or* whose weight is greater than 1.2%

■ SOLUTION

```
filter(acwi.short,  Market.Cap > 1700000 | Weight > 1.2)
```

```
  Ticker Weight Market.Cap              GICS.Sector
1   AAPL   3.47    2193582 Information Technology
2   MSFT   2.81    1899924 Information Technology
3   AMZN   2.29    1744112 Consumer Discretionary
4     FB   1.21     926153 Communication Services
5 BJTM27   0.03    1886985                   Energy
        Country   ROE Return another.mega.cap.column
1 United States 73.69  78.98                    TRUE
2 United States 40.14  40.72                    TRUE
3 United States 27.44  40.15                    TRUE
4 United States 25.42  58.80                   FALSE
5  Saudi Arabia 18.26  11.99                    TRUE
```

■ END OF SOLUTION

6.11.2 Performance by Sector

Based on the **MSCI All Country World Index** data set, calculate the average return of index constituents grouped by GICS sector.

■ SOLUTION

```
acwi %>%
  na.omit() %>%
  group_by(GICS.Sector) %>%
  summarize(sector.performance = mean(Return))
```

```
# A tibble: 11 x 2
   GICS.Sector             sector.performance
   <chr>                                <dbl>
 1 Communication Services                34.9
 2 Consumer Discretionary                69.7
 3 Consumer Staples                      29.6
 4 Energy                                32.5
 5 Financials                            40.0
 6 Health Care                           39.6
 7 Industrials                           51.7
 8 Information Technology                59.1
 9 Materials                             80.9
10 Real Estate                           24.8
11 Utilities                             19.9
```

■ END OF SOLUTION

6.11.3 Ordering and Plotting Returns

Sort the constituents of the ACWI index by decreasing return order and plot the returns in that order.

■ SOLUTION

```
sorted = acwi.short %>%
  na.omit() %>%
  arrange(desc(Return))

plot(sorted$Return)
```

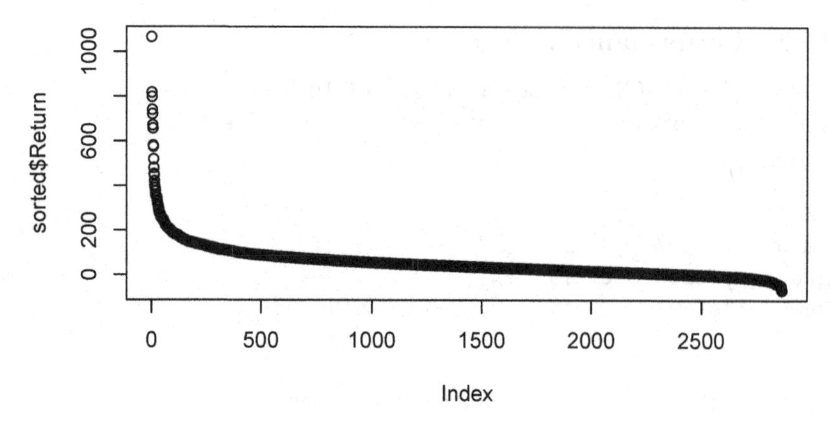

■ End of solution

6.11.4 Removing NAs

On page 102, we calculated the average return for the ACWI index as follows:

```
acwi.short %>%
  summarize(average.weight = mean(Return, na.rm=TRUE))
```

```
  average.weight
1          52.22
```

We needed to remove `NA`'s that are present in the **Return** column of the data, and we did so by setting the optional **na.rm=** parameter of **mean()** to TRUE. But that parameter is also available in **summarize()**. So, could we have equivalently written our code as follows? Why?

```
acwi.short %>%
  summarize(average.weight = mean(Return), na.rm=TRUE)
```

6.11.5 Removing Outliers

Assume we want to remove NIO from our computation of return statistics on the ACWI constituents because the stock was a non-representative and average-distorting out-performer. Use **filter()** and **summarize()** to calculate the average return of the ACWI benchmark, excluding that stock.

■ Solution

```
acwi.short %>%
  na.omit() %>%
  filter(Return < 1068) %>%
  summarize(index.return = mean(Return))
```

```
  index.return
```

```
1          51.99
```

■ END OF SOLUTION

6.11.6 Deutsche Bank's Long-Term Debt

Come back to the Deutsche Bank example seen on page 27 using the **Nasdaq** **Quandl** package and extract the historical amount of long-term debt carried by the bank.

■ SOLUTION

```
db = deutschebank %>%
  filter(indicator == "Long Term Debt")
```

```
plot(db$reportdate, db$amount, type = "l")
```

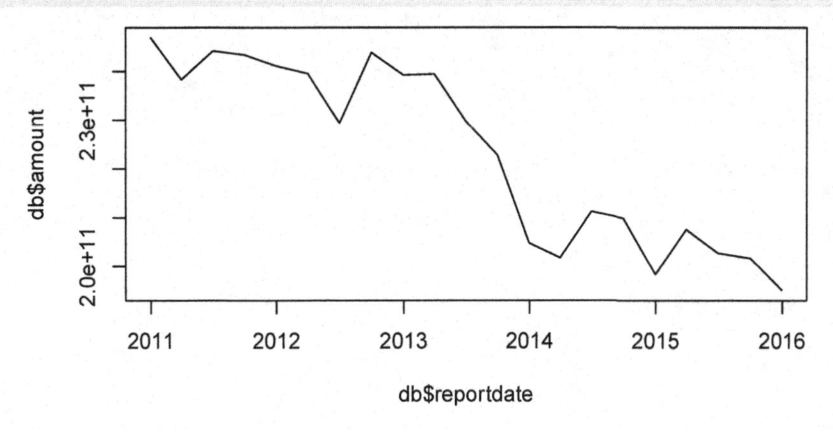

■ END OF SOLUTION

7

Merging Data Sets

Merging tables or data sets is very frequent in finance if only because different data providers have different strengths. Also, the internal systems of financial institutions often produce separate reports that sometimes need to be combined, if only for ad-hoc analyses.

We will illustrate how to perform such merging on two small data sets extracted from the **MSCI All Country World Index**. These two files are:

```
financials = read.csv(file = 'MSCI AC World - Financial Data.csv')
sectors    = read.csv(file = 'MSCI AC World - Sectors.csv')
```

The headers of these datasets are:

```
head(financials, 2)
```

```
  Ticker            Name Dividend.Yield
1   AAPL      Apple Inc.           0.62
2   MSFT Microsoft Corp.           0.85
```

and:

```
head(sectors, 2)
```

```
  Ticker Market.Cap.            GICS.Sector
1   AAPL   2,193,582 Information Technology
2   MSFT   1,899,924 Information Technology
```

Note that there is only one column, **Ticker**, that is shared between the two tables. That column will be the **key** according to which the records will be merged: The two tables are merged by looking for records in both tables that share the same key. This sounds easy enough, but it gets more complicated when we need to specify what to do with records where the key can't be found in one of the two tables.

DOI: 10.1201/9781003320555-7

7.1 Inner Join

The **inner join** of x and y returns all rows from x where there are matching values in y, with all columns from x and y. If there are multiple matches between x and y, all combination of the matches are returned. This is performed as follows on our two datasets using the `inner_join()` function:

```
inner_join(financials, sectors, by = "Ticker")
```

```
  Ticker                 Name Dividend.Yield Market.Cap.
1   AAPL           Apple Inc.           0.62   2,193,582
2   MSFT      Microsoft Corp.           0.85   1,899,924
3   AMZN Amazon.com, Inc.                 --   1,744,112
4     FB       Facebook, Inc.             --     926,153
5   GOOG        Alphabet Inc.             --   1,617,417
6   TSLA           Tesla Inc              --     683,191
7    JPM      JPMorgan Chase            2.34     465,598
             GICS.Sector
1 Information Technology
2 Information Technology
3 Consumer Discretionary
4 Communication Services
5 Communication Services
6 Consumer Discretionary
7             Financials
```

The key was provided explicitly using `by=`, but since `Ticker` was the only column shared by the two dataframes, we could also write:

```
inner_join(financials, sectors)
```

Note also that R offers an alternative function, `merge()`:

```
merge(financials, sectors, by = "Ticker")
```

Or simply:

```
merge(financials, sectors)
```

7.2 Left Join

The **left join** of x and y returns all rows from x, with all columns from x and y. If there are multiple matches between x and y, all combination of the matches are returned:

```
left_join(financials, sectors, by = "Ticker")
```

```
  Ticker                Name Dividend.Yield Market.Cap.
1   AAPL          Apple Inc.           0.62   2,193,582
2   MSFT     Microsoft Corp.           0.85   1,899,924
3   AMZN Amazon.com, Inc.                --   1,744,112
4     FB     Facebook, Inc.              --     926,153
5   GOOG       Alphabet Inc.             --   1,617,417
6   TSLA           Tesla Inc             --     683,191
7 688910          TSMC, Ltd.           1.67        <NA>
8    JPM      JPMorgan Chase           2.34     465,598
             GICS.Sector
1 Information Technology
2 Information Technology
3 Consumer Discretionary
4 Communication Services
5 Communication Services
6 Consumer Discretionary
7                   <NA>
8             Financials
```

Note that the definition of the inner join does say *all* rows from x, the first data frame. Since TSMC is in the dataset called **financials**, its record is kept into the output. Columns that cannot be completed however (because there is no entry for TSMC in **sector**) are filled in with NA's.

The **merge()** equivalent is:

```
merge(financials, sectors, all.x=TRUE)
```

```
  Ticker                Name Dividend.Yield Market.Cap.
1 688910          TSMC, Ltd.           1.67        <NA>
2   AAPL          Apple Inc.           0.62   2,193,582
3   AMZN Amazon.com, Inc.                --   1,744,112
4     FB     Facebook, Inc.              --     926,153
5   GOOG       Alphabet Inc.             --   1,617,417
6    JPM      JPMorgan Chase           2.34     465,598
7   MSFT     Microsoft Corp.           0.85   1,899,924
8   TSLA           Tesla Inc             --     683,191
```

```
                GICS.Sector
1                      <NA>
2 Information Technology
3 Consumer Discretionary
4 Communication Services
5 Communication Services
6                Financials
7 Information Technology
8 Consumer Discretionary
```

Note that the result is the same dataframe – the *order* in which the records were joined is just different.

7.3 Right Join

The **right join** of x and y returns all rows from y, with all columns from x and y. If there are multiple matches between x and y, all combination of the matches are returned.

```
right_join(financials, sectors, by = "Ticker")
```

```
  Ticker                Name Dividend.Yield Market.Cap.
1   AAPL          Apple Inc.           0.62   2,193,582
2   MSFT     Microsoft Corp.           0.85   1,899,924
3   AMZN Amazon.com, Inc.                --   1,744,112
4     FB     Facebook, Inc.              --     926,153
5   GOOG       Alphabet Inc.             --   1,617,417
6   TSLA           Tesla Inc             --     683,191
7    JPM      JPMorgan Chase           2.34     465,598
8    JNJ                <NA>           <NA>     428,389
                GICS.Sector
1 Information Technology
2 Information Technology
3 Consumer Discretionary
4 Communication Services
5 Communication Services
6 Consumer Discretionary
7                Financials
8               Health Care
```

This is in a sense the symmetrical situation: *all* rows from y, the second dataframe, get included in the merged dataframe even if there is no match in x. JNJ is an example of an entry in **sectors** whose key was not found in **financials**. Its record is included however, with missing entries for the columns of x filled in with NA.

The merge equivalent of right join is:

```
merge(financials, sectors, all.y = TRUE)
```

```
  Ticker                Name Dividend.Yield Market.Cap.
1   AAPL          Apple Inc.           0.62   2,193,582
2   AMZN Amazon.com, Inc.               --   1,744,112
3     FB    Facebook, Inc.               --     926,153
4   GOOG     Alphabet Inc.               --   1,617,417
5    JNJ                <NA>           <NA>     428,389
6    JPM    JPMorgan Chase           2.34     465,598
7   MSFT   Microsoft Corp.           0.85   1,899,924
8   TSLA           Tesla Inc             --     683,191
            GICS.Sector
1 Information Technology
2 Consumer Discretionary
3 Communication Services
4 Communication Services
5            Health Care
6             Financials
7 Information Technology
8 Consumer Discretionary
```

Here as well, the resulting dataframe is the same but the rows are stored in a different order.

7.4 Full Join (a.k.a. Outer Join)

The **full join**, also known as **outer join**, of x and y returns all rows and all columns from both x and y. Where there are not matching values, returns NA for the one missing.

```
full_join(financials, sectors, by = "Ticker")
```

```
  Ticker                Name Dividend.Yield Market.Cap.
1   AAPL          Apple Inc.           0.62   2,193,582
2   MSFT   Microsoft Corp.           0.85   1,899,924
3   AMZN Amazon.com, Inc.               --   1,744,112
4     FB    Facebook, Inc.               --     926,153
5   GOOG     Alphabet Inc.               --   1,617,417
6   TSLA           Tesla Inc             --     683,191
```

```
7 688910         TSMC, Ltd.           1.67       <NA>
8    JPM    JPMorgan Chase           2.34    465,598
9    JNJ               <NA>           <NA>    428,389
             GICS.Sector
1 Information Technology
2 Information Technology
3 Consumer Discretionary
4 Communication Services
5 Communication Services
6 Consumer Discretionary
7                   <NA>
8             Financials
9            Health Care
```

Per the definition of a full join, we see records of **financials** that were not matched in **sectors**, such as TSMC, and entries in the opposite situation, such as JNJ.

The **merge()** equivalent is:

```
merge(financials, sectors, all = TRUE)
```

```
  Ticker               Name Dividend.Yield Market.Cap.
1 688910         TSMC, Ltd.           1.67       <NA>
2   AAPL         Apple Inc.           0.62  2,193,582
3   AMZN Amazon.com, Inc.              --  1,744,112
4     FB     Facebook, Inc.             --    926,153
5   GOOG      Alphabet Inc.             --  1,617,417
6    JNJ               <NA>           <NA>    428,389
7    JPM    JPMorgan Chase           2.34    465,598
8   MSFT    Microsoft Corp.           0.85  1,899,924
9   TSLA          Tesla Inc             --    683,191
             GICS.Sector
1                   <NA>
2 Information Technology
3 Consumer Discretionary
4 Communication Services
5 Communication Services
6            Health Care
7             Financials
8 Information Technology
9 Consumer Discretionary
```

where the **all=** named argument makes the semantics quite clear.

7.5 Merging Nasdaq Datasets

Merging datasets pulled from the internet is equally easy. For example, let's pull data on employee ("full time" or "ft," hence the column name, **emp_ft_cnt**), total revenue and gross **profits and losses** (often abbreviated **P&L**) from the **ZACKS/FC** database provided by Nasdaq through the **Quandl** package discussed on page 25:

```
zacks.data =
  Quandl.datatable("ZACKS/FC",
    qopts.columns=c("ticker", "comp_name", "comp_name_2",
                    "per_type", "per_cal_year", "per_cal_qtr",
                    "emp_ft_cnt", "tot_revnu", "gross_profit"))
```

(**Zacks** is a data provider with multiple datasets available through **Quandl**, and details can be found at https://data.nasdaq.com/databases/ZFA/documentation. **FC** is one of the datasets, and the exercise on page 123 makes use of their **EE** dataset.)

We limit our example to annual reports (**per**iod **type** is "A") issued in the 4th **cal**endar quarter:

```
zacks.data = zacks.data %>%
  filter(per_cal_qtr == "4", per_type == "A")
```

The obtained dataset is not very rich but plenty for our purposes:

```
unique(zacks.data$ticker)
```

```
 [1] "AXP"  "BA"   "CAT"  "CVX"  "DD"   "GE"   "GS"
 [8] "IBM"  "INTC" "JNJ"  "JPM"  "KO"   "MCD"  "MMM"
[15] "MRK"  "PFE"  "TRV"  "UNH"  "VZ"   "XOM"
```

Separately, let's pull data from the **Sharadar** database, also provided through Nasdaq:

```
shara.data = Quandl.datatable('SHARADAR/SEP',date.gte='2018-12-31')
```

We can verify that the firms it reports on overlaps with the Zacks dataset above:

```
unique(shara.data$ticker)
```

```
 [1] "XOM"  "WMT"  "VZ"   "V"    "UNH"  "TSLA" "TRV"
 [8] "PG"   "PFE"  "NKE"  "MSFT" "MRK"  "MMM"  "MCD"
[15] "KO"   "JPM"  "JNJ"  "INTC" "IBM"  "HD"   "GS"
[22] "GE"   "DIS"  "DD"   "CVX"  "CSCO" "CAT"  "BA"
[29] "AXP"  "AAPL"
```

We can now merge the datasets using **merge()**, and remove the NA's thanks to **na.omit()**:

```
merged.data = merge(zacks.data, shara.data, by="ticker")
merged.data = na.omit(merged.data)
```

The Sharader database provided information such trading volume, and the Zachs repository contained gross profit. Is there a relationship between the two? Let's plot the line of best fit, in Figure 7.1.

```
plot(merged.data$gross_profit, merged.data$volume)
linearRegr = lm(merged.data$volume ~ merged.data$gross_profit)
abline(linearRegr)
legend("topright",
       legend=paste("R-squared = ",
                    format(summary(linearRegr)$adj.r.squared,
                           digits=2)))
```

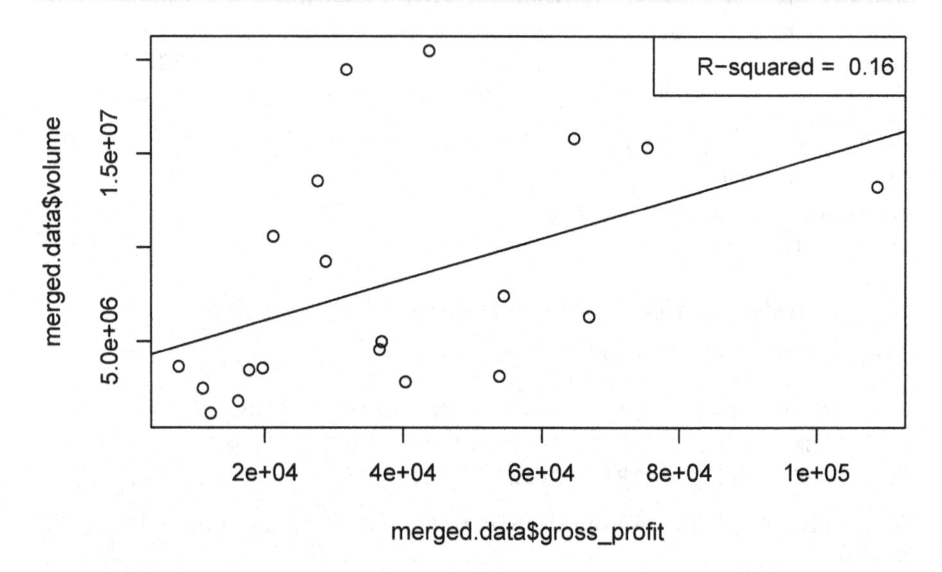

FIGURE 7.1 Plotting two statistics, trading volume and gross profit, obtained from merged datasets

Figure 7.1 indicates that there isn't much of a correlation.

7.6 Chapter-End Summary

Data is frequently aggregated from different tables pulled from different sources. As long as the tables agree on at least one unambiguous code to refer to the same objects, data can be crossed or merged. We saw that the most subtle issue is what to do with object that don't have an entry in one of the tables: we can either drop these objects, i.e., prefer to lose the information provided by one of the data sets, in an operation called **inner join**; or we need to treat objects and records differently depending on which of the two tables they are in, using either a **left join** or a **right join**; or we keep all the data we have using a **full join**, also known as **outer join**.

7.7 Exercises

7.7.1 The Zacks EE Dataset

The Zacks data provider also offers the `EE` database. Use that database to pull from annual reports estimates on earnings per share (**EPS**, field `eps_mean_est`) for 2023 (next year, as of the time of this writing). Merge this dataset with the `FC` data seen on page 121 to investigate if total revenues (field `tot_revnu` from `FC`) have a linear relationship with EPS forecasts.

■ SOLUTION

```
ee.data = Quandl.datatable("ZACKS/EE")
ee.data = ee.data %>% filter(per_cal_year == 2023, per_type == "A")
```

```
z.data = merge(zacks.data, ee.data, by="ticker")
z.data = na.omit(z.data)
```

```
plot(z.data$tot_revnu, z.data$eps_mean_est)
```

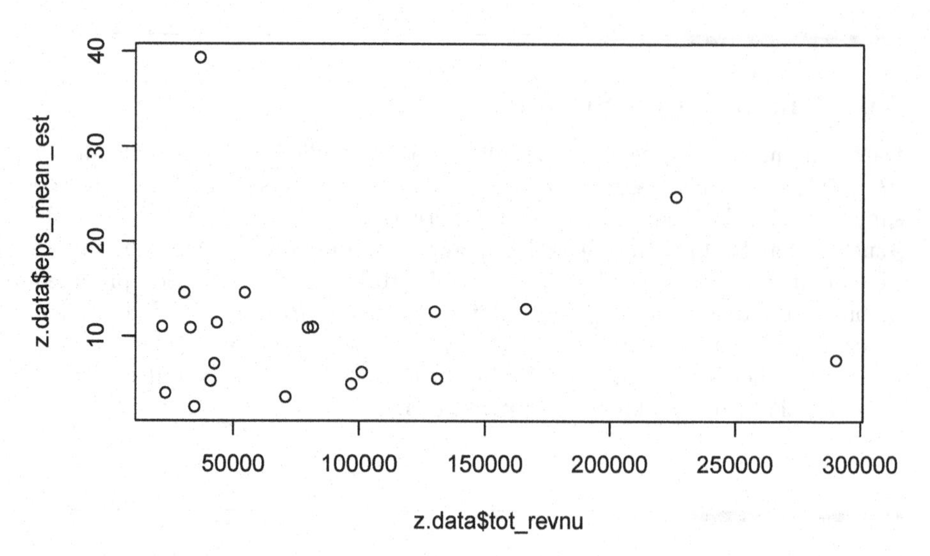

■ END OF SOLUTION

7.7.2 Merging Dividend and Split Data

We saw how to pull dividend and stock split data, including the dates at which they happen. Get all dividend payments made by Apple in 2020 and all splits the stock had in the same year, then merge the two datasets.

■ SOLUTION

```
aapl.splits = tq_get("AAPL",
                     get  = "splits",
                     from = "2020-01-01",
                     to   ="2020-12-31")
aapl.div = getDividends("AAPL",
                        from = "2020-01-01",
                        to   = "2020-12-31")
```

```
aapl.div2 = data.frame(date=index(aapl.div),
                       dividend = coredata(aapl.div))
colnames(aapl.div2) = c("date", "dividend")
```

```
aapl.div2
```

```
        date dividend
1 2020-02-07  0.04813
2 2020-05-08  0.05125
3 2020-08-07  0.05125
4 2020-11-06  0.20500
```

```
merge(aapl.splits, aapl.div2,  all=TRUE)
```

```
        date symbol value dividend
1 2020-02-07   <NA>    NA  0.04813
2 2020-05-08   <NA>    NA  0.05125
3 2020-08-07   <NA>    NA  0.05125
4 2020-08-31   AAPL  0.25       NA
5 2020-11-06   <NA>    NA  0.20500
```

■ End of solution

8

Graphing Using Ggplot

The **ggplot2** package is a versatile graphing package and its main function is **ggplot()**. "GG" stands for Grammar of Graphics[1] and the grammar of **ggplot()** commands allows us to independently specify elements of a graph and to combine them quite freely using a consistent syntax. These elements include: the data of course, the horribly named **aesthetic** mappings, the so-called **geometry**, and of course specifications for the title, axes and axis legends. **Aesthetic** here means "something you can see." Each aesthetic is a mapping between a visual element and a variable. Examples of visual elements include position (i.e., on the x and y axes), outline and fill colors, shapes (for example, of markers), line types, and sizes and widths.

8.1 The Grammar of Ggplot Commands

The **ggplot()** command is typically followed by options, such as a **geometric** layer. The **ggplot()** command initiates a plot and specifies general attributes, such as which data set is being used and which columns of this data set are used. This mapping of data set columns to, typically, the x and y axes of the graph is specified using a so-called **aesthetic**, abbreviated as **aes()**. So a typical **ggplot()** command will usually have the following syntax:

```
ggplot([data=] DATA, [mapping=] aes(MAPPING)) +
  GEOM_FUNCTION([mapping=] aes(MAPPING))
```

Above, square brackets indicate optional **named arguments** (see page 70) and **UPPERCASE** text represents parts that you need to fill in depending on your intent For example, **DATA** would typically be the name of the dataframe containing data you want to graph, and **GEOM_FUNCTION** can be **geom_point()** that adds a layer of points to create a scatter plot. The **ggplot2** package comes with many **GEOM_FUNCTION**s that each offers (adds) a different type of layer to a plot.

[1] Wilkinson, Leland (2005). The Grammar of Graphics. Springer. ISBN 978-0-387-98774-3.

DOI: 10.1201/9781003320555-8

Note that the **aes()** mapping specification can be in the **ggplot()** function (we will call it **global** for reasons explained later) or specific to one geom layer (we will call it **local**).

8.2 Geometric Objects

A **geom** is the geometrical object that a plot uses to represent data. People often describe plots by the type of geom that the plot uses, so for example, bar charts use "bar geoms," line charts use "line geoms," boxplots use "boxplot geoms," and so on. Scatter plots break the pattern though: the corresponding geom is, as we alluded to earlier, **geom_point()**.

Let's try our first example. We will again use the data set of the constituents of the **MSCI All Country World Index**. Please refer to page 7 to read the input file and convert specific columns to numerical values.

Then of course, you need to install the **ggplot2** package (we recommend using the "Tools" drop-down menu of RStudio) and to load it using the **library()** command:

```
library(ggplot2)
```

Then, following the syntax sketched earlier, we would call **ggplot()** on our **acwi** dataframe, followed (using the **+** operator) by **geom_point()**. This **geom_point()** only needs to know what goes to the x axis and what goes to the y axis. This "what goes where" is the **aes**thetic mapping we mentioned earlier. Here's our first example, whose output appears in Figure 8.1.

```
ggplot(acwi) +
  geom_point(aes(x = Market.Cap, y = Weight)) +
  ggtitle("ACWI Weights vs. Market Capitalization")
```

Note how easy it was to add a feature like a title: we used the **+** operator to add a new specification – in our case, the **ggtitle()** function.

As always, one should pause for a moment to see what we can learn from the graphs we produce. Clearly, constituents of the index tend to plot along a line, so weights in the index tend to be proportional to the company's market capitalization – and indeed, the MSCI ACWI is essentially market cap-weighted. We notice a few outliers though, and in particular one dot in the bottom right corner, and two stocks with market caps exceeding 1.5 trillion dollars but with weights barely greater than 1%.

Likewise, plotting Price-to-Cash-Flow versus Price-to-Sales also turns out to be a bit surprising, resulting in Figure 8.2.

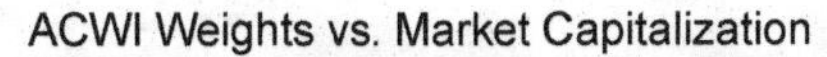

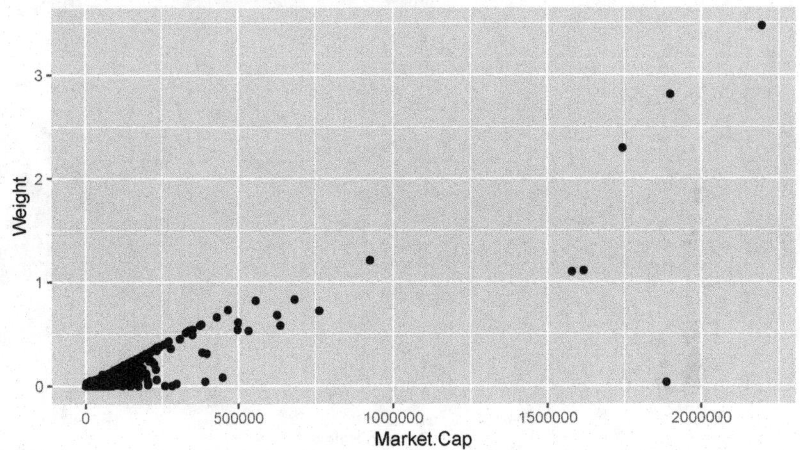

FIGURE 8.1 Using the geom-point() function

```
ggplot(acwi) +
  geom_point(aes(x = Price.to.Sales, y = Price.to.Cash.Flow))
```

For one, the scale of the x-axis goes all the way to 4,000, whereas most of the points plot at Prices-to-Sales less than 500. This is when our discussion on page 10 comes handy: at least one outlier is distorting the scale and makes the range larger than it should probably be.

But we can also notice that there is no point on the graph with Price-to-Sales around 4,000! So how come ggplot stretched the x-axis so far? The answer is that the data point that had a Price-to-Sales around 4,000 had no entry in the Price-to-Cash-Flow column and therefore could not be plotted. We can check that as follows:

```
acwi %>%
  filter(Price.to.Sales > 4000)
```

```
  Ticker                                   Name Weight
1 BJKDJS CanSino Biologics, Inc. Class H   0.01
  Market.Cap Dividend.Yield Price.to.Cash.Flow
1      12393             NA                 NA
  Price.to.Book Price.to.Sales GICS.Sector Country
1         13.37           4102 Health Care   China
     ROE Return
1 -10.08    136
```

Such outliers are going to complicate any analysis you may want to perform so you will probably have to remove these outliers or handle them in a case-specific

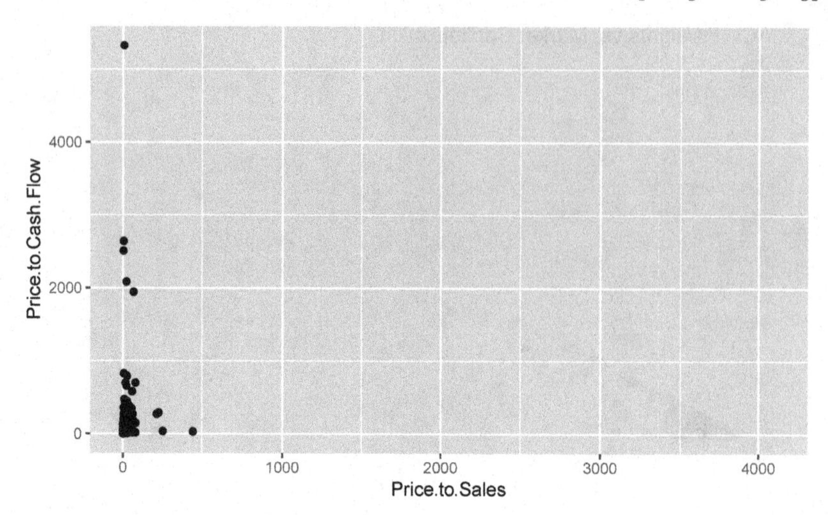

FIGURE 8.2 Plotting Price-to-Cash-Flow versus Price-to-Sales

way. We will discuss this throughout this material. For now, one solution could be to **filter()** this one outlier away:

```
acwi %>%
  filter(Price.to.Sales < 4000) %>%
  ggplot() +
    geom_point(aes(x = Price.to.Sales, y = Price.to.Cash.Flow))
```

However, this kind of one-by-one thinking quickly shows its limits: executing the code above would produce a graph with the same problem, because *another* stock with abnormally high Price-to-Sales has `NA` for its Price-to-Cash-Flow. So a more comprehensive strategy to handle `NA`'s and outliers is needed. One tool is to remove *all* rows in the dataframe that contain an `NA`, using the **na.omit()** function. The code below does that and produces Figure 8.3.

```
acwi %>%
  na.omit() %>%
  ggplot() +
    geom_point(aes(x = Price.to.Sales, y = Price.to.Cash.Flow))
```

However, this method removes data points even if the `NA` is *not* Price-to-Sales or Price-to-Cash-Flow, so those data points would be needlessly deleted from our analysis of Price-to-Cash-Flow vs Price-to-Sales.

Alternatively, since we are focusing on just plotting, we can add (using the **+** operator again) instructions to limit the scale of the x-axis thanks to the **xlim()** function. This results in Figure 8.4.

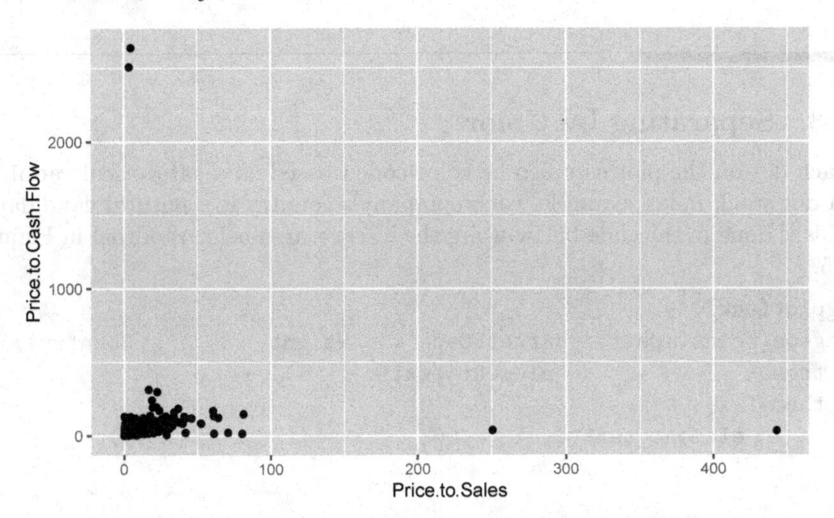

FIGURE 8.3 Removing NA's before plotting

```
ggplot(acwi) +
  geom_point(aes(x = Price.to.Sales, y = Price.to.Cash.Flow)) +
  xlim(0,500) +
  theme(aspect.ratio=1/2)
```

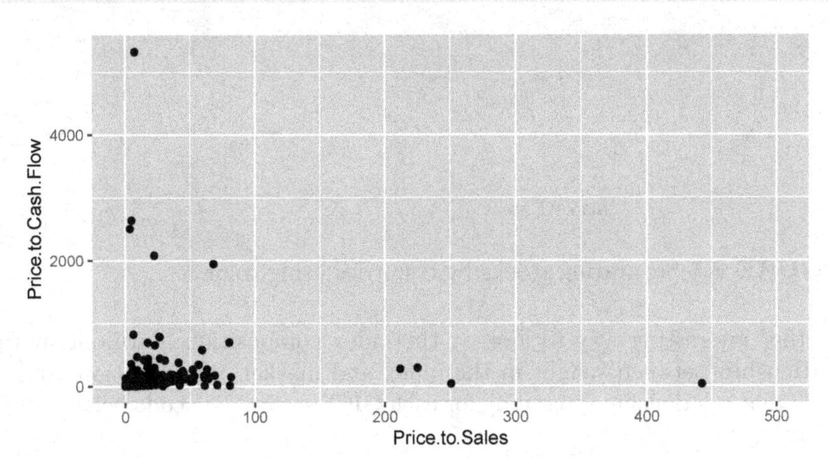

FIGURE 8.4 Using aspect.ratio to resize a graph

Note also the addition, using the **+** operator, of the **theme()** function with the **aspect.ratio** named argument.

8.3 Separating by Color

Each dot on the plot can also be color-coded based on a categorical variable. In our stock index example, each company's country is a natural candidate. This is done in the code below using the **color=** argument, resulting in Figure 8.5.

```
ggplot(acwi) +
  geom_point(aes(x = Market.Cap, y = Weight, color = Country)) +
  theme(legend.text = element_text(size=4)) +
  theme(axis.text.x =
        element_text(angle = 90, vjust = 0.5, hjust=1))
```

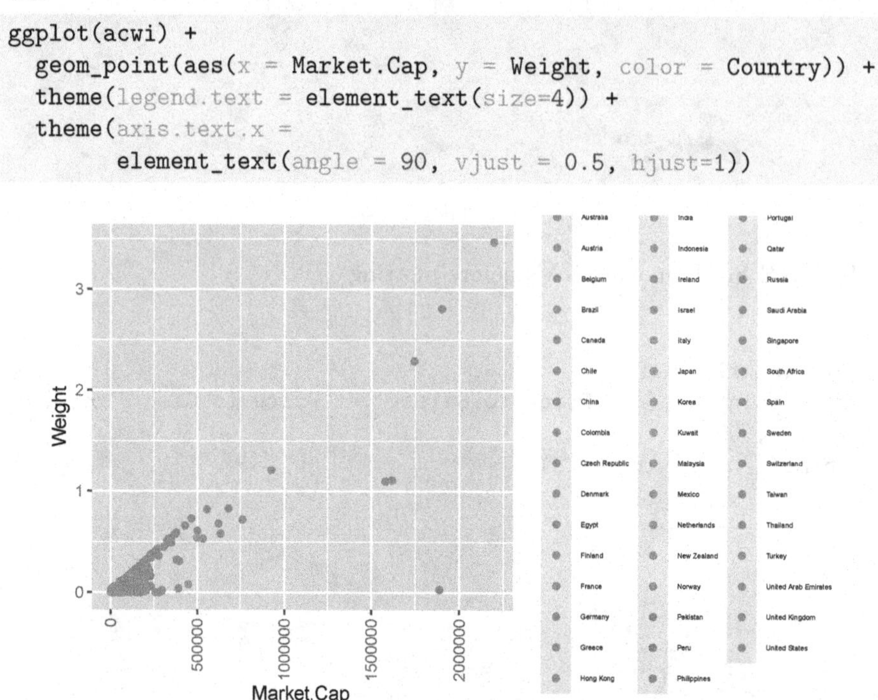

FIGURE 8.5 Separating stocks by countries using color

Another interesting way to look at the index composition is to look at the relationship between weight in the index and market capitalization, and to study that relationship separately for each GICS sector. The code below results in Figure 8.6.

```
ggplot(acwi) +
  geom_point(aes(x = Market.Cap, y = Weight, color = GICS.Sector)) +
  theme(aspect.ratio=2/3)
```

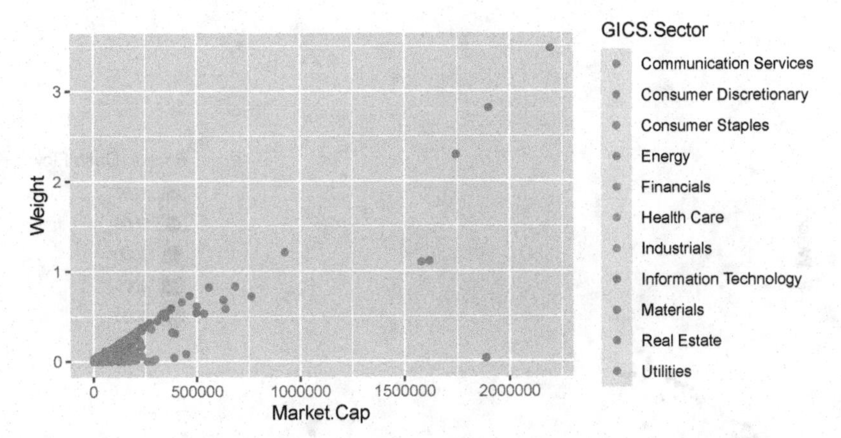

FIGURE 8.6 Separating stocks by sectors using color

8.4 Separating by Size

One key feature of **ggplot** that we haven't highlighted yet is that it is fully compatible with the pipe operator **%>%**. What this means is that, once you have sliced and diced your data set in a way you like, you can directly pipe the result into a **ggplot()** command.

Let's illustrate this while making the size of each dot proportional to one of the numeric fields in your data set. Note that we now start with the name of the data frame, in effect asking R to use its content:

```
acwi %>%
  ggplot() +
  geom_point(aes(x = Market.Cap,
                 y = Weight,
                 size = Price.to.Cash.Flow))
```

8.5 Separating by Shape

For qualitative data, changing the shape of each dot can provide a clearer view:

```
acwi %>%
  ggplot() +
  geom_point(aes(x = Market.Cap,
```

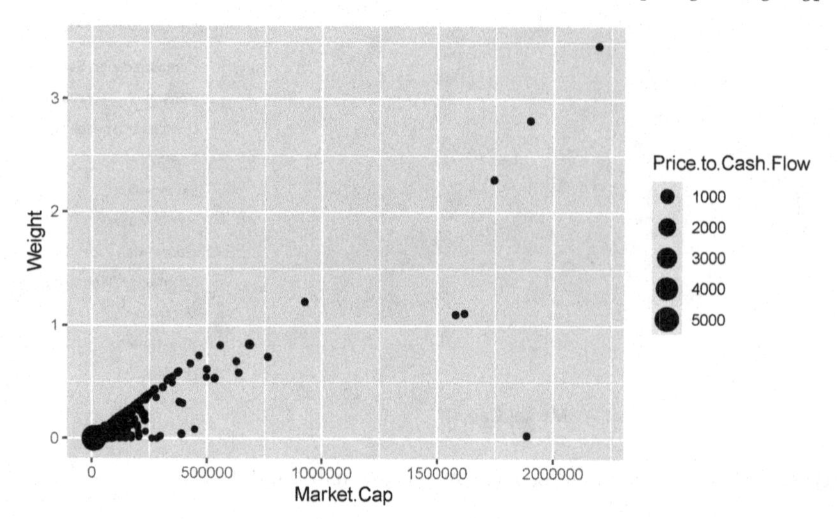

FIGURE 8.7 Using the size of markers to convey information in a scatter plot

```
              y = Weight,
              shape = GICS.Sector))
```

```
Warning: The shape palette can deal with a maximum of 6
discrete values because more than 6 becomes
difficult to discriminate; you have 11. Consider
specifying shapes manually if you must have
them.

Warning: Removed 1342 rows containing missing values
(geom_point).
```

If you observe closely the resulting graph, shown in Figure 8.8, you will note that some sectors disappeared: **ggplot2** only uses up to six shapes at a time. By default, additional groups will not be plotted when you use the **shape** to separate more than 6 categories. Hence the warning message we got when running the command.

8.6 Curves of Best Fit

A frequent goal of data analysis is to detect any trend in the data. To do that, the most widely tool is linear regression, which is the subject of an entire

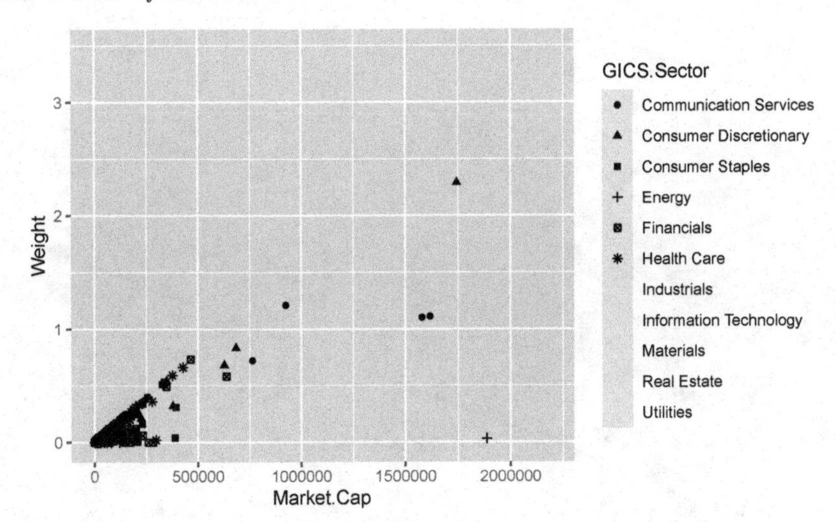

FIGURE 8.8 Using the shape of markers to convey information in a scatter plot

chapter. For now, we use it to plot the line of best fit using the **geom_smooth()** layer and the linear model **lm**. The output of this code appears in Figure 8.9.

```
ggplot(acwi, aes(x = Market.Cap, y = Weight)) +
  geom_point() +
  geom_smooth(method='lm')
```

Sometimes a curve (i.e., a polynomial) of best fit is more appropriate than a straight line. By default (that is, if the method of fitting is not specified), a polynomial curve of best fit is calculated and graphed by **geom_smooth()** using **loess**, a form of polynomial regression also known as **locally estimated scatterplot smoothing**. The code below plots Price-to-Cash-Flow against ROE together wit a loess curve of best fit, resulting in Figure 8.10.

```
ggplot(acwi) +
  geom_point(aes(x = ROE,
                 y = Price.to.Cash.Flow))+
  geom_smooth(aes(x = ROE,
                  y = Price.to.Cash.Flow)) +
  ylim(0, 500)
```

Global Mappings

As the previous example shows, aesthetics risk being repeated when multiple geometries are used. Fortunately, you can avoid such repetition by passing an aesthetic mapping *to* **ggplot()** instead of adding it to all the geometries.

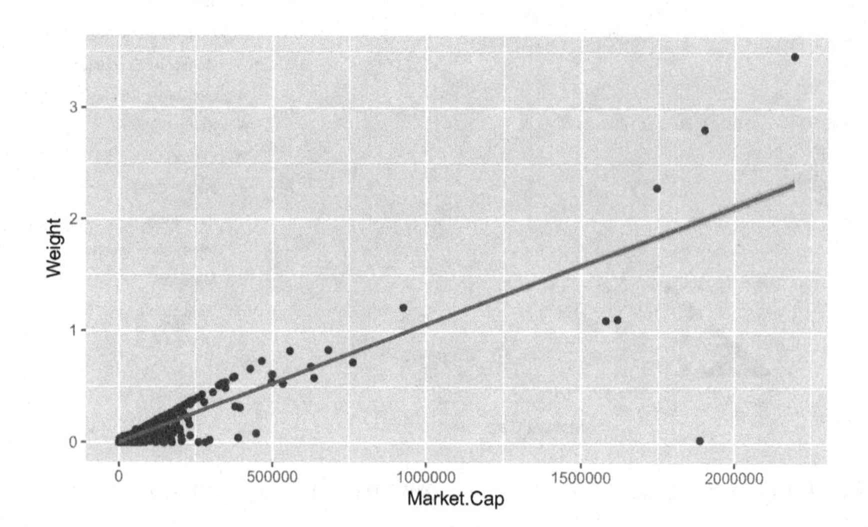

FIGURE 8.9 Adding a line of best fit to a scatter plot

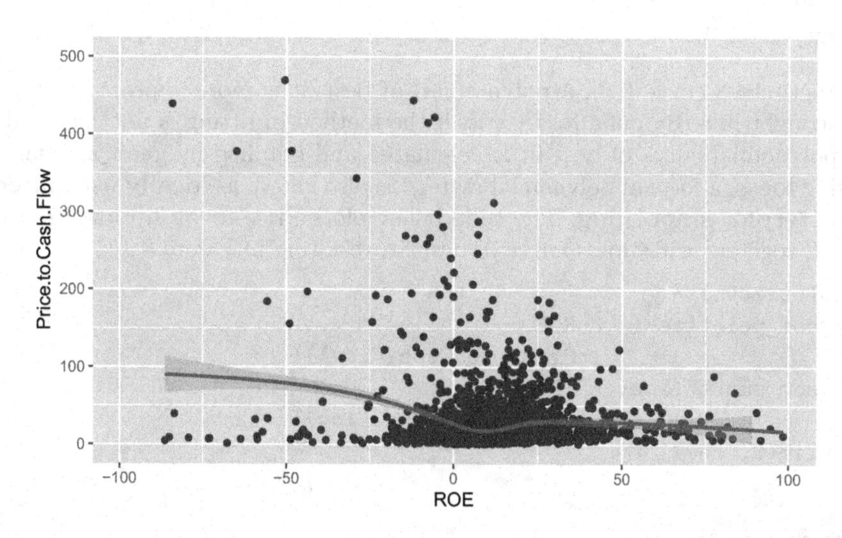

FIGURE 8.10 Polynomial fitting using locally estimated scatterplot smoothing

ggplot2 treats this mapping as a **global** (i.e., common) mapping that applies to each geom in the graph. In other words, the code below produces the exact same graph as Figure 8.10.

```
ggplot(acwi, aes(x = ROE,
                 y = Price.to.Cash.Flow)) +
  geom_point() +
  geom_smooth() +
  ylim(0, 500)
```

Local Mappings

If you place mappings in a geom function, **ggplot2** will treat them as local mappings for that layer. I.e., **ggplot2** will use these mappings to extend or overwrite the global mappings for that layer only. This makes it possible to display different aesthetics in different layers. For example, the code below generates the plot in Figure 8.11. Note that the color scheme is applied only to the scatter plot.

```
ggplot(acwi, aes(x = Market.Cap, y = Weight)) +
  geom_point(aes(color = GICS.Sector)) +
  geom_smooth(method='lm')
```

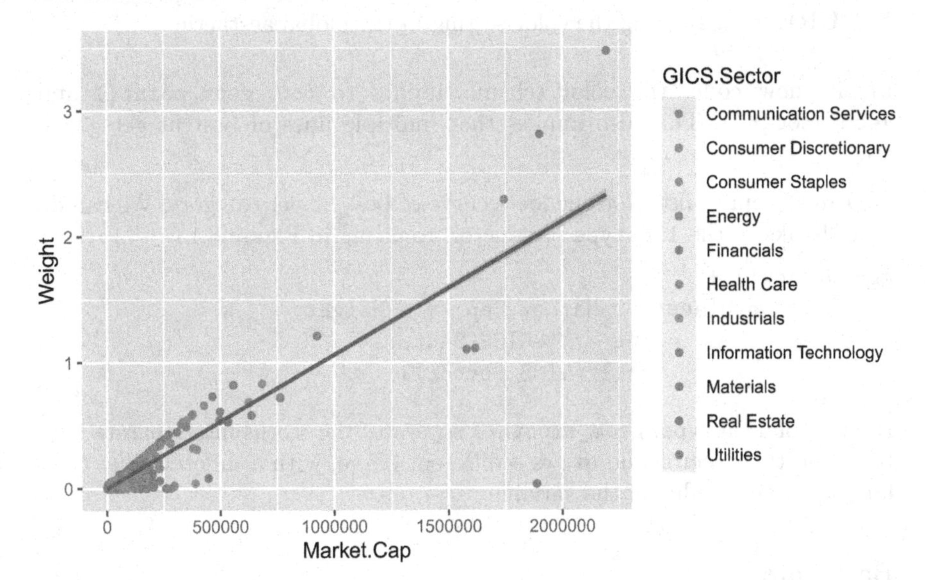

FIGURE 8.11 Adding a local setting (here, color) to only one of the local mapping (here, for the scatter plot)

To contrast local vs global mappings, compare the previous plot (Figure 8.11) with Figure 8.12 produced by the code below:

```
ggplot(acwi,
       aes(x = Market.Cap, y = Weight, color = GICS.Sector)) +
  geom_point() +
  geom_smooth(method='lm')
```

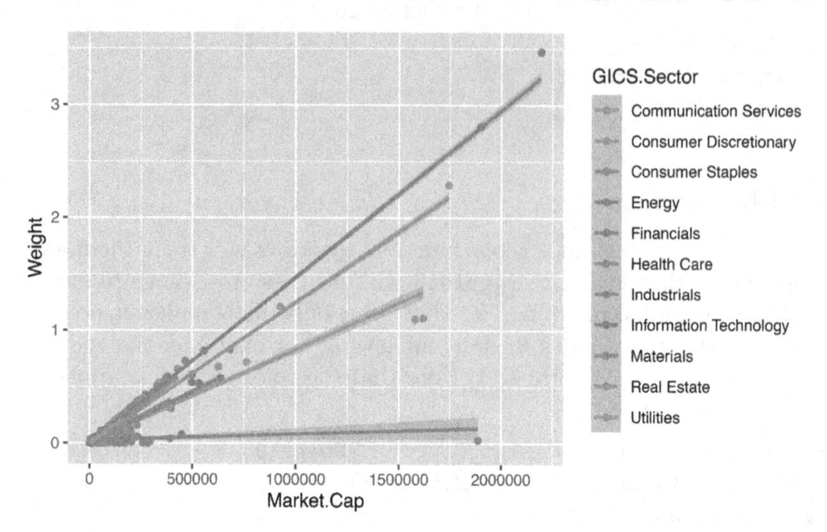

FIGURE 8.12 Putting the color setting in the global aesthetic

In the new code, the color scheme applies to *both* **geom_point()** and **geom_smooth()**. This also implies that multiple lines of best fit get generated, one per sector.

We can of course also plot one loess curve of best fit per category. We can do that thanks to the **linetype** parameter, resulting in Figure 8.13.

```
ggplot(acwi) +
  geom_smooth(aes(x = Market.Cap, y = Weight,
                  linetype=GICS.Sector,
                  color=GICS.Sector))
```

Because of **linetype**, **geom_smooth()** separates the stocks into separate lines based on their sector and draws a different curve, with a different line type, for each unique value of the variable.

Bar chart

A **bar graph** typically comes to mind as a way to visualize how a value changes over different buckets or groups. Staying with our ACWI example, one metric we might be interested in is the number of index constituents in each of the GICS sectors. We've learned to do that easily – for example:

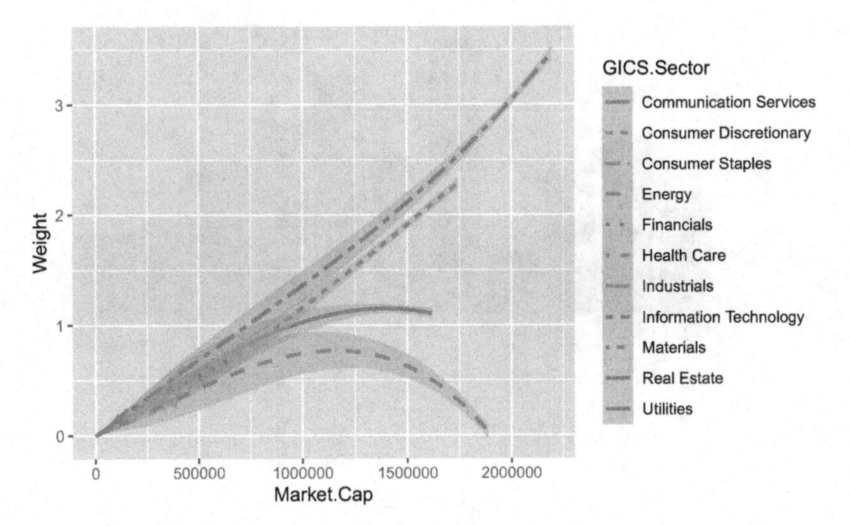

FIGURE 8.13 Curves of best fit by sector

```
acwi.grouped =
  acwi %>%
  group_by(GICS.Sector) %>%
  summarize(count = n())
```

Now that we created a new column called **count**, we will plot its values for each of the GICS sector. In the example below, the **acwi.grouped** dataframe is piped into a **ggplot** command. That command is itself built of the addition of a **ggplot()** function to start the plot and of **geom_col()** that contains the **aes()**thetic mapping: that mapping says that the x-axis corresponds to the **GICS.Sector** and the y-axis corresponds to **count** (both implicitly contained in the dataframe injected at the entry of the pipe, i.e. **acwi**). The code below results in the bar chart of Figure 8.14.

```
acwi.grouped %>%
  ggplot() +
  geom_col(aes(x=GICS.Sector, y=count)) +
  theme(axis.text.x =
        element_text(angle = 90, vjust = 0.5, hjust=1))
```

Note the three parameters in **element_text()** for, respectively, the text angle and the vertical and horizontal justifications.

By default, the columns appear along order in alphabetical order if the values are characters, or the factor order if the values are factors. But another improvement we might want is to sort the columns in increasing or decreasing order. To enforce an order, we use the **reorder()** function that takes two

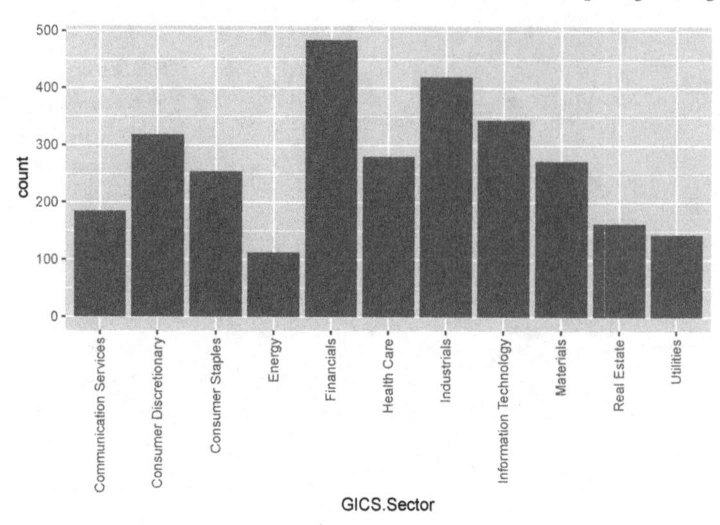

FIGURE 8.14 Producing a bar chart

arguments: the data frame column to be ordered and the data frame column whose values are as the order. By default, bars in the plot are placed in increasing order as we can see in Figure 8.15.

```
acwi.grouped %>%
  ggplot() +
  geom_col(aes(x=reorder(GICS.Sector, count), y=count)) +
  ggtitle("Number of ACWI Consitutents by GICS Sector") +
  xlab("GICS Sectors") +
  theme(axis.text.x =
        element_text(angle = 90, vjust = 0.5, hjust=1))
```

To show the bars in decreasing order, the trick is more a math trick than a weird R notation: Since 1, 2 and 3 are increasing order but $-1, -2$ and -3 are in decreasing order, and since **count** contains values, we can specify the order to be the *negative* of **count**; so the **reorder()** command would look like this, with a negative sign added:

```
aes(x=reorder(GICS.Sector, -count)
```

The **ggplot** package offers another function to produce bar plots, and given its name, **geom_bar()** could look like the function of choice. Indeed, it can allow for more concise code, such as the example below that results in Figure 8.16:

```
acwi %>%
  ggplot(aes(x=GICS.Sector)) +
    geom_bar() +
    ggtitle("Number of ACWI Constituents by GICS Sector") +
```

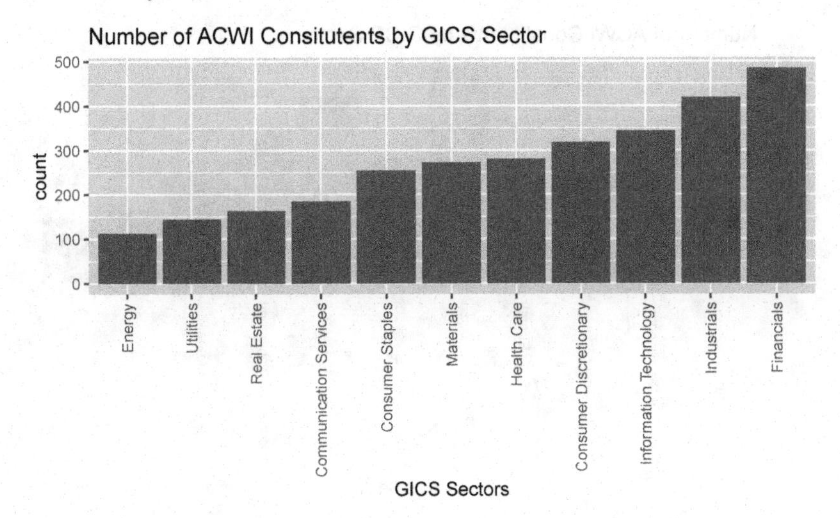

FIGURE 8.15 Displaying bars in increasing order

```
    xlab("GICS Sectors") +
    theme(axis.text.x =
              element_text(angle = 90, vjust = 0.5, hjust=1))
```

One subtle aspect of **geom_bar()** is that, if you look at the code closely, we didn't have to specify what the vertical bars are plotting: by default, **geom_bar()** assumes you want to plot the *count* of data points in the category specified by the **x** variable of the **aes**thetic. In fact, we didn't even have to create the **acwi.grouped** variable! The starting point of the pipeline is plain **acwi**.

Facets

Another way to plot the same graph for different categories is to separate these graphs, each smaller graph being one tile (or **facet**) of the larger figure. The function to perform this tiling is **facet_wrap()**. Its first argument is the categorical variable, introduced with the $\sim$ symbol, across which the different facets will be created; an optional argument, **nrow=**, is the number of rows you want tiles to be spread over.

In our running example on the ACWI stock index, GICS sectors come naturally as an example: it makes sense to compare the relationship between benchmark weight and market capitalization across sectors. We suspect that the slope of the relationship should be approximately the same across sectors, but some sectors are expected to include "mega caps," while others can be expected to

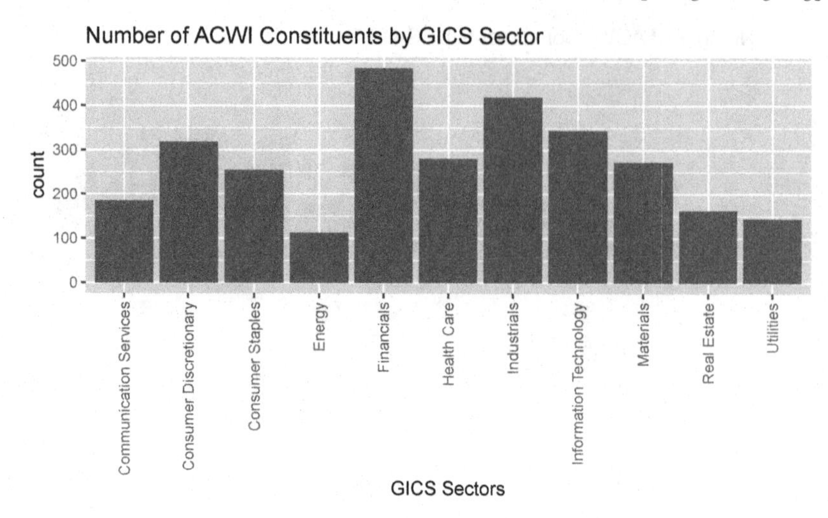

FIGURE 8.16 Bar chart produced using the geom-bar() function

not even include large caps. We are going to use **facet_wrap()** to examine one plot for each of the sectors.

```
ggplot(acwi) +
  geom_point(aes(x = Market.Cap,
                 y = Weight)) +
  facet_wrap(~ GICS.Sector,
             nrow = 3) +
  theme(axis.text.x =
          element_text(angle = 90,
                       vjust = 0.5,
                       hjust=1))
```

Note that **facet_wrap()** comes as a function added on top of the "geometry" specification and that, as said earlier, the categorical variable used to separate the data is introduced with the tilde ∼ symbol. The result appears in Figure 8.17.

Note also the use of **theme()** to rotate the axis labels by **angle=90** degrees. (And please see page 145 for another example of using **theme()** to beautify axis labels.)

As we suspected, Figure 8.17 shows that some sectors (Materials, Real Estate and Utilities for example) do not include very large companies. However, in each sector the relationship between weight and market capitalization seems to hold and to be approximately linear.

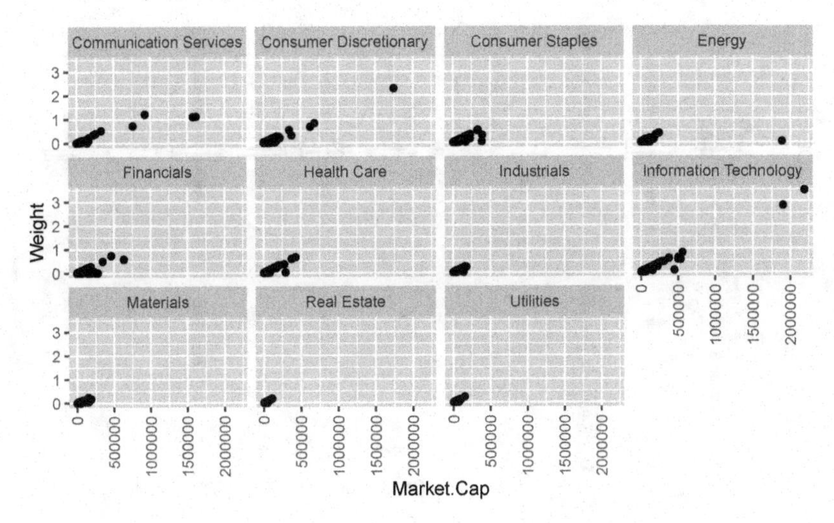

FIGURE 8.17 Using facet-wrap() to display one plot per value of a categorical variable

Boxplots

In a **boxplot**, a.k.a. **box-and-whiskers**, the box stretches from the 25th percentile to the 75th percentile, a distance known as the **interquartile range (IQR)**. In the middle of the box, a line displays the median, i.e. the 50th percentile. These three lines give you a sense of the spread of the distribution and whether or not the distribution is symmetric about the median or skewed to one side. Dots mark outliers that fall more than 1.5 times the IQR from either edge of the box. A "whisker" line extends from each end of the box and goes to the farthest non-outlier point in the distribution.

The **geom_boxplot()** geometry allows us to plot such box-and-whiskers plot, illustrating for example the dispersion of returns on equity across sectors. The result appears in Figure 8.18:

```
acwi %>%
  na.omit() %>%
  ggplot(aes(x = GICS.Sector, y = ROE)) +
    geom_boxplot() +
    theme(axis.text.x = element_text(angle = 90,
                                     vjust = 0.5, hjust=1))
```

Note that, in the case of time series of returns of different investments, a boxplot chart can more easily be plotted using the **chart.Boxplot()** function of **PerformanceAnalytics** discussed on page 180.

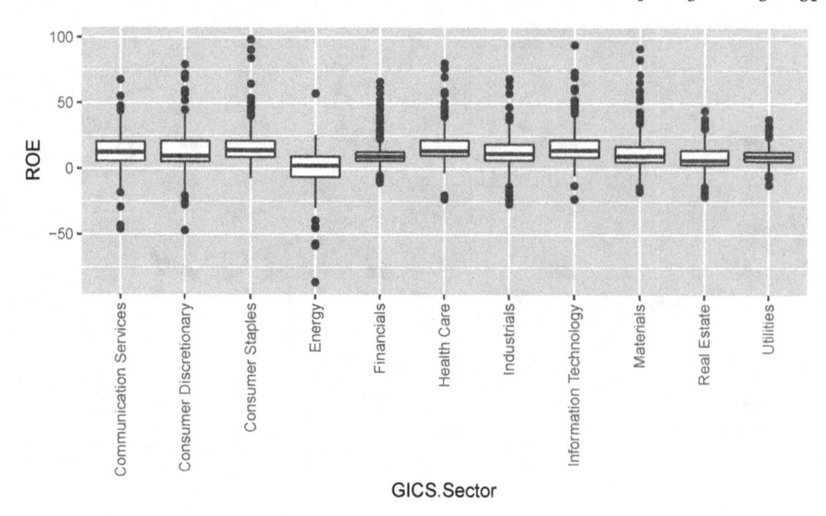

FIGURE 8.18 Box and whiskers plot

8.7 Case Study: The House Price Dataset

This case study is adapted from:

`https://colinreimerdawson.com/stat209/labs/03-ggplot.html`

and the dataset was made available on **github** by Harvard's **Institute for Quantitative Data Sciences** and contains prices on houses and land across the US.

```
f = "https://raw.githubusercontent.com/IQSS/dss-workshops-
    archived/master/R/Rgraphics/dataSets/landdata-states.csv"
housing = read.csv(f)
```

A quick look at the data set shows that the Dates can be quarters denoted in a fractional format. So Q1'13 for example is recorded as 2013.25. Now, assume we are interested in studying the buildings' construction costs as a function of land value. To start doing that, we want to plot one against the other. First, to make the data set more manageable, we will extract data for Q1'13 only.

```
hp2013Q1 =
  housing %>%
  filter(Date == 2013.25)
```

We then prepare the plot in the R statement below:

```
base_plot = ggplot(hp2013Q1,
                   aes(y = Structure.Cost, x = Land.Value))
```

Note that nothing gets printed yet! We only created an object that contains some information the plot will need – specifically, the dataset, and the mapping from dataset columns to the x and y of the plot. We plot the scatter plot using the **geom_point()** function seen earlier plus elements to format monetary numbers cleanly. This code results in Figure 8.19.

```
base_plot +
  geom_point() +
  theme(axis.text=element_text(size=14),
        axis.title=element_text(size=14,face="bold")) +
  scale_x_continuous(labels = scales::dollar_format()) +
  scale_y_continuous(labels = scales::dollar_format())
```

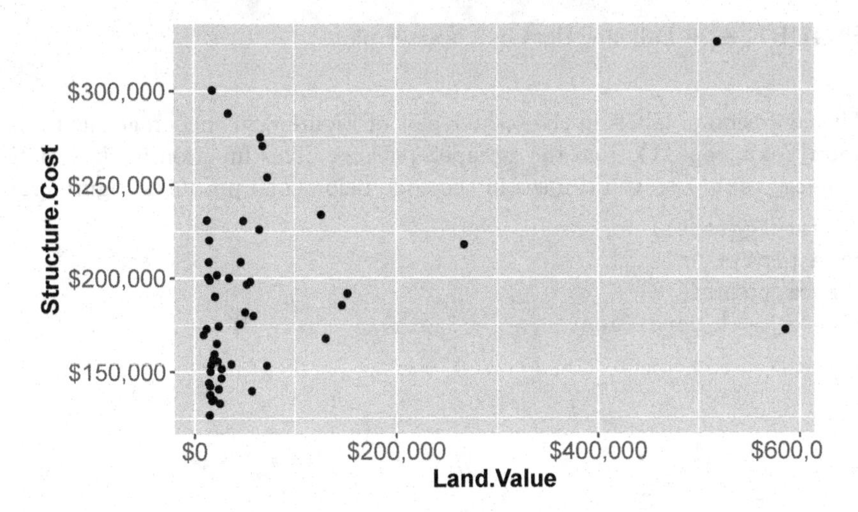

FIGURE 8.19 Improving the scale labels

Note the addition of **theme()** to specify the font size in **axis.text** and **axis.title**. Without it, the axis fonts are too small. (Please see page 142 for another example of using **theme()** to polish axis labels.) Note also that the dollar amounts are shown using their format thanks to **scales::dollar_format()**.

In this case study, labeling each dot can be useful. We do that using the **geom_text()** function, as shown below. The resulting plot appears in Figure 8.20.

```
base_plot +
  geom_point() +
  geom_text(aes(label = State), size = 3)
```

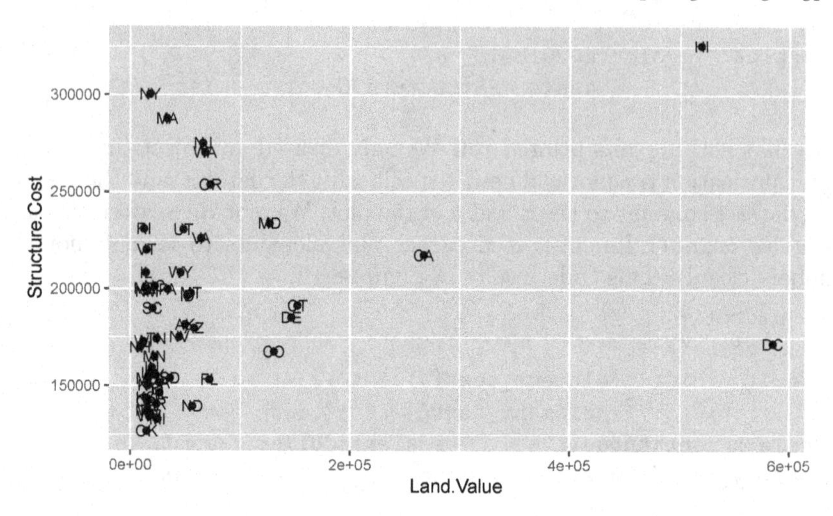

FIGURE 8.20 Labeling dots in a scatter plot

The confusion of labels in the scatter plot of Figure 8.20 encourages us to use **geom_text_repel()** from the **ggrepel** package. This function works exactly as **geom_text()** as you can see in the code below, and produces Figure 8.21.

```
library(ggrepel)
base_plot +
  geom_point() +
  geom_text_repel(aes(label = State), size = 3)
```

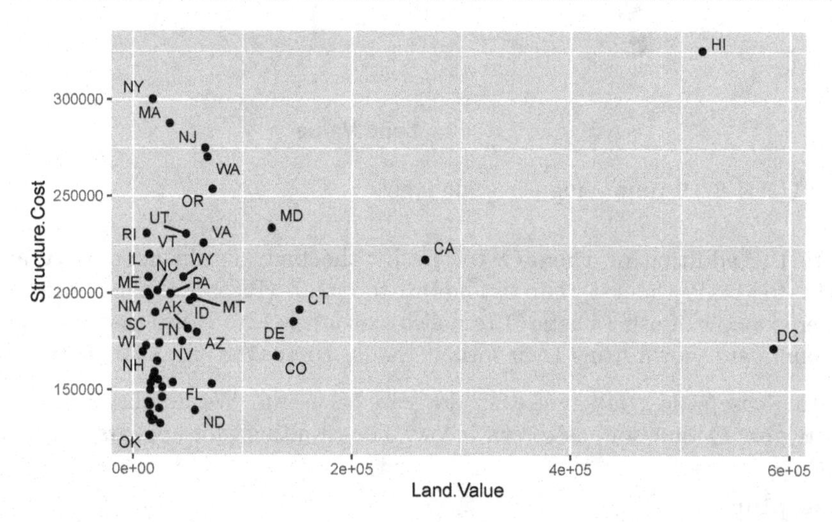

FIGURE 8.21 Labeling dots in a scatter plot, with the repel version

The scatter plots we have seen so far plotted continuous variables on both axes. But variables can also be discrete, which can create elegant graphs. For example, let's plot changes in home prices for each State. The code below results in Figure 8.22.

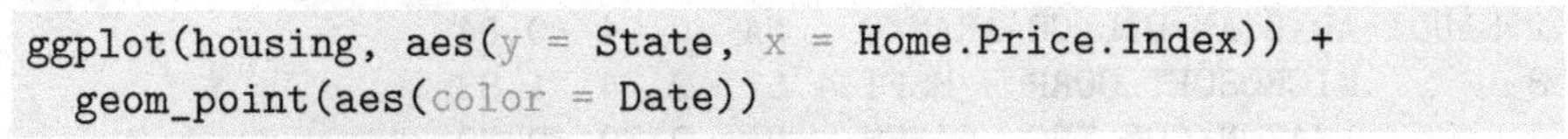

```
ggplot(housing, aes(y = State, x = Home.Price.Index)) +
  geom_point(aes(color = Date))
```

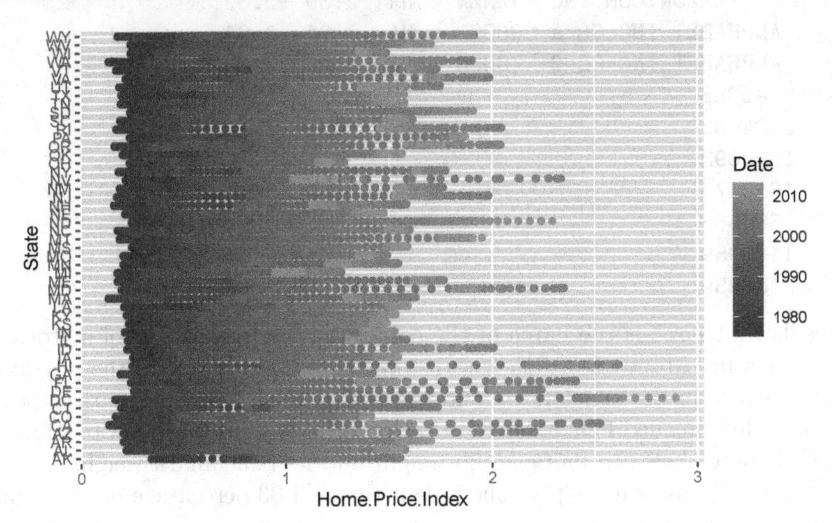

FIGURE 8.22 House price per state

8.8 Case Study: The Ocean Portfolio

Uploading the Portfolio

This is a stock portfolio called Ocean compared to a broad benchmark. For each stock, its weight in the portfolio and/or in the benchmark is provided, plus its market capitalization.

```
library(openxlsx)
```

```
ocean = read.xlsx(xlsxFile = "OceanMarketCaps.xlsx")
```

Even if we don't always show these commands and their results, we strongly recommend you get into the habit of immediately inspect any new dataset using the **head()** function to have a first sense of the data and column names, **str()** to see if some columns need to be converted to numerical value, and **summary()** to look for NA's and for outliers. For example:

```
head(ocean)
```

```
                    Name Ticker Port Bench    AW
1              APPLE INC   AAPL   NA  3.67 -3.67
2 SAUDI ARABIAN OIL CO ARAMCO   NA  0.04 -0.04
3         MICROSOFT CORP   MSFT 4.57  2.74  1.83
4         AMAZON.COM INC   AMZN   NA  2.39 -2.39
5      ALPHABET INC-CL A  GOOGL   NA  0.94 -0.94
6      ALPHABET INC-CL C   GOOG   NA  0.95 -0.95
  MarketCap
1   2024065
2   1919693
3   1618475
4   1589568
5   1188959
6   1188959
```

reveals that **Port** is the name of the column giving the weight of a stock in the portfolio. Many show NA's because those stocks are not in the portfolio, but they are in the benchmark, and their weights in the benchmark appears in the column called **Bench**. The **active weight** of a position in a portfolio is the difference between its portfolio weight and its benchmark weight. In this portfolio, Microsoft has a portfolio weight that is 1.83 percentage points higher than in the benchmark.

Likewise, **summary()** reveals interesting facts about the portfolio:

```
summary(ocean)
```

```
     Name               Ticker                 Port     
 Length:3009        Length:3009        Min.   :0.4  
 Class :character   Class :character   1st Qu.:0.5  
 Mode  :character   Mode  :character   Median :0.6  
                                       Mean   :1.2  
                                       3rd Qu.:1.6  
                                       Max.   :4.8  
                                       NA's   :2929 
     Bench             AW           MarketCap      
 Min.   :0.000   Min.   :-3.67   Min.   :   1142  
 1st Qu.:0.000   1st Qu.:-0.03   1st Qu.:   5676  
 Median :0.010   Median :-0.01   Median :  10834  
 Mean   :0.033   Mean   : 0.00   Mean   :  27960  
 3rd Qu.:0.030   3rd Qu.: 0.00   3rd Qu.:  24398  
 Max.   :3.670   Max.   : 4.69   Max.   :2024065  
 NA's   :20                                       
```

We see 2,929 `NA`'s in the portfolio weights, which just means that most of the stocks in the very large benchmark are not in our portfolio. Reciprocally, we see 20 `NA`'s in the benchmark weights. Assuming all the records in the file are for stocks that are either in the portfolio or the benchmark or both, this indicates that 20 stocks in the Ocean portfolio are **out-of-benchmark**. We also note that active weights range from negative 3.7 percentage points (which possibly means that a benchmark heavyweight is passed on in the portfolio) to positive 4.7 percentage points, indicating that the portfolio manager has a strong conviction in the corresponding stock. Finally, we can note the wide dispersion in and long right tail of market capitalizations, with a mean much larger than its median.

Stocks are typically grouped as small caps (less than $2bn), mid caps (between $2 and $10bn), large caps (between $10 and $200bn) and mega caps (above $200bn). The question is how much of the ocean portfolio is in each of these capitalization categories – in number of stocks, but also what percentage of the investment is placed in each.

To do that, we introduce the **`cut()`** function, with specific thresholds specified with **`breaks`**:

```
ocean %>%
  group_by(capitalizations =
             cut(MarketCap,
                 breaks=c(0,2000, 10000, 200000,2000000000))) %>%
  count()
```

```
# A tibble: 4 x 2
# Groups:   capitalizations [4]
  capitalizations     n
  <fct>           <int>
1 (0,2e+03]          46
2 (2e+03,1e+04]    1385
3 (1e+04,2e+05]    1530
4 (2e+05,2e+09]      48
```

However, we see that we are counting all the stocks, not those in the portfolio. Also, we have the number of stocks, but we also want the total weight allocated to each category.

A refinement, therefore, is to group the stocks by the value of their weight **`Port`** and by market capitalization:

```
ocean %>%
  group_by(isInPort = (Port>0),
       capitalizations=
             cut(MarketCap,
                 breaks=c(0,2000, 10000, 200000,2000000000))) %>%
  summarize(Count=n(), Weight=sum(Port))
```

```
# A tibble: 8 x 4
# Groups:   isInPort [2]
  isInPort capitalizations Count Weight
  <lgl>    <fct>           <int>  <dbl>
1 TRUE     (0,2e+03]           2   0.94
2 TRUE     (2e+03,1e+04]      30  15.9
3 TRUE     (1e+04,2e+05]      47  78.6
4 TRUE     (2e+05,2e+09]       1   4.57
5 NA       (0,2e+03]          44  NA
6 NA       (2e+03,1e+04]    1355  NA
7 NA       (1e+04,2e+05]    1483  NA
8 NA       (2e+05,2e+09]      47  NA
```

We notice that the bottom 4 rows report on stocks that are not portfolio holdings. We can clean that up when we do the same for the benchmark, by keeping only those stocks whose weight in the benchmark is strictly greater than 0:

```
ocean %>%
  group_by(isInBench=(Bench>0),
          capitalizations=
            cut(MarketCap,
                  breaks=c(0,2000, 10000, 200000,2000000000))) %>%
  summarize(Count=n(), Weight=sum(Bench))%>%
  filter(isInBench == TRUE)
```

```
# A tibble: 7 x 4
# Groups:   isInBench [2]
  isInBench capitalizations Count Weight
  <lgl>     <fct>           <int>  <dbl>
2 TRUE      (2e+03,1e+04]    1368   6.32
3 TRUE      (1e+04,2e+05]    1528  63.3
4 TRUE      (2e+05,2e+09]      48  29.0
```

We can plot a bar graph of portfolio holdings after having extracted the raw values:

```
portcap = ocean %>%
  group_by(isInPort = (Port>0),
          capitalizations =
            cut(MarketCap,
                  breaks=c(0,2000, 10000, 200000,2000000000))) %>%
  summarize(Count=n(), Weight=sum(Port))
```

The bar graph itself is produced by **barplot()** and appears in Figure 8.23.

```
barplot(portcap$Weight[1:4])
```

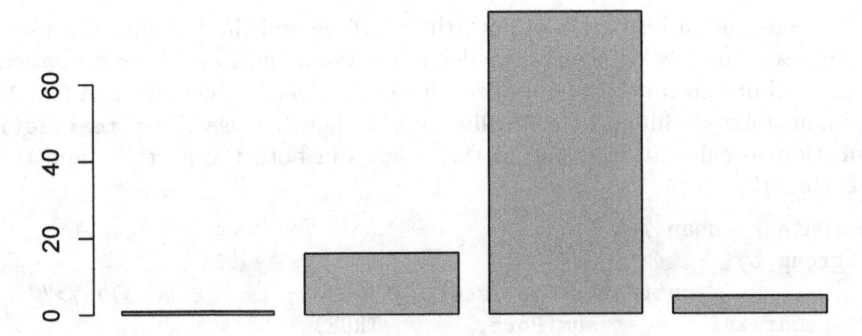

FIGURE 8.23 Bar graph of the holdings in the Ocean portfolio

But the separation into Small/Mid/Large/Mega is arbitrary. Let's instead plot the histogram of capitalizations for the portfolio. The histogram appears in Figure 8.24.

```
portfolio = filter(ocean, Port>0)
Portfolio_Capitalizations = portfolio$MarketCap
hist(Portfolio_Capitalizations, breaks=50)
```

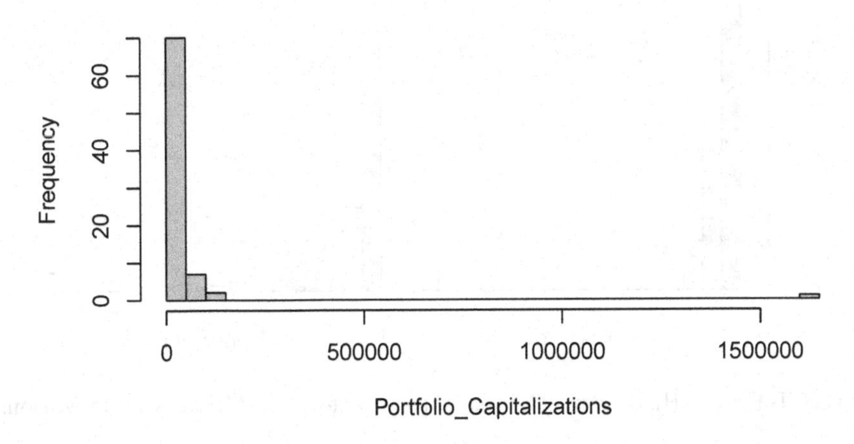

FIGURE 8.24 Histogram of Market Capitalizations

Figure 8.24 shows the frequency, in percentage, of each bucket. 70% of the portfolio holdings belong to the smaller capitalization bucket, but we see a bump on the far right side – what is sometimes called a **barbell** distribution. These percentages correspond to the count of stocks in each category, but don't reflect the weight allocated to each capitalization bucket.

Let's construct a histogram of both the portfolio and the benchmark on one graph. As a further refinement, we define not the number of the capitalization buckets but the range of capitalizations each cover, in increment of 50,000 million dollars. But most critically, observe how we use the **summarize()** function to calculate the sum of the weights in both the portfolio and the benchmark.

```
mktData = ocean %>%
  group_by(capitalizations =
             cut(MarketCap, seq(0, 1700000, by = 50000))) %>%
  summarize(PortWgt=sum(Port, na.rm=TRUE),
            BenchWgt=sum(Bench, na.rm=TRUE))
```

Finally, we plot the histogram of market capitalizations for both the portfolio and the benchmark, side by side using the optional **beside=** argument. The data of interest are in the second and third columns of the dataframe. **barplot()** accepts vectors of data, for a single histogram as shown in all previous example, or multiple vectors in the form of a matrix – hence the need to explicitly convert the dataframe using the **as.matrix()** function. The result appears in Figure 8.25.

```
barplot(as.matrix(mktData[,2:3]), beside=TRUE)
```

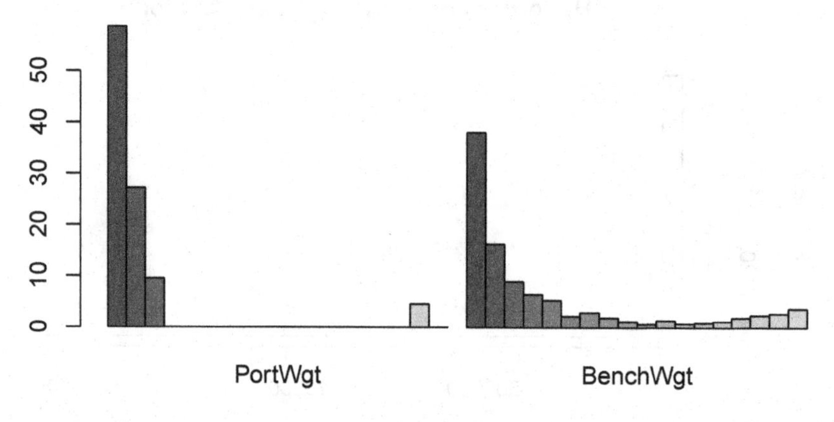

FIGURE 8.25 Histogram of Stocks in the Portfolio, by Market Capitalizations

Even taking weights into account, Figure 8.25 shows that the portfolio's distribution of market capitalizations is more barbelled than that of the benchmark.

8.9 Exercises

8.9.1 Change the Marker Shape by Region

Coming back to the House Price dataset, plot the structure cost as a function of land value with a different market shape for each region.

■ Solution

```
base_plot +
  geom_point(aes(shape = region))
```

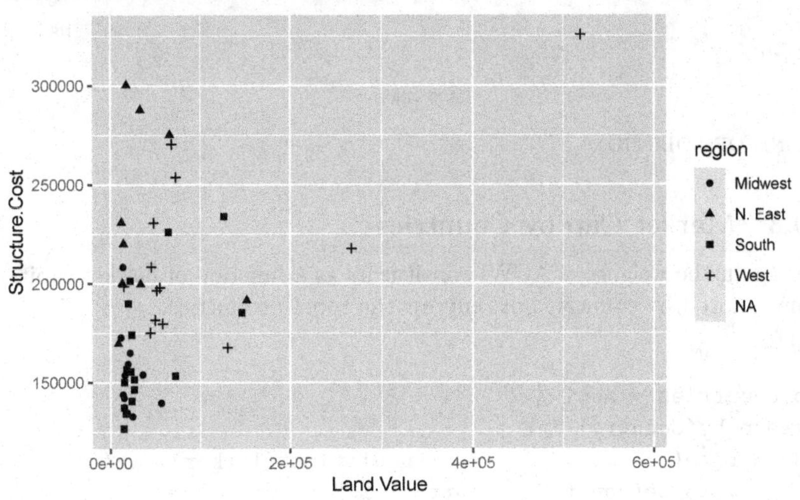

■ End of solution

8.9.2 Change the Marker Color by Price

On the same plot, set the market color to reflect the value of the house.

■ Solution

```
base_plot +
  geom_point(aes(shape = region, color = Home.Value))
```

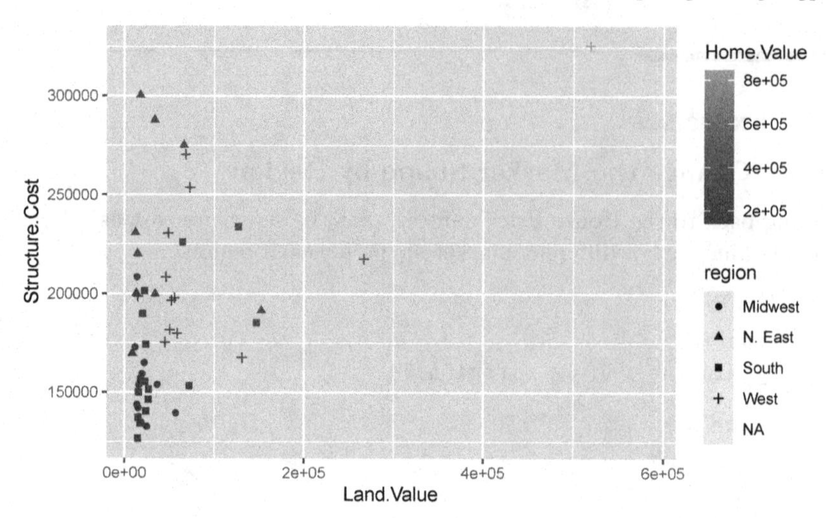

■ END OF SOLUTION

8.9.3 Market Cap by Countries

Plot again the weights of ACWI consituents as a function of market capitalizations, colored by country, but only for the top 10 countries.

■ SOLUTION

```
top.countries = acwi %>%
  group_by(Country) %>%
  summarize(count.by.country = n_distinct(Ticker)) %>%
  arrange(desc(count.by.country)) %>%
  head(10) %>%
  select(Country)
```

```
acwi %>%
  filter(Country %in% unlist(top.countries)) %>%
  group_by(Country) %>%
  ggplot() +
    geom_point(aes(x = Market.Cap, y = Weight, color =
                     Country)) +
    theme(legend.text = element_text(size=10)) +
    theme(axis.text.x =
        element_text(angle = 90, vjust = 0.5, hjust=1))
```

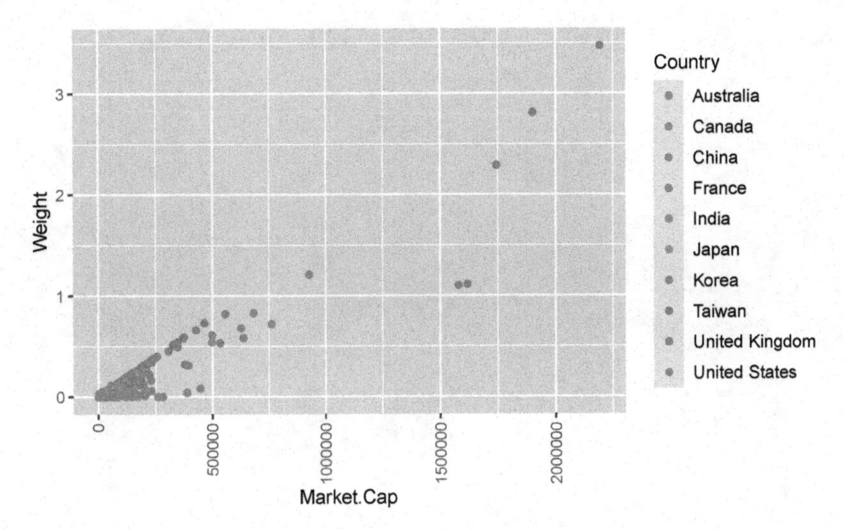

■ END OF SOLUTION

9

Returns and Returns-based Statistics

Most of the data we discussed so far are either a description of the investment (market capitalization of a company, for example), or are price histories. But as we hinted to already, the price of an asset does not matter much, but your **profit or loss** does: you need to take into account how much you paid for that asset.

But that's not enough: Imagine you paid \$1 for one asset and you sell it for \$2, thus pocketing \$1, and another asset required an investment of \$10,000 and is now worth \$15,000, which one is the best investment? You may argue that making \$1 on the first trade is hardly a life changer, whereas making \$5,000 on the second investment will fund some nice vacation. But I would retort that investing ten thousand times in the first asset would have made you an even nicer 10 grands for the same initial outlay.

In conclusion, what matters is the *percentage increase* in the value of your investment. That may sound obvious, but investors are not always rational. That's why **stock splits** are still a thing even though they in no way change the value of a company and that most trade brokers now allow us to buy **fractional shares**, i.e., a portion of a stock, if one share is too costly. So let's focus on definining and calculating returns.

9.1 Single-Period Returns

Let P_t be the price (in, say, dollars) of an asset at time t. Assuming no additional payments (such as dividends) are received from your investment, the profit (or loss) from time $t-1$ to time t is simply the difference in value, $P_t - P_{t-1}$, also expressed in dollars. But profits expressed in dollars are not comparable, because they depend on how much was invested in the first place, as said earlier. In contrast, returns are profits normalized as percentages.

The simplest form of return is called net return in some financial engineering books but for practitioners, "net" and "gross" refer to with or without fees, so we will eschew these terms in this book. Instead, we simply call **return** the

DOI: 10.1201/9781003320555-9

profit or loss divided by the price at the beginning of the period:

$$R_t = \frac{P_t - P_{t-1}}{P_{t-1}}$$

We immediately note that

$$\frac{P_t - P_{t-1}}{P_{t-1}} = \frac{P_t}{P_{t-1}} - 1.$$

The **ratio return** (also called "gross return" in some financial engineering books, but we will not follow that terminology) is defined as current price divided by the price at the end of the previous period:

$$\text{ratio return} = \frac{P_t}{P_{t-1}}.$$

We also observe immediately that $\frac{P_t}{P_{t-1}} = 1 + R_t$.

For example, if $P_{t-1} = 10$ and $P_t = 10.3$ then the ratio return is

$$1 + R_t = \frac{P_t}{P_{t-1}} = \frac{10.3}{10} = 1.03$$

and the return is $R_t = 0.03$ or $R_t = 3\%$

Dividing P_t by P_{t-1} doesn't make a lot of practical sense, but it's the property that the ratio returns equal $1 + R_t$ that makes them important – because $1 + R_t$ is the basis of **geometric compounding** discussed on page 159.

The ratio return can also be flipped on its head: if you know what value to expect at t for an asset, the ratio return can also be seen as telling you how much you should be willing to pay at $t - 1$ if you require a certain return:

$$P_{t-1} = \text{ratio return}^{-1} P_t$$

What we will later call a **discount factor** follows the same logic. Finance, because of its siloed sub-fields, tends to have multiple names for the same concepts.

Note also that returns are scale-free, meaning that they do not depend on monetary units (dollars, cents, etc.), but that they depend on the time frame: is a return for a day, a month or a year? A 1% return you can earn every day is better than 100% in a year!

Another point to keep in mind is that, if you invest \$1, what you can earn is unbounded, so the return and the ratio return are unbounded. However, the most you can lose is \$1: The value of your investment can't go below 0. So $\frac{P_t}{P_{t-1}} \geq 0$, and returns have a lower bound of $R_t \geq -1$. This is a big problem for modeling, because the domain (support) of either the ratio return or the regular return is not symmetric. We'll come back to this point.

9.2 Multiple Periods

Once we have returns for short periods, such as days, we typically need to calculate the return over a longer periods, such as a month or a year. Starting at time step $t = 0$, the return at time t after t periods can be written as:

$$R_{t,t} = \frac{P_t - P_0}{P_0} = \frac{P_t}{P_0} - 1$$

More generally, we denote with $R_{t,k}$ the return at time t over k periods $t-k, t-k+1, \ldots, t$[1]. The most accurate way to compound returns, then, is **geometric compounding**:

$$1 + R_{t,t} = \frac{P_t}{P_0} = \frac{P_t}{P_{t-1}} \frac{P_{t-1}}{P_{t-2}} \cdots \frac{P_1}{P_0} = (1 + R_t)(1 + R_{t-1}) \ldots (1 + R_1)$$

This leads to the very useful Equation (9.1):

$$R_{t,t} = \frac{P_t - P_0}{P_0} = (1 + R_t)(1 + R_{t-1}) \ldots (1 + R_1) \; - \; 1 \qquad (9.1)$$

Let's work out an example. Assume the value of an investment on 4 consecutive dates (with the initial day being Day 0) are $P_0 = \$200$, $P_1 = \$210$, $P_2 = \$206$ ad $P_3 = \$212$. The different $R_{t,k}$ returns are as follows: There are three $R_{t,1}$'s: $R_{1,1}$, $R_{2,1}$ and $R_{3,1}$. Their values are, respectively: 0.05, -0.019 and 0.0291. There are thus also three $1+R_{t,1}$'s: $1+R_{1,1}$, $1+R_{2,1}$ and $1+R_{3,1}$. Their values are, respectively: 210/200 = 1.05, 206/210 = 0.981 and 212/206 = 1.0291[2]. There are two $1 + R_{t,2}$'s: 206/200 = 1.03 and 212/210 = 1.0095. And there is only one $1 + R_{t,3}$, $1 + R_{3,3}$: 212/200 = 1.06.

This can be summed up in the following table:

time	0	1	2	3
P	200	210	206	212
1+ *Rt*1		1.05	0.981	1.03
1+ *Rt*2			1.03	1.01
1+ *Rt*3				1.06

So $R_{3,3} = 1.05 \times 0.981 \times 1.0291 - 1 = 0.06$, or 6%.

Note that, in R, a simple way to implement the geometric compounding of the $R_{t,1}$'s, per Equation (9.1), is to add 1 to a vector of returns then to use **`prod()`** to calculate their product:

[1] There are indeed k periods, not $k + 1$.

[2] You may wonder why I typed all these decimals? I did not. More on this on page 376.

```
R.t.1 = c(0.05, -0.01905, 0.0291)
one.plus.R.t.1 = 1 + R.t.1
(compounded.return = prod(one.plus.R.t.1) - 1)
```

```
[1] 0.05997
```

This of course matches the 6% we calculated a moment ago.

If you need each of the successive geometrically compounded returns, i.e. the sequence of $R_{1,1}, R_{2,2}, \ldots, R_{t,t}$, then **cumprod()** is here to help you:

```
cumprod(one.plus.R.t.1) - 1
```

```
[1] 0.05000 0.03000 0.05997
```

The results are $R_{1,1}$, $R_{2,2}$ and $R_{3,3}$, respectively. We can verify of course that the last value, 6%, is the same value as the one computed immediately earlier.

The average of these returns (i.e. an average *over multiple incremental periods*) should not be calculated using the usual arithmetic mean but using the geometric mean, i.e.:

$$\text{Average Return} = \sqrt[t]{(1 + R_t)(1 + R_{t-1}) \ldots (1 + R_1)} - 1$$

From that expression, it follows that a daily return can be annualized as:

$$R_{annualized} = (1 + R_{daily})^{TradingDays} - 1$$

Reciprocally, a return over N days can be averaged into a daily return as follows:

$$R_{daily} = \sqrt[N]{(1 + R_{N\,days})} - 1$$

9.3 Prices and Adjusted Prices

A stock's price doesn't reveal all we need to know about how much money we make investing in it. For example, if a stock's price was $3 at markets' close yesterday and is $2.50 today, it doesn't necessarily mean you lost money: the company could have paid out a dividend of $1 this morning, in which case you now have $3.50 – a profit of 50 cents and a 16.67% **total return**, including the dividend. Or the stock could have had a 3-to-1 split, and your 1 share valued at $3 yesterday became today 3 shares each valued $2.50, so your return is $\frac{3 \times 2.50 - 1 \times 3}{1 \times 3} = 1.5$ or 150%.

One way to account for this is to adjust yesterday's price so we can better compare it to today's price. So the **adjusted price** is calculated backward in time, starting from today's price and back. Today's price is not adjusted, so today's "adjusted price" is the price currently quoted. Then, the multiplier applied to yesterday's price is:

$$\frac{\text{Today's price}}{\text{Today's price} + \text{dividend}},$$

so the adjusted price is:

$$\text{Yesterday's adjusted price} = \frac{\text{Today's price}}{\text{Today's price} + \text{dividend}} \text{Yesterday's price}.$$

You can verify that the total return calculated from non-adjusted prices,

$$\frac{\text{Today's price} + \text{dividend} - \text{Yesterday's price}}{\text{Yesterday's price}},$$

equals that calculated from adjusted levels, or

$$\frac{\text{Today's price} - \text{Yesterday's adjusted price}}{\text{Yesterday's adjusted price}}.$$

In our first example, yesterday's close price is $\frac{2.5}{2.5+1} \times 3 = 2.1429$, and the total return is indeed $\frac{2.1429}{2.5} - 1 = 16.67\%$.

This backward-calculated adjusted return is what is reported by, among other, **Yahoo Finance**. For example, pulling Yahoo data through the **quantmod** package as of February 7, 2022, the command **getSymbols()** produces the following data frame (some columns were omitted for clarity):

```
getSymbols("AAPL", from = "2022-02-01", to ="2022-02-07")
AAPL
```

```
           AAPL.Close  AAPL.Adjusted
2022-02-01     174.6          174.4
2022-02-02     175.8          175.6
2022-02-03     172.9          172.7
2022-02-04     172.4          172.4
2022-02-07     171.7          171.7
```

You'll notice that the adjusted prices are not the same as the **AAPL.Close** column. More precisely, the **AAPL.Close** and **AAPL.Adjusted** columns differ only until February 3.

The reason we had to specify an as-of day is that these adjusted prices are, as explained earlier, calculated backward from the last (current) day. So, as of close on February 7, the price of an Apple share was $171.70, and there's no need for adjustment. But going back we see that an adjustment of close price was needed on Feb 3. We can verify that a dividend was paid recently using the **getDividends()** function:

```
(getDividends("AAPL", from = "2022-02-01", to = "2022-02-07"))
```

Since the ex-dividend date was February 4, an adjustment was indeed required on February 3. (Remember, adjustment calculations go backward in time.)

The **getDividends()** function also has side-effects in that it creates a variable called with the name of the ticker, here **AAPL**, followed by **.div**.

Before we move on, let's come back to **getSymbols()** and the different fields it provides. These fields report the open, close, high, low prices for each trading session, and the trading volume and adjusted price for that day. Because the result is a data frame, each column can be retrieved using the **$** notation. However, package **quantmod** offers a series of helper shorthands named, respectively, **Op()**, **Cl()**, **Hi()**, **Lo()**, **Vo()** and **Ad()**.

Another way to pull the same data uses **tq_get()** what we saw on page 21:

```
library(tidyquant)
(aapl.tq = tq_get('AAPL',
                  from = "2022-02-01",
                  to   = "2022-02-07",
                  get  = "stock.prices"))
```

```
# A tibble: 6 x 8
date         open  high   low close    volume adjusted
<date>      <dbl> <dbl> <dbl> <dbl>     <dbl>    <dbl>
2022-02-01  174.0    174.8    172.3    174.6    86213900    174.4
2022-02-02  174.8    175.9    173.3    175.8    84914300    175.6
2022-02-03  174.5    176.2    172.1    172.9    89418100    172.7
2022-02-04  171.7    174.1    170.7    172.4    82465400    172.4
```

Again, the results look slightly different, but they are the same after we account for rounding. This is not a surprise: both pull their data from the same source, **Yahoo Finance**.

Finally, note that **tq_get()** can also allow pull historical prices of multiple stocks at the same time. For example:

```
basket = tq_get(c("AAPL", "GOOG", "NFLX"),
                from = "2019-04-01",
                to   = "2021-04-01")
head(basket, 3)
```

```
# A tibble: 3 x 8
  symbol date        open  high   low close    volume
  <chr>  <date>     <dbl> <dbl> <dbl> <dbl>     <dbl>
```

```
1 AAPL   2019-04-01  47.9  47.9  47.1  47.8 111448000
2 AAPL   2019-04-02  47.8  48.6  47.8  48.5  91062800
3 AAPL   2019-04-03  48.3  49.1  48.3  48.8  93087200
# ... with 1 more variable: adjusted <dbl>
```

9.4 Returns

A bit earlier, we saw how to pull **adjusted prices** for Apple using **getSymbols**. For instance:

```
getSymbols("AAPL", from = "2021-11-24", to = "2021-11-26")
```

Remember that adjusted prices already contain the effects of dividends stocks is **stock splits**. We can extract adjusted prices as follows:

```
(aapl.adjusted = as.numeric(AAPL$AAPL.Adjusted))
```

```
[1] 161.7 156.6
```

or equivalently, using function **Ad()** seen on page 162:

```
(aapl.adjusted2 = Ad(AAPL))
```

```
           AAPL.Adjusted
2021-11-24         161.7
2021-11-26         156.6
```

From these adjusted prices, daily total returns can be calculated as follows:

```
aapl.adjusted[-1]/aapl.adjusted[-length(aapl.adjusted)]-1
```

```
[1] -0.03168
```

The **periodReturn()** function in the **quantmod** package calculates that directly:

```
periodReturn(AAPL, period="daily")
```

```
           daily.returns
2021-11-24      0.007403
2021-11-26     -0.031678
```

The main difference is that **periodReturn()** calculated an additional data point, the daily return for November 24 based on the close on November 23 that we did not get when we used **getSymbols()**.

Another way to calculate returns from successive prices uses the **Return.calculate()** function of the **PerformanceAnalytics** package:

```
library(PerformanceAnalytics)
(aapl.ret = Return.calculate(aapl.adjusted2))
```

```
           AAPL.Adjusted
2021-11-24            NA
2021-11-26      -0.03168
```

The two functions, **periodReturn()** and **Return.calculate()**, are equivalent. But whereas **Return.calculate()** assumes regular price data and will calculate returns for these regular price intervals (e.g., daily returns in the example above), the **periodReturn()** function also allows us to calculate returns over periods different from the input data, such as annually or quarterly based on daily data. For example, the two commands below show calendar-year returns, plus the **year-to-date** performance:

```
getSymbols("AAPL", from="2018-01-01")
```

```
[1] "AAPL"
```

```
periodReturn(AAPL, period="yearly")
```

```
           yearly.returns
2018-12-31       -0.07299
2019-12-31        0.86161
2020-12-31        0.80746
2021-12-31        0.33823
2022-04-14       -0.06916
```

So **periodReturn()** allows us to aggregate returns with any standard calendar periodicity. But another function, **table.CalendarReturns()** of the **PerformanceAnalytics** package provides a similar functionality that you can find useful:

```
aapl.monthly = periodReturn(AAPL, period="monthly")
table.CalendarReturns(aapl.monthly)
```

```
      Jan   Feb  Mar  Apr   May  Jun  Jul  Aug   Sep
2018 -1.6   6.4 -5.8 -1.5  13.1 -0.9  2.8 19.6  -0.8
2019  5.5   4.0  9.7  5.6 -12.8 13.1  7.6 -2.0   7.3
2020  5.4 -11.7 -7.0 15.5   8.2 14.7 16.5 21.4 -10.3
2021 -0.6  -8.1  0.7  7.6  -5.2  9.9  6.5  4.1  -6.8
2022 -1.6  -5.5  5.7 -5.3    NA   NA   NA   NA    NA
      Oct   Nov   Dec monthly.returns
2018 -3.0 -18.4 -11.7            -7.3
2019 11.1   7.4   9.9            86.2
2020 -6.0   9.4  11.5            80.7
2021  5.9  10.3   7.4            33.8
2022   NA    NA    NA            -6.9
```

The annual returns, expressed as percentages in the rightmost column, are of course equal to those calculated earlier using **periodReturn()** with **period="yearly"**. Be careful, however, that **table.CalendarReturns()** currently assumes monthly returns as an input and does not handle other time scales.

Rolling Returns

The functions above divide time in calendar periods, which are usual but arbitrary. In contrast, plots of **rolling performance** are often more informative. For example, we can plot annualized returns over rolling windows of 12 months with the code below, producing the plot in Figure 9.1.

```
chart.RollingPerformance(R = aapl.monthly,
           width = 12,
           main  = "Rolling 12-mo Annualized Return",
           FUN   = "Return.annualized")
```

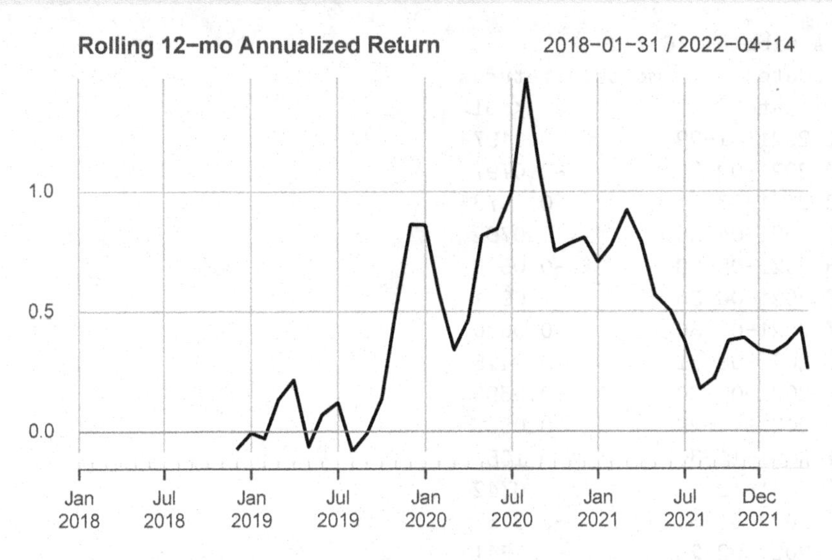

FIGURE 9.1 Rolling 12-month Annualized Performance of Apple

Technical Note on Mixing Quantmod and Tidyquant

As a side-note, be careful that **quantmod** and **tidyquant** functions do not always mix well. For example, let's pull the daily prices of Apple again using **tq_get()**, this time on a longer period to derive a recent history of monthly returns:

```
aapl.tq = tq_get('AAPL', from = "2021-01-01", to = "2022-04-12")
```

This table of daily prices cannot be used by **periodReturn()**, at least not directly, to produce monthly returns:

```
periodReturn(aapl.tq, period="monthly")
```

```
Error in try.xts(x) :
  Error in as.POSIXlt.character(x, tz, ...) : character string
  is not in a standard unambiguous format
```

To fix this, we can call **periodReturn()** within **tq_transmute()**, which arranges things for us:

```
aapl.tq %>%
    tq_transmute(select      = adjusted,
                 mutate_fun = periodReturn,
                 period     = "monthly")
```

```
# A tibble: 16 x 2
   date       monthly.returns
   <date>               <dbl>
 1 2021-01-29         0.0197
 2 2021-02-26        -0.0797
 3 2021-03-31         0.00734
 4 2021-04-30         0.0762
 5 2021-05-28        -0.0505
 6 2021-06-30         0.0991
 7 2021-07-30         0.0650
 8 2021-08-31         0.0425
 9 2021-09-30        -0.0680
10 2021-10-29         0.0587
11 2021-11-30         0.105
12 2021-12-31         0.0742
13 2022-01-31        -0.0157
14 2022-02-28        -0.0541
15 2022-03-31         0.0575
16 2022-04-11        -0.0507
```

This output can then be piped directly into a variety of additional functions. One of them is a general performance statistics function called **tq_performance()**. That function in turn calls one of multiple predefined functions specified with the **performance_fun=** parameter:

```
aapl.tq %>%
     tq_transmute(select      = adjusted,
                  mutate_fun = periodReturn,
                  period     = "monthly") %>%
```

```
tq_performance(Ra = monthly.returns,
               Rb = NULL,
               performance_fun = table.AnnualizedReturns)
```

```
# A tibble: 1 x 3
  AnnualizedReturn `AnnualizedSharpe(~ AnnualizedStdDev
             <dbl>                <dbl>            <dbl>
1            0.211                0.965            0.218
```

The leftmost entry in the table is the average return for the period, as we requested. The rightmost entry in the table is the annualized standard deviation, or annualized volatility, over the entire period. We are going to come back to this concept in the next section. The entry in the middle, the annualized Sharpe, is the topic of the section after next, Section 9.6. Finally, the reader is referred to page 317 for other uses of the **tq_performance()** function, such as the calculation of Value at Risk.

9.5 Volatility

A key measure of the riskiness of an investment is the **volatility** of its returns. Volatility is the standard deviation of returns. If returns are daily, then their standard deviation is a daily number. However, this number can be scaled to the desired time frame. Scaling a daily volatility to a monthly volatility involves multiplying it by the square root of number of trading days in a month, typically 22. Similarly, **annualized volatility** equals daily volatility multiplied by the square root of the number of trading days per year, typically 252. Likewise, a monthly volatility can be annualized by multiplying it by the square root of 12.

How many past data points should be taken into account to calculate volatility is another parameter. Practitioners often calculate annual standard deviations from 12 monthly returns, but 12 is too low. More data points are recommended, for example using weekly returns and multiplying their standard deviation by the square root of 52.

Moreover, irrespective of the choice of the parameters, volatility numbers change every day: there is not *one* annualized volatility for Apple's stock, for example: every day, a new data point is added, and the oldest daily return rolls off. In the example below, we calculate standard deviation over a **rolling window** of 22 days, scale it to an annualized volatility, and plot the graph.

```
getSymbols("AAPL", from = "2021-05-01", to="2022-01-13")
```

```
[1] "AAPL"
```

```
aapl.rets = periodReturn(AAPL, period = "daily")
```

Moreover, to save time and reduce the risk of calculation errors (and get a nice plot, too!) we can use the **chart.RollingPerformance()** function of the **PerformanceAnalytics** analytics package that we introduced on page 5. The result appears in Figure 9.2.

```
chart.RollingPerformance(R=aapl.rets,
                         width=22,
                         FUN="sd.annualized",
                         scale=252,
                         main="Rolling 1-month volatility")
```

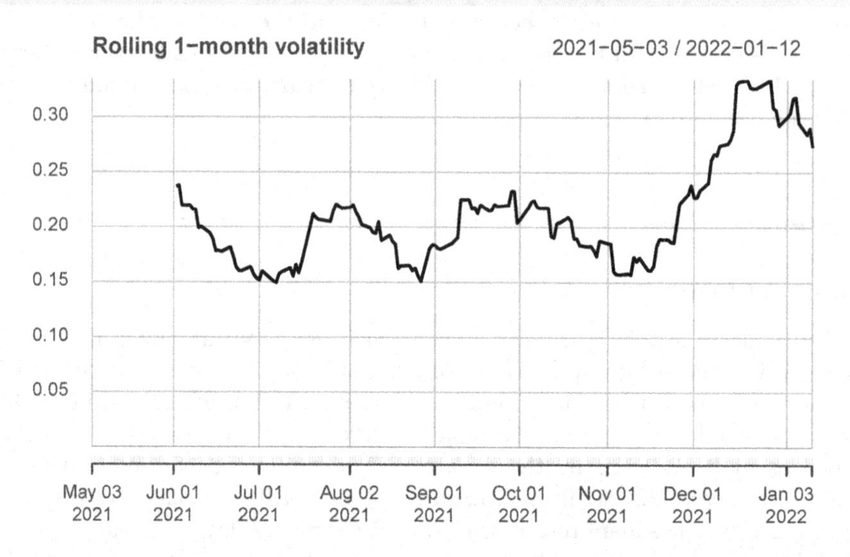

FIGURE 9.2 Rolling Volatility, annualized from rolling windows of 22 trading days

Figure 9.2 clearly shows the ups and downs of volatility, which can be due to events at the companies (such as the introduction of a new product; **earning calls** organized by company management, typically quarterly, can also have an impact) but can also be due to changes in the market environment.

9.6 Sharpe

So we know how to calculate the average return of an investment over a period and we saw how to calculate volatility, which is the way to measure the riskiness of that investment. So what is best: more return, or less risk?

If two investments have the same return, of course you'll want the one with the lower risk, and if they have the same risk, of course you'll favor the one with the highest return. There's a trade-off here, and what you probably want is the investment that offers the higher return *per unit of risk*. That is, you want the highest ratio of return divided by volatility. That idea is essentially the idea of what is called the **Sharpe ratio**.

There is just one subtlety left. Let's assume that cash deposits at a bank earn a return of, say, 3% per year. Cash deposits at a bank are essentially risk-free – in fact, deposits of $250,000 are automatically insured by the **FDIC**, or **Federal Deposit Insurance Corporation**, so they really are risk-less. A stock investment should thus be judged on its return *beyond* a "floor" of 3%. In fact, when you buy a company's stock, that company could take your money, do nothing but deposit it in a savings account, get the 3% interest, and buy back your share. Would that be a good investment?

The **Sharpe ratio** of an investment is thus the average of the periodic returns that investment achieved in excess of the risk-free rate, divided by the volatility of the excess returns, measured at the same periodicity. More formally, if R is the series of successive periodic returns and r_f is the series of prevailing interest rates available at the bank on the same periods, then the Sharpe ratio is:

$$Sharpe = \frac{\mathbf{E}[R - r_f]}{\text{StdDev}(R - r_f)} \tag{9.2}$$

Keep in mind that, in Equation (9.2), **E**[..] is the **expected value** function, which is the probability-weighted average but is, in many cases included here, equivalent to the simple arithmetic average.

In practice, and in particular when interest rates are as low as they have been from 2009 until the time of this writing (2022), people often simply divide the average return realized on an investment by the volatility of its return. Otherwise, the calculations require a little bit more care. Fortunately, functions like **`table.AnnualizedReturns()`** does the calculation for us. The risk-free rate is specified with the same periodicity as the returns, in this case monthly – so the rate of 0.3% compounds to about 3.6%, annualized.

```
aapl.monthly = periodReturn(AAPL, period="monthly")
table.AnnualizedReturns(aapl.monthly, Rf = 0.003)
```

```
                              monthly.returns
Annualized Return                      0.4617
Annualized Std Dev                     0.2222
Annualized Sharpe (Rf=3.6%)            1.8513
```

A back-of-the-envelope approximation would be to take the volatility of the investment as the denominator instead of the standard deviation of the differences between R and r_f. That approximation gives us:

$$\frac{0.4617 - 0.036}{0.2222} = 1.916$$

which is not too far off from the exact calculation provided by **`table.AnnualizedReturns()`**.

9.7 Drawdowns

A **drawdown** is the loss in the value of your investment from its last peak. To calculate its value on any day (or at any time), find the maximum value this investment has had in the past, and subtract today's price from that peak price: The difference is the current dollar loss, per share. A better (because it is comparable) metric is to divide that loss by the latest peak price, resulting in the drawdown, a unit-less percentage that is always less than or equal to zero.

Here again, the **`PerformanceAnalytics`** comes to our help to calculate this metric thanks to the **`chart.Drawdown()`** function, whose output appears in Figure 9.3. Note that the drawdown is reset to zero as soon as the stock reaches a new peak.

```
chart.Drawdown(aapl.rets)
```

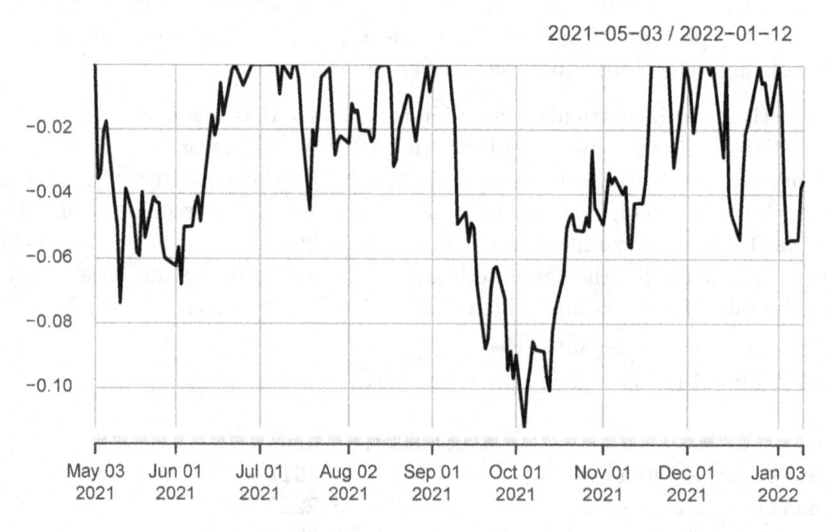

FIGURE 9.3 Drawdown of Apple's Stock in the period under consideration.

As can be read on this graph, Apple's stock reached a peak in early September 2021 but then lost some of its ground until early October. The dip of the graph means that an investor unfortunate enough to have bought the stock at its peak in September would have lost about 11% if we eyeball the graph. On the right side of the plot, we also see that the latest data point is also below the top line, indicating that the latest day in the time series also is in drawdown (relative to the highest-ever peak reached by the stock).

The five worst drawdowns in the period included in a time series of returns can also be tabulated using the **table.Drawdowns()** function:

```
table.Drawdowns(aapl.rets)
```

```
        From     Trough         To   Depth Length
1 2021-09-08 2021-10-04 2021-11-18 -0.1120     52
2 2021-05-04 2021-05-12 2021-06-22 -0.0737     35
3 2022-01-04 2022-01-06       <NA> -0.0550      8
4 2021-12-13 2021-12-20 2021-12-27 -0.0541     10
5 2021-07-15 2021-07-19 2021-08-16 -0.0449     23
  To Trough Recovery
1        19       33
2         7       28
3         3       NA
4         6        4
5         3       20
```

The drawdowns are sorted by decreasing depth. This table confirms that a drawdown of 11.2% was observed from September 8 to October 4, and that investors had to wait until November 18 to break even. The third worst drawdown in the period started on Jan 4, 2022 and saw its trough on Jan 6. The `NA` in the `To` column indicates that this drawdown is still ongoing as of Jan 13, 2022.

9.8 Benchmark-Relative Performance and Risk

Performance and risk are often assessed relative to a benchmark. Let's take a longer history for Apple's stock as an example, and the corresponding history for the **S&P 500** index, here proxied by the **VOO** ETF:

```
aapl.monthly =
  tq_get("AAPL", from = "2017-01-01", to="2021-01-01") %>%
  tq_transmute(select      = adjusted,
               mutate_fun = periodReturn,
               period     = "monthly")
benchmark =
```

```
  tq_get("VOO", from = "2017-01-01", to="2021-01-01") %>%
  tq_transmute(select      = adjusted,
               mutate_fun = periodReturn,
               period     = "monthly")
```

The functions we saw earlier are able to do benchmark-relative analysis but the benchmark returns have to be side-by-side with that of the stock or portfolio. One possibility, if you are absolutely sure that the dates in both time series are aligned, is to use `cbind()`. A much safer and cleaner way to bring the two return series into a single table is to perform a **join** using for example the `merge()` function:

```
stock.and.benchmark = merge(aapl.monthly, benchmark, by="date")
```

See also the section on time series on page 58 and more details on `merge()` on page 61.

Remember that, because the two tables being merged both have a column called "date" and a column called "monthly.returns," we must specify which column should be used as a **key** for the merge.

After the merge, the resulting table has 3 columns: the dates, and the monthly returns suffixed with ".x" and ".y" to identify the two series. The table can then be passed to `tq_performance()`:

```
stock.and.benchmark %>%
  tq_performance(monthly.returns.x,
                 monthly.returns.y,
                 performance_fun = table.CAPM)
```

```
# A tibble: 1 x 12
  ActivePremium  Alpha AnnualizedAlpha  Beta `Beta-`
          <dbl>  <dbl>           <dbl> <dbl>   <dbl>
1         0.325 0.0204           0.275  1.26   0.979
# ... with 7 more variables: Beta+ <dbl>,
#   Correlation <dbl>, Correlationp-value <dbl>,
#   InformationRatio <dbl>, R-squared <dbl>,
#   TrackingError <dbl>, TreynorRatio <dbl>
```

The **alpha** and **beta** statistics correspond to the intercept and slope, respectively, of the line of best fit when regressing the stock's returns to that of the benchmark. This can be checked quickly:

```
lm(stock.and.benchmark$monthly.returns.x ~
     stock.and.benchmark$monthly.returns.y)$coefficients
```

```
(Intercept)
    0.02043
```

```
stock.and.benchmark$monthly.returns.y
                              1.25920
```

The intercept is for monthly returns, so the annualized alpha is indeed:

```
(1+0.02043)^12 - 1
```

```
[1] 0.2747
```

Observe that, because we know the structure of the object returned by `lm()`, we can use the `$` operator to pull the **`coefficients`** directly from `lm()`.

Beta is an important measure of risk because it captures how sensitive an investment is to the gyrations, up or down, of the broader market. (Most frequently, beta refers to the sensitivity of a stock to the movements of a broad index, like the **S&P 500** or the **Russell 1000** indices.) It is the slope of the line of best fit, and that is equal to the correlation $\rho_{r,b}$ of the investment's returns to that of the index *times* the ratio of that investment's volatility σ_r over that of the benchmark, σ_b. In equation form, the beta of an investment relative to a benchmark b is:

$$\beta_{r,b} = \rho_{r,b} \frac{\sigma_r}{\sigma_b} \tag{9.3}$$

If you don't trust your ability at performing linear regressions, you also call the **`CAMP.beta()`** function of the **`PortfolioAnalytics`** package. One requirement of that function, however, is that the returns be in the form of time series. Constructing a time series object is done using **`ts()`** function. That function's syntax is **`ts(vector, start=, end=, frequency=)`** where **`vector`** is the set of values in historical order, **`start`** and **`end`** are the times of the first and last observations, and **`frequency`** is the number of observations per year (so 1 = annual, 4 = quarterly, 12 = monthly, etc.)

```
library(PortfolioAnalytics)
apple.as.ts = ts(stock.and.benchmark$monthly.returns.x,
                 start = c(2017,1),
                 frequency = 12)
sp500.as.ts = ts(stock.and.benchmark$monthly.returns.y,
                 start = c(2017,1),
                 frequency = 12)
```

We can now call **`CAMP.beta()`** to get the beta of Apple to the S&P 500:

```
CAPM.beta(apple.as.ts, sp500.as.ts)
```

```
[1] 1.259
```

This is of course the same value as the slope calculated earlier.

The next column in the table produced by `tq_performance()` was `Beta-`, which deserves some explanation.

Investors look at how an investment reacts to up and down markets. Standard measures include the **up-capture** and **down-capture**, which are the average returns of the investment on months (or other periods) when the reference benchmark was up and down, respectively. Similarly, we can separate beta when the market is up (called **beta+**, or at times the **bull beta**) from when the market is down (**beta-** or **bear beta**). The `Beta-` number reported by `tq_performance()` can be verified using `filter()` to select only the periods when the stock market's return was negative:

```
sp.down = filter(stock.and.benchmark, monthly.returns.y<0)
lm.down = lm(sp.down$monthly.returns.x ~
                       sp.down$monthly.returns.y)
lm.down$coefficients
```

```
            (Intercept) sp.down$monthly.returns.y
              -0.001357                  0.979412
```

The 0.979 number matches what we saw earlier, and the same can be done for `Beta+`:

```
sp.up = filter(stock.and.benchmark, monthly.returns.y>=0)
lm(sp.up$monthly.returns.x ~
     sp.up$monthly.returns.y)$coefficients
```

```
           (Intercept) sp.up$monthly.returns.y
               0.02592                 1.15790
```

Yet another way to get the same statistics is to use the `CAPM.beta.bull()` and `CAPM.beta.bear()` functions of the `PortfolioAnalytics` package. To get the **beta-** or **bear beta**, we would write:

```
CAPM.beta.bear(apple.as.ts, sp500.as.ts)
```

```
[1] 0.9794
```

Interestingly, the overall beta is higher than in either the "Up" or "Down" market environments. That's because the relationship in the period has some curvature, as can be seen on the plot in Figure 9.4. (But see page 264.)

```
plot(stock.and.benchmark$monthly.returns.x ~
    stock.and.benchmark$monthly.returns.y,
     xlab="Returns on the SP500",
     ylab="Returns on Apple")
```

But even more convincing is to plot the three market environments and their regression lines separately. First, we add a column denoting the relevant environment:

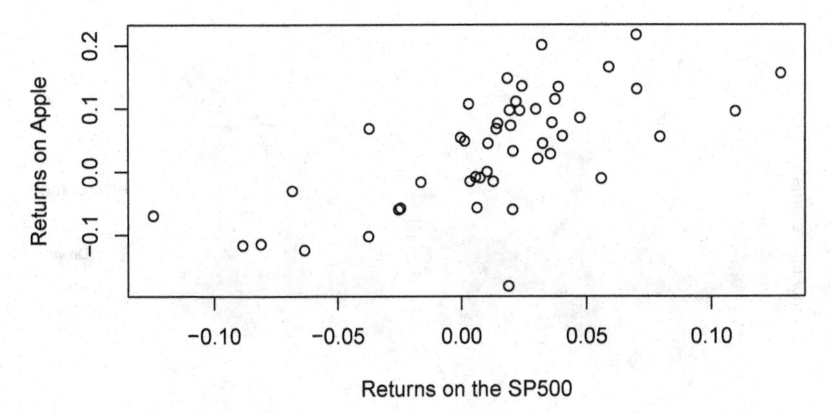

FIGURE 9.4 Return on Apple's Stock against that of a Stock Index

```
stock.and.benchmark$environment = "All"
sp.up$environment = "Up"
sp.down$environment = "Down"
```

Then we bind the dataframes by rows thanks to the function **rbind()** introduced on page 55:

```
data = rbind(stock.and.benchmark, sp.up)
data = rbind(data, sp.down)

ggplot (data, aes(x=monthly.returns.y,
                  y=monthly.returns.x,
                      shape = environment,
                      color = environment,
                      fill= environment)) +
  geom_point(aes(col=environment), size=1) +
  geom_smooth(method="lm")
```

As we can clearly see in Figure 9.5, the regression line that covers both environments (shown in red when you try the code) has a slightly steeper slope than either of the other two regression lines (green and blue, if your version of this book is in color).

Calculating and graphing cumulative performance is offered by the **charts.PerformanceSummary()** function of the **PerformanceAnalytics** package: If you pull the returns of one investment and that of one or multiple comparisons such as benchmarks, the function allows us to graph their cumulative performance and drawdowns over the period. In the code below, note that **^GSPC** is the symbol for the **S&P 500** index.

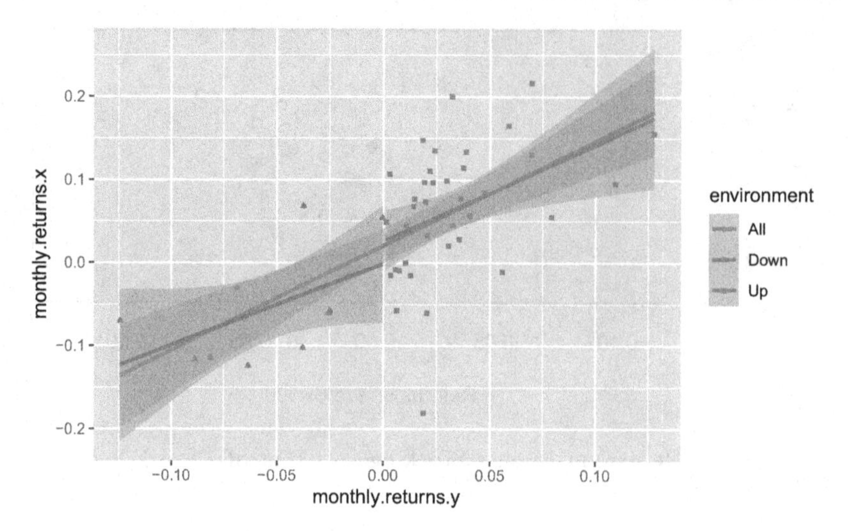

FIGURE 9.5 Different sensitivity of a stock in down and up markets

```
getSymbols("AAPL", from = "2021-05-01", to = "2021-11-01")
```

```
[1] "AAPL"
```

```
aapl.rets = periodReturn(AAPL, period = "daily")
getSymbols("^GSPC",  from = "2021-05-01", to = "2021-11-01")
```

```
[1] "^GSPC"
```

```
sp500.rets = periodReturn(GSPC,  period = "daily")
```

We can now call the **charts.PerformanceSummary()** function. Note that "charts" is plural.

```
all.returns = cbind(aapl.rets, sp500.rets)
colnames(all.returns) = c("Apple", "S&P 500")
charts.PerformanceSummary(all.returns)
```

More than two investments can be plotted in **charts.PerformanceSummary()** but the chart can become crowded. Note also that a different look-and-feel can be obtained by using **ggplot** as the plot engine. The code below, for example, produces the charts in Figure 9.7.

```
charts.PerformanceSummary(all.returns,
                          plot.engine="ggplot2")
```

The charts in Figure 9.7 emphasize drawdown as a measure of risk, but of course volatility, or standard deviation, is also a standard metric. To compare the risk and reward of the two investments, volatility should also be considered. Rolling

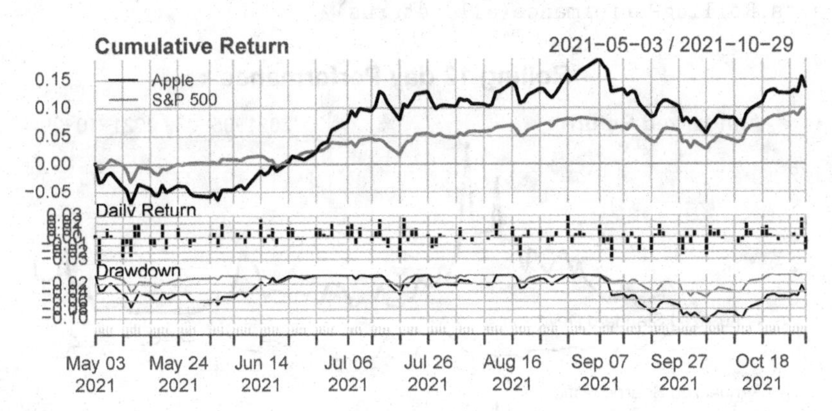

FIGURE 9.6 Cumulative return, drawdown and periodic returns all reported in one chart

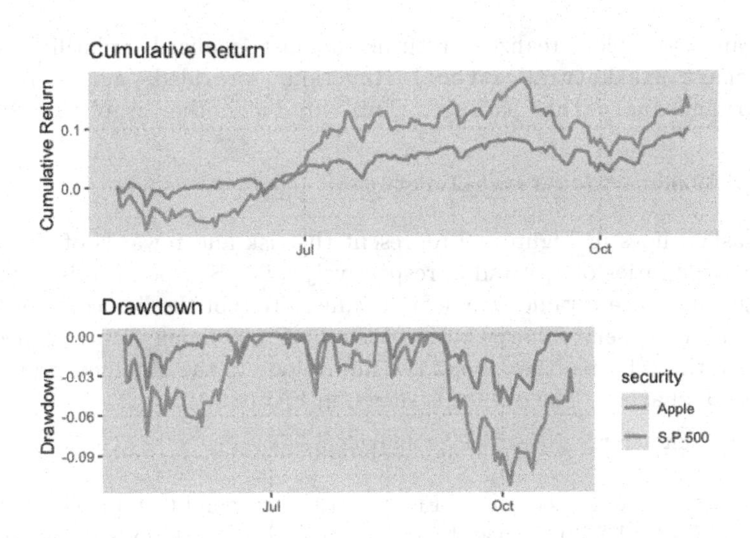

FIGURE 9.7 Illustrating charts.PerformanceSummary() on Apple Stock

performance and rolling standard deviations can be plotted easily thanks to **charts.RollingPerformance()**. The output of the command below appears in Figure 9.8.

```
charts.RollingPerformance(all.returns)
```

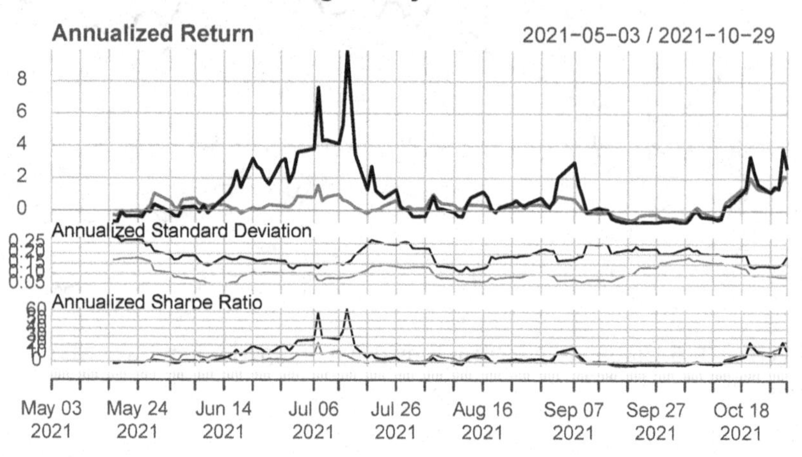

FIGURE 9.8 charts.RollingPerformance() plots rolling performance, volatility and Sharpe ratio

We can also plot realized returns against realized volatility using the **chart.RiskReturnScatter()** function, provided again by the **PerformanceAnalytics** package. This produces the graph in Figure 9.9.

```
chart.RiskReturnScatter(all.returns)
```

The dashed lines in Figure 9.9 represent the risk and rewards of portfolios with Sharpe ratios of 1, 2 and 3, respectively. (The Sharpe ratio is discussed on page 168.) These values are default values. To plot the line corresponding to the actual observe Sharpe ratio of the S&P 500 index, you first need to calculate that Sharpe, then to set the line's slope to the obtained value ratio using **add.sharpe=**. The output is shown in Figure 9.10.

```
chart.RiskReturnScatter(all.returns, add.sharpe = c(1.83))
```

The points on the dashed line in Figure 9.10 correspond to portfolios with an equal Sharpe of 1.83. One way to construct such a portfolio is to blend cash and the S&P 500 in varying proportions. (Of course, any other investment can happen to have the same Sharpe ratio.) Each point on a line to the left of the circle represent portfolios with no more than 100% in the index. If we use **leverage** however, that is if we borrow from a bank to invest more into

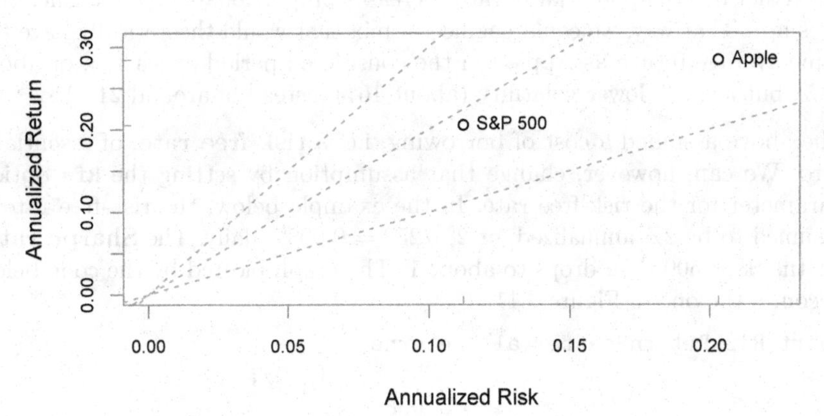

FIGURE 9.9 Risk and Return of Apple and the SP 500 with lines representing investments with Sharpe ratios of 1, 2 and 3

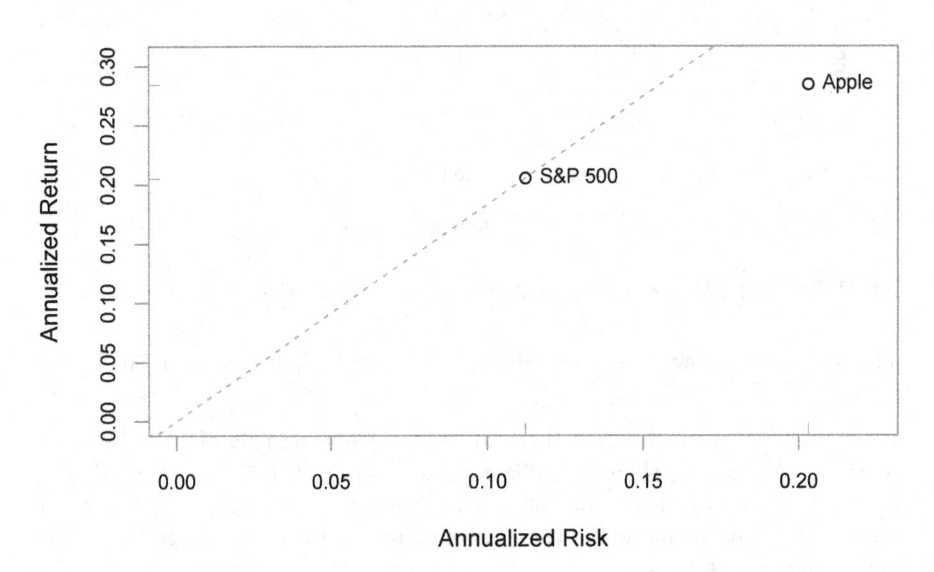

FIGURE 9.10 Risk and Return of Apple and the SP 500, with lines representing investment with a Sharpe ratio of 1.83

the S&P 500, the corresponding portfolio would still plot on the dashed line, but to the right of the circle: the investment becomes more risky (more to the right) but the potential reward increases proportionately (the dashed line goes up). That way, we could engineer a mix that would theoretically have the same average return as Apple (on the considered period at least..), or about 29%, but a much lower volatility (about 16% instead of around 21%).

The above assumed a cost of borrowing, i.e. a **risk-free rate**, of essentially zero. We can, however, change that assumption by setting the `Rf=` option parameter for the risk-free rate. In the example below, the risk-free rate is assumed to be 2% annualized, or 2%/252 = 0.008% daily. The **Sharpe ratio** for the S&P 500 then drops to about 1. The graph plotted by the code below becomes the one in Figure 9.11.

```
chart.RiskReturnScatter(all.returns,
                        add.sharpe = c(1.0),
                        Rf = 0.008)
```

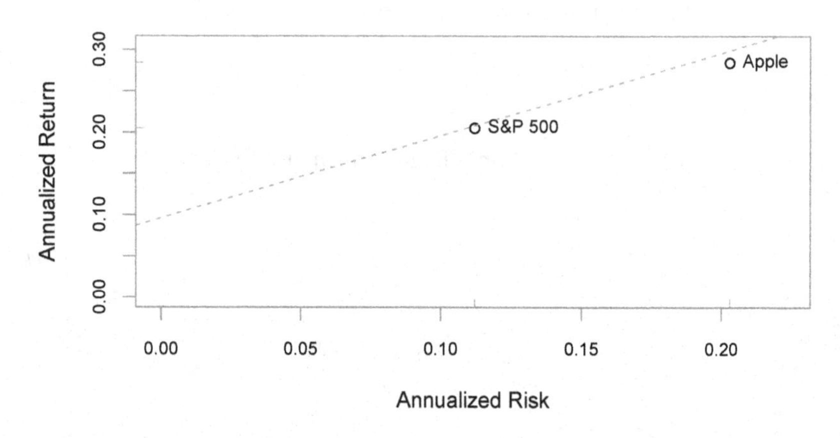

FIGURE 9.11 Risk/Return tradeoffs with a higher risk-free rate

The dashed line now goes essentially through Apple, indicating similar Sharpe ratios.

When comparing two or more investments, consideration should also be given to the dispersion of their periodic returns. One way to gauge that dispersion is to plot **boxplot**. One easy way to produce that graph is to use the **`chart.Boxplot()`** function of the **`PerformanceAnalytics`** package. Note that here, "chart" is singular.

```
chart.Boxplot(all.returns)
```

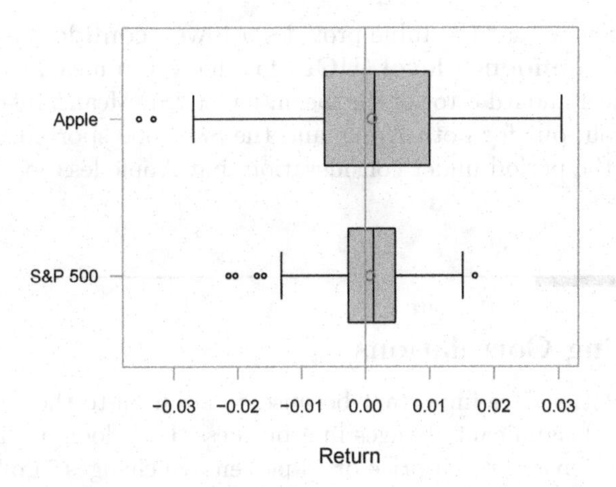

FIGURE 9.12 Box plot of return dispersions for Apple and the SP500 index

Notice that boxplot charts could also be produced, but with some effort, using the **geom_boxplot()** function of **ggplot2** seen on page 143. **chart.Boxplot()** makes it easy: In particular, note that the names of the investments appearing in Figure 9.12 are filled in for us from the head row of **all.returns** – we didn't have to specify these names in the above command.

Of course, such graphs are not enough to appreciate differences in the risk and potential reward of two investments. A table of quantitative properties is often a better tool, and the **table.Stats()** function of the **PerformanceAnalytics** package conveniently provides this service:

```
table.Stats(all.returns)
```

```
                    Apple  S&P 500
Observations     127.0000 127.0000
NAs                0.0000   0.0000
Minimum           -0.0354  -0.0214
Quartile 1        -0.0063  -0.0027
Median             0.0015   0.0013
Arithmetic Mean    0.0011   0.0008
Geometric Mean     0.0010   0.0007
Quartile 3         0.0100   0.0047
Maximum            0.0304   0.0171
SE Mean            0.0011   0.0006
LCL Mean (0.95)   -0.0012  -0.0005
UCL Mean (0.95)    0.0033   0.0020
Variance           0.0002   0.0000
Stdev              0.0128   0.0071
Skewness          -0.3475  -0.4586
```

```
Kurtosis              0.1362    0.6752
```

Note in particular that the table provides a **lower confidence level (LCL)** and a **upper confidence level (UCL)** for the return means. The table also reports on the standard error of the mean, under "SE Mean." Interestingly, the return distributions for both Apple and the S&P 500 sported very negative skew during the period under consideration, but Apple less so.

9.9 Rolling Correlations

Rolling correlations are important because, in addition to their average value, they can reveal significant changes in a business. Let's look at airlines' stocks and their relationship to oil price or to percentage changes in oil price.

```
getSymbols("JBLU", from = "2019-01-01", to = "2021-12-31")
```

```
[1] "JBLU"
```

```
jetblue.rets = periodReturn(JBLU, period = "weekly")
getSymbols("LUV",  from = "2019-01-01", to = "2021-12-31")
```

```
[1] "LUV"
```

```
southwest.rets = periodReturn(LUV,  period = "weekly")
getSymbols("AAL",  from = "2019-01-01", to = "2021-12-31")
```

```
[1] "AAL"
```

```
aal.rets = periodReturn(AAL,  period = "weekly")
getSymbols("DCOILWTICO",  src="FRED",
           from = "2019-01-01", to = "2021-12-31")
```

```
[1] "DCOILWTICO"
```

```
wti.rets = periodReturn(DCOILWTICO,  period = "weekly")
```

As you might remember, the price of oil dropped dramatically in the first half of 2020 due to COVID-19, to the point that **futures** (which are derivative contracts) had a negative price at some point, as shown in Figure 9.13. Note the square-bracket syntax seen on page 58, allowed by **xts** to truncate the time series:

```
plot(DCOILWTICO["2019/2022"])
```

We can chart, in Figure 9.14 the rolling correlation of the weekly returns of these stocks using **chart.RollingCorrelation()** from **PerformanceAnalytics**:

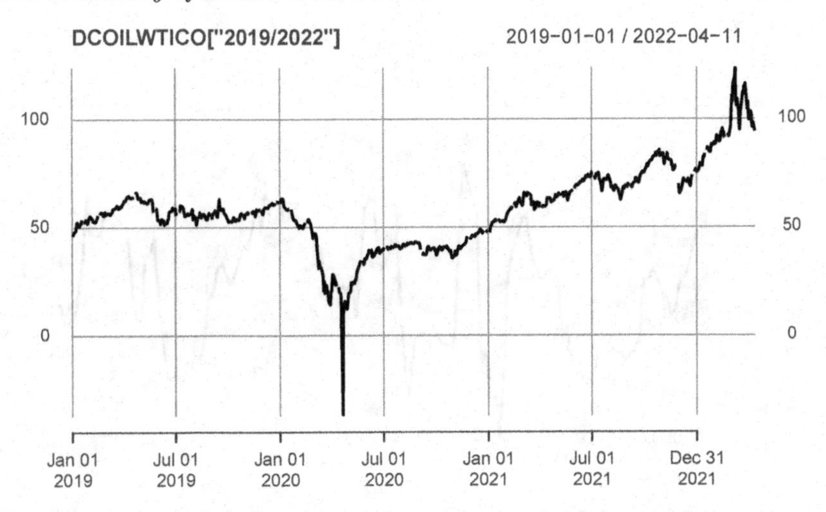

FIGURE 9.13 Plotting a subset of a time series

```
airlines = merge(jetblue.rets, southwest.rets, aal.rets)
colnames(airlines) = c("JBLU", "Southwest", "American")
colnames(wti.rets) = "WTI Oil"
chart.RollingCorrelation(airlines, wti.rets,
                         legend.loc = "bottomleft")
```

We can also investigate the correlation of weekly stock returns with the absolute level of oil price of course, as shown in Figure 9.15:

```
colnames(DCOILWTICO) = "WTI Oil"
chart.RollingCorrelation(airlines, DCOILWTICO,
                         legend.loc = "bottomleft")
```

Despite the sensitivity of airlines' business to oil price, the correlation is hovering around zero. This may be due to their hedging programs that makes their operating costs quite predictable, hence mitigating the impact of oil price on their profits and therefore reducing the correlation of their stock price with oil.

9.10 Normality of Return Distributions

Data analysis would be fancy hand-waiving without statistics, and in particular no forecasting could be anchored in a solid framework. This is very true in

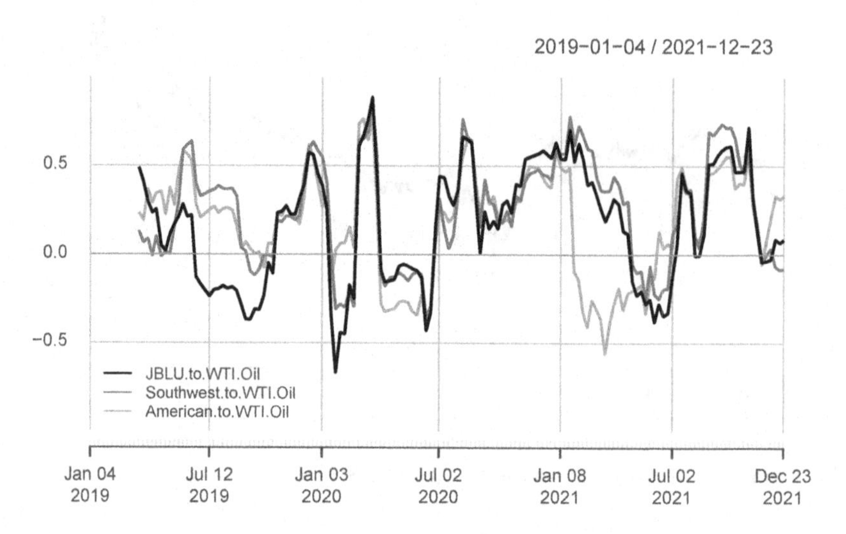

FIGURE 9.14 Rolling correlation of some airline stocks to changes in WTI prices

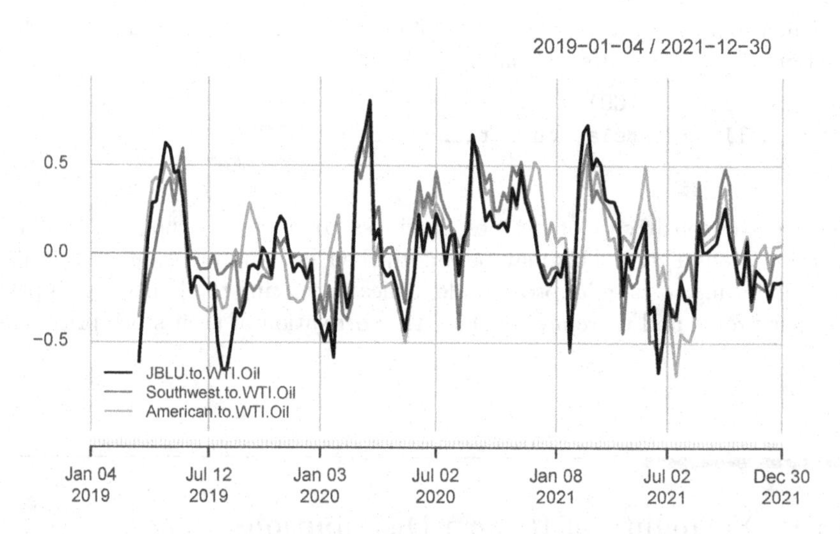

FIGURE 9.15 Using chartRollingCorrelation() again, but this time with absolute oil prices

finance, where probably the most desired forecast is a down-to-earth question: how much money can I make or lose from an investment?

There are, of course, many types of investments, but the best known one, or at least the one that attracts most popular interest, is the stock market. But how reliable is the stock market in creating wealth? By that I mean, how consistent is the increase in wealth – can we really make money on most days from the stock market? That question is essentially a statistics question, and more precisely: what is the nature of the distribution of daily stock-market returns?

Here is a quote from **Myron Scholes**: "Compound returns, not average returns or performance relative to a benchmark, should be our major focus. They are enhanced most by mitigation of tail losses and participation in tail gains, each period of time. [..] The realized return of U.S. equities can be explained by the extreme tail gains or extreme tail losses. Take out the extreme tail gains, and realized return falls to almost zero. Take out the extreme tail losses, and realized return nearly doubles. All the little moves up or down don't matter."

What the Nobel laureate was saying was, as we discussed earlier, that **skew** (and, to a lesser extent, **kurtosis**) is critical in finance: if you lose 50%, you have to make 100% to just break even, so the possibility of a left-tail event needs to be compensated for by an equal probability of an even more extreme right-tail event. In other words, if your investment has a high kurtosis (steep peak in the middle and/or fat tails) then that tail had better be on the right side (positive skew).

What does that translate into in practice for a stock like Apple's? Let's load the `moments` package:

```
library(moments)
```

The skew and kurtosis are, respectively:

```
skewness(aapl.monthly$monthly.returns)
```

```
[1] -0.3015
```

```
kurtosis(aapl.monthly$monthly.returns)
```

```
[1] -0.4181
```

What we do see is that, in that period at least, the monthly returns of Apple exhibit a skew and kurtosis that contrast with the hypothesis that Apple's returns can be modeled by a normal distribution. How wrong is that assumption? To answer that question, we need statistical tests of normality.

Tests of Normality

Finance often assumes that returns are normally distributed without spending a second to check if that assumption is valid (or even caring, frankly). However, there are multiple ways to check for normality, and to do it fast:

- QQ-plots
- Jarque-Bera
- Kolmogorov-Smirnov
- Shapiro-Wilk

We're going to review them one by one.

QQ-Plots

A **quantile-quantile plot**, or **QQ-plot**, is a plot of the quantiles of one sample or distribution against the quantiles of a second sample or distribution. The intuitive appeal of a QQ-plot is that deviation of the plot from a straight line is evidence of non-normality.

A QQ-plot is done in R using the function **qqnorm()**, as illustrated in the code below on a variable **x1** containing 20 data points sampled from the log-normal distribution of standard deviation 1. (You can find on page 258 another example of using the **qqnorm()** function.)

```
qqnorm(x1, datax=TRUE, main="n=20, sigma=1")
```

The output of this code can be found below. We overlaid on this graph a couple of annotations to make interpreting it easier.

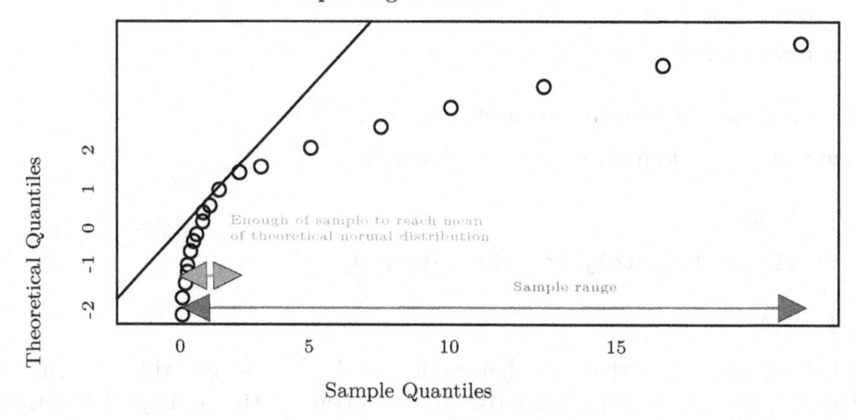

When interpreting a normal plot with a nonlinear pattern, it is essential to know which axis contains the data. But interpreting it is actually easy. Statistical software packages differ about whether the data are on the x-axis (horizontal axis) and the theoretical (reference) quantiles on the y-axis (vertical axis) or vice versa. The **qqnorm()** function in R allows the data to be on either axis depending on the choice of a parameter called **datax**, and by default this parameter is set to **FALSE**.

We remember that **x1** contains 150 data points sampled from the log-normal distribution of mean 0 and standard deviation 1. Obviously the distribution is not normal, with a long right tail. The QQ-plot we just saw corresponds to the same distribution, with the theoretical (reference) quantiles on the y-axis.

What the QQ-plot tells us, and what the small arrow (shown in green depending on your book version) highlights, is that the first few quantiles of the sample distribution were enough to reach a cumulative count that would be the mean of a normal distribution (the horizontal line that goes through $y = 0$). That green arrow, moreover, is small compared to the full range of the sample's values. In other words, a lot of the sample's distribution is "packed" in the lowest values. In contrast, larger values in the sample are spread over what is left of the sample range, and they are spread so thin that the data points diverge farther and farther away from the line that shows what a normal distribution should look like. The **x1** distribution then, clearly, has a long right tail and is far from normality.

Another way to produce **QQ-plots** is to use the **chart.QQPlot()** function of the **PerformanceAnalytics** function. For instance, let's pull the historical values of **Bitcoin** over the past three years, at the time of this writing, using **tq_get()**, then convert them to returns using **tq_transmute()** and **periodReturn()**, and check whether the quantiles of the daily returns of Bitcoin match those of a normal distribution:

```
bitcoin = tq_get("CBBTCUSD",
                 from = "2019-02-11",
                 to = "2022-02-11",
                 get = "economic.data")
```

As discussed on page 166, converting levels obtained by **tq_get()** to returns is most easily achieved using **tq_transmute()**:

```
bit.returns = bitcoin %>%
  tq_transmute(mutate_fun = periodReturn,
               period = "daily")
```

And finally, we call **chart.QQPlot()**, whose output appears in Figure 9.16.

```
chart.QQPlot(bit.returns$daily.returns)
```

Clearly, the distribution of daily returns for Bitcoin exhibits many more extreme points, positive and negative, than a normal distribution would predict. Note also that the sample quartiles are always shown on the y-axis when using **chart.QQPlot()**.

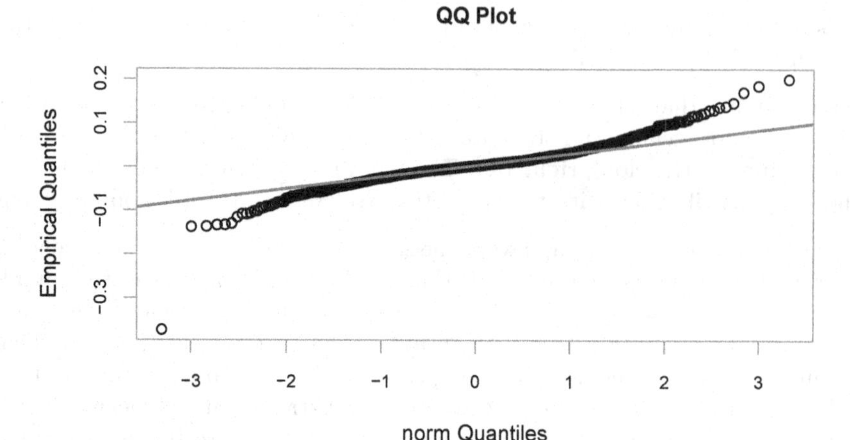

FIGURE 9.16 QQ-plot of Bitcoin's daily returns using the chart.QQPlot() function

Jarque-Bera

Implementations for this test can be found in the **`normtest`** package (function **`jb.norm.test()`**) or the **`tseries`** package (function **`jarque.bera.test()`**).

The **Jarque-Bera test** is another test of normality, based on the skewness and kurtosis coefficient.

Statistics Note: The Jarque-Bera expression is:

$$JB = n(\hat{SK}^2/6 + (\hat{Kur} - 3)^2/24)$$

This test implicitly compares the sample **skewness** and **kurtosis** to 0 and 3 respectively, which are their respective values under normality. More precisely, Jarque-Bera's null hypothesis is the joint hypothesis of the skewness being zero and the kurtosis being 3. JB converges in distribution to a chi-square with 2 degrees of freedom when the sample size tends to infinity, so a large-sample approximation is used to compute the p-value. That p-value is $P(\chi^2_2 > JB)$.

Using function **`jarque.bera.test()`** in the **`tseries`** package, we get:

```
jarque.bera.test(aapl.monthly$monthly.returns)
```

```
    Jarque Bera Test

data:  aapl.monthly$monthly.returns
X-squared = 1.1, df = 2, p-value = 0.6
```

The p-value can be read from the output above or extracted explicitly using a field called **p.value**. How can we know that? The trick is to look at the **summary()** of the result:

```
summary(jarque.bera.test(aapl.monthly$monthly.returns))
```

```
          Length Class  Mode
statistic 1      -none- numeric
parameter 1      -none- numeric
p.value   1      -none- numeric
method    1      -none- character
data.name 1      -none- character
```

So now we know we can write the following:

```
jarque.bera.test(aapl.monthly$monthly.returns)$p.value
```

```
[1] 0.5836
```

As Wikipedia reminds us, "the p-value is the probability of obtaining the observed sample results when the null hypothesis is actually true. if a p-value is very small, usually less than or equal to a threshold value previously chosen called the significance level (traditionally 5% or 1%), it suggests that the observed data is inconsistent with the assumption that the null hypothesis is true." Therefore, a p-value of 0.58 indicates we can't reject the hypothesis that the daily returns of Apple are normally distributed.

Other tests that we are going to see next compare the observed cumulative distribution function $\hat{F}(x)$ with what a the cumulative distribution of a normal distribution centered at μ and with standard deviation σ, $\Phi((x-\hat{\mu})/\hat{\sigma})$ would indicate.

Kolmogorov-Smirnov

The **Kolmogorov-Smirnov test** is based on maximum divergence between the experimental cumulative distribution of returns, which we saw on page 86 in section 5.7, and the cumulative distribution of a normal distribution.

Statistics Note: More precisely, Kolmogorov-Smirnov looks at:

$$\sup_x |\hat{F}(x) - \Phi((x-\hat{\mu})/\hat{\sigma})|$$

which is he maximum distance between $\hat{F}(x)$ with $\Phi((x-\hat{\mu})/\hat{\sigma})$

Intuitively, Kolmogorov-Smirnov tests whether the largest difference between ECDF and normal CDF is statistically significant. For example, Figure 9.17 reproduces Figure 5.8 with an additional arrow (in red, depending on your book version) showing the largest vertical dispersion between the ECDF of

Google's returns and the CDF of the normal distribution with same mean and standard deviation.

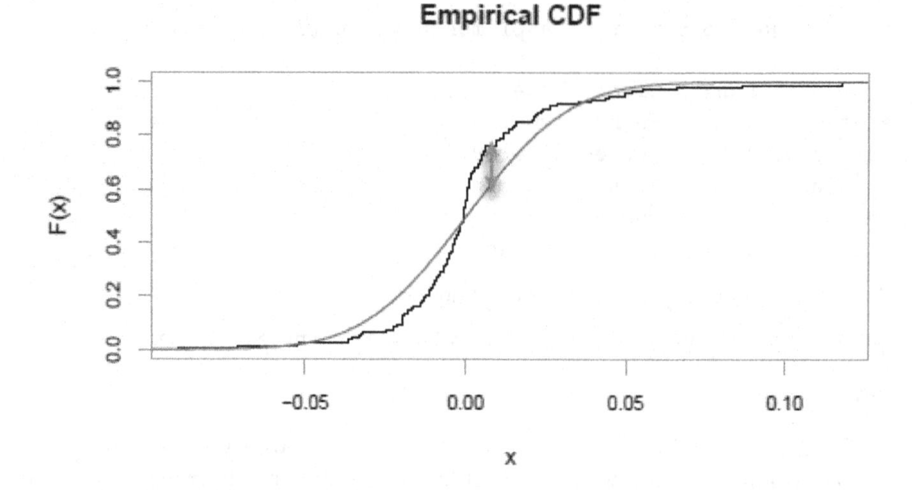

FIGURE 9.17 Illustrating the Kolmogorov-Smirnov test

Note that Kolmogorov-Smirnov requires we provide μ and σ, the mean and standard deviation of the experimental distribution, respectively. These values are calculated using the **mean()** and **sd()** functions, respectively.

```
distribution = aapl.monthly$monthly.returns
(ks = ks.test(distribution,
              "pnorm",
              mean = mean(distribution),
              sd = sd(distribution)))
```

```
    One-sample Kolmogorov-Smirnov test

data:  distribution
D = 0.096, p-value = 0.7
alternative hypothesis: two-sided
```

Since the p-value is large, we cannot reject the null hypothesis and we can consider that Apple's returns are normally distributed.

Shapiro-Wilk

Another test of normality is Shapiro-Wilk. Its mathematical basis is beyond the scope of this book but we should know that, unlike Kolmogorov-Smirnov, this test takes the correlation of the sample with the standard normal distribution into account.

We can apply the easily thanks to the **shapiro.test()** function of the **stats** package:

```
shapiro.test(aapl.monthly$monthly.returns)
```

```
    Shapiro-Wilk normality test

data:  aapl.monthly$monthly.returns
W = 0.98, p-value = 0.7
```

Here again, the null hypothesis is that the returns are normally distributed, and the high p-value means we cannot reject that null hypothesis.

9.11 Fitting A Distribution

Once we decided whether a distribution is normal, how can we decide which parameter values provide the best fit? This is the purpose of **distribution fitting**, which can be done using the **fitdistr()** function of the **MASS** package:

```
library(MASS)
(f = fitdistr(distribution, "normal"))
```

```
      mean          sd
  0.037234    0.088859
 (0.012826) (0.009069)
```

In this example, we ask **fitdistr()** to fit a **normal** distribution, but we will see other cases shortly. The output are the mean and the standard deviation of the normal distribution that best fits the data.

Remember that you can typically find the different fields inside objects returned by R functions using the **str()** or **names()** functions:

```
names(f)
```

```
[1] "estimate" "sd"       "vcov"     "n"
[5] "loglik"
```

Printing **f$estimate** shows this field is a **named vector**. This vector has two entries, **mean** and **sd**, so the mean can be extracted as follows:

```
f$estimate["mean"]
```

```
   mean
0.03723
```

We can verify that the mean 0.0372 of the fitted distribution is indistinguishably close to that of the empirical distribution:

```
mean(distribution)
```

```
[1] 0.03723
```

Likewise, we could fit a **t-distribution**:

```
stocks = read.csv(file = 'stocks5.csv')
```

```
df=data.frame(stocks)
amzn = df$AMZN
fitdistr(amzn,"t")
```

```
       m           s           df
  0.03652     0.06194     4.20758
 (0.01319)   (0.01492)   (3.57958)
```

For a t distribution, the fitted parameters are the **location** **`m`**, the **scale** **`s`**, and the **degree of freedom**. For a t distribution, the location equals the mean, and the mean is close to what we got when fitting a normal distribution. The scale parameter is not the standard deviation, but it is related (the variance equals $s \times \frac{df}{df-2}$). The relatively low degree of freedom confirms the data have fat tails – at least fatter than what a normal distribution would predict.

Which of the distribution, normal or t, has the best fit? Of course, the more complex the distribution is, the better fit we might be able to achieve, but an exceedingly complex model will be over-fitting the empirical data. We can answer that question using the **log-likelihood** L of the fit and, based on that, using the **AIC** (**Akaike's Information Criterion**), defined as:

$$\mathrm{AIC} = 2k - 2\ln(L),$$

where k is the number of parameters in the fitted distribution.

The AIC criterion offers a trade-off between fit and complexity as it depends on the number of parameters used by the type of distribution: only 2 for a normal distribution (mean and standard deviation), whereas the t-distribution adds a third parameter. The lower the value of the AIC, the better the fit:

```
fitted = fitdistr(amzn,"normal")
loglikelihood = fitted$loglik
(AIC = -2 * loglikelihood + 2 * 2)
```

```
[1] -63.06
```

```
fitted = fitdistr(amzn,"t")
loglikelihood = fitted$loglik
(AIC = -2 * loglikelihood + 2 * 3)
```

```
[1] -63.03
```

In this example of Amazon's returns, we can conclude that both fits are equally good.

9.12 Are Differences in Returns Significant?

Another frequent question in finance is to decide whether the returns of two investments are statistically different. We can illustrate that question on one of our running examples, the `stocks5.csv` data set:

```
stocks = read.csv("stocks5.csv")
head(stocks, 2)
```

```
    GM    F   JPM BAML GOOG AMZN AAPL  CRM
1 0.05 0.04 -0.01 0.02 0.03 0.10 0.05 0.16
2 0.01 0.01  0.07 0.09 0.03 0.03 0.13 0.03
```

We see that the returns on GM's and Ford's stocks are in the first and second columns, respectively. Now the question is: on this (very short) period, with what confidence can we say that their returns are different – or more precisely, that the observed returns are samples of distributions with different means? One way to answer this question is to apply a **t-test** using the **t.test()** function:

```
(gmf = t.test(stocks[1], stocks[2]))
```

```
    Welch Two Sample t-test

data:  stocks[1] and stocks[2]
t = 0.5, df = 60, p-value = 0.6
alternative hypothesis: true difference in means is not
equal to 0
95 percent confidence interval:
 -0.02917  0.04853
sample estimates:
mean of x mean of y
  0.01097   0.00129
```

So the t-test, or more precisely here a variation called the Welch t-test, shows a high p-value. So we cannot reject the null hypothesis the returns on the two stocks have equal mean.

This t-test reports a lot of information, which we can extract, one-by-one, if we know the names of each of these fields. We can get that information using the **names()** function:

```
names(gmf)
```

```
 [1] "statistic"   "parameter"   "p.value"
 [4] "conf.int"    "estimate"    "null.value"
 [7] "stderr"      "alternative" "method"
[10] "data.name"
```

```
gmf$p.value
```

```
[1] 0.6201
```

```
gmf$conf.int
```

```
[1] -0.02917  0.04853
attr(,"conf.level")
[1] 0.95
```

Clearly, this result could come from our using too short of a period. You can easily pull data from the web for any period you like, and perform the test again. For instance, here is a test of equality of means over 5 years of daily data:

```
getSymbols("GM", from = "2016-05-01", to = "2021-05-01")
```

```
[1] "GM"
```

```
getSymbols("F", from = "2016-05-01", to = "2021-05-01")
```

```
[1] "F"
```

```
gm.returns = periodReturn(GM, period="daily")
f.returns  = periodReturn(F, period="daily")
```

And we apply the same t-test again:

```
t.test(gm.returns$daily.returns, f.returns$daily.returns)
```

```
    Welch Two Sample t-test

data:  gm.returns$daily.returns and f.returns$daily.returns
t = 0.7, df = 2507, p-value = 0.5
alternative hypothesis: true difference in means is not
equal to 0
95 percent confidence interval:
 -0.001144  0.002401
sample estimates:
mean of x mean of y
0.0007328 0.0001045
```

The result is essentially the same and the high p-value confirms we have to assume that the average return of the two stocks are not statistically different.

This would be consistent with the observation that, in Figure 9.18 produced by the code below, the histogram of returns appear almost undistinguishable:

```
breakpoints = seq(-0.25, 0.25, by=0.002)
hist(gm.returns,
    breaks=breakpoints,
    xlab="",
    main="Daily Returns of Ford (red) and GM (gray)")
hist(f.returns,
    breaks=breakpoints,
    add=TRUE,
    col="red",
    main="")
```

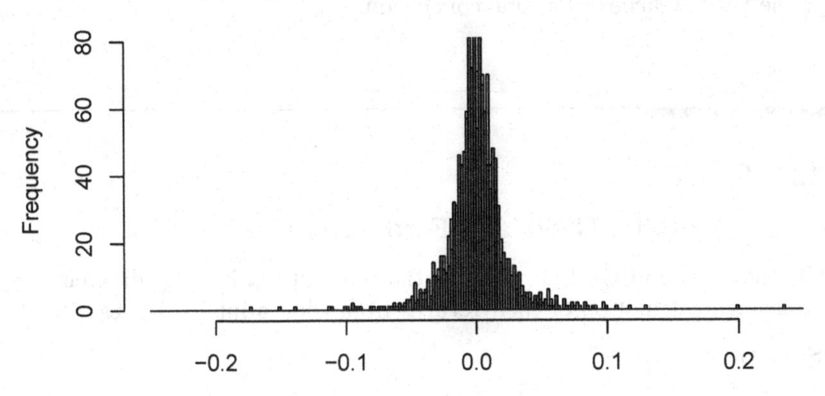

FIGURE 9.18 Histogram of the daily returns of GM and Ford

If we calculate the overall return over the entire period, however, we get a surprise. The compounded returns over the entire period are:

```
(gm.5year.return =
   (as.numeric(GM$GM.Adjusted[1259])/
    as.numeric(GM$GM.Adjusted[1])) - 1)
```

```
[1] 1.135
```

and:

```
(f.5year.return =
    (as.numeric(F$F.Adjusted[1259])/
     as.numeric(F$F.Adjusted[1]))    - 1)
```

```
[1] 0.06461
```

Ford's stock gained a paltry 6.5% over these 5 years, whereas General Motors more than doubled, with a return of 113% over the same period. Despite that, the distributions of their returns have statistically indistinguishable means! This illustrates Myron Scholes' words that the frequent small returns do not make a difference, it is the presence of returns on either tail of the distribution that does.

Technical Detail Note that the adjusted prices come as **`xts`** objects, on which arithmetic calculations don't immediately apply. (That's per design, since the **eXtensible Time Series** package ensures dates are aligned when operating on time series. We don't need to worry about that here.) We used **`as.numeric()`** to extract the raw values of the **`xts`** and drop its dates.

Exercise 9.13.1 on page 196 asks you to verify that the returns constructed by **`periodReturn()`** match those you construct from the adjusted prices, and that the t-test reaches the same conclusion.

9.13 Exercises

9.13.1 Verifying GM's and Ford's Returns

Verify that GM and Ford have means that are not statistically different based on your own calculation of their returns from their adjusted prices.

■ Solution

The returns for General Motors can be constructed from the adjusted prices as follows. Note that we first extract the raw values out of the **`xts`** object, then calculate returns as usual:

```
gm.adjusted.prices = as.numeric(GM$GM.Adjusted)
gm.returns.from.adjusted.prices =
  gm.adjusted.prices[-1] /
    gm.adjusted.prices[-length(gm.adjusted.prices)] - 1
```

We can inspect the two time series we constructred for GM:

```
head(gm.returns)
```

```
           daily.returns
2016-05-02     -0.009360
2016-05-03     -0.015748
2016-05-04     -0.021120
2016-05-05     -0.001634
2016-05-06      0.013752
2016-05-09     -0.005814
```

```
head(gm.returns.from.adjusted.prices)
```

```
[1] -0.015748 -0.021120 -0.001635  0.013752 -0.005814
[6]  0.018519
```

Now, Ford:

```
f.adjusted.prices = as.numeric(F$F.Adjusted)
f.returns.from.adjusted.prices =
  f.adjusted.prices[-1] /
  f.adjusted.prices[-length(f.adjusted.prices)]  - 1
```

We can do a t-test on the newly constructed time series, and reach the same conclusion:

```
t.test(gm.returns.from.adjusted.prices, f.returns.from.adjusted.prices)
```

```
    Welch Two Sample t-test

data: gm.returns.from.adjusted.prices and f.returns.from.adjusted.prices
t = 0.65, df = 2504, p-value = 0.5
alternative hypothesis: true difference in means is not equal to 0
95 percent confidence interval:
 -0.001181  0.002356
sample estimates:
mean of x mean of y
0.0008743 0.0002872
```

■ END OF SOLUTION

9.13.2 Computing Monthly Percentage Changes of Oil Prices

On page 89, we calculated daily percentage changes in the price of oil:

```
oil.returns =
  oil.prices[-1]/oil.prices[-length(oil.prices)]-1
```

Convert these changes to monthly returns.

■ SOLUTION

```
oil = data.frame(oil.dates[-1], oil.returns)
colnames(oil) = c("date", "return")
```

```
oil.monthly =
  oil %>%
  tq_transmute(select     = return,
               mutate_fun = periodReturn,
               period     = "monthly")
```

Please see page 166 if you tried and failed to use **periodReturn()**.

■ END OF SOLUTION

9.13.3 Comparing Returns and Log-Returns

In Exercise 5.9.5, we compared the mean and standard deviations of the returns and log-returns of Google. Now compare their skew and kurtosis.

■ SOLUTION

```
google.close.prices = google.data[, 2]
google.returns =
  google.close.prices[-1] /
  google.close.prices[-length(google.close.prices)] -1
google.logreturns = diff(log(google.close.prices))
```

```
skewness(google.returns)
```

```
[1] 1.076
```

```
skewness(google.logreturns)
```

```
[1] 0.8214
```

```
kurtosis(google.returns)
```

```
[1] 5.962
```

```
kurtosis(google.logreturns)
```

```
[1] 5.427
```

■ END OF SOLUTION

9.13.4 Worst and Best Days for Bitcoin

Between March 1, 2019 and March 1, 2022, what were the worst and best days for Bitcoin? How much did it lose and gain, respectively, on these two days?

■ SOLUTION

We first pull the daily prices:

```
bitcoin = tq_get("CBBTCUSD",
                 from = "2019-03-01",
                 to = "2022-03-01",
                 get = "economic.data")
```

We convert them into returns:

```
bit.returns = bitcoin %>%
  tq_transmute(mutate_fun = periodReturn,
               period = "daily")
```

The obtained dataframe looks like this:

```
head(bit.returns, 3)
```

```
# A tibble: 3 x 2
  date       daily.returns
  <date>             <dbl>
1 2019-03-01      0
2 2019-03-02      0.000927
3 2019-03-03     -0.00598
```

In the dataframe, the row indexes of the minimum and maximum returns are, respectively:

```
which.min(bit.returns$daily.returns)
```

```
[1] 378
```

```
which.max(bit.returns$daily.returns)
```

```
[1] 710
```

We use these indexes to refer to the corresponding dates:

```
bit.returns$date[which.min(bit.returns$daily.returns)]
```

```
[1] "2020-03-12"
```

```
bit.returns$date[which.max(bit.returns$daily.returns)]
```

```
[1] "2021-02-08"
```

And finally we extract the corresponding returns:

```
bit.returns$daily.returns[which.min(bit.returns$daily.returns)]
```

```
[1] -0.3741
```

```
bit.returns$daily.returns[which.max(bit.returns$daily.returns)]
```

```
[1] 0.2002
```

So Bitcoin lost 37% in one day and gained 20% in a single trading session.

■ END OF SOLUTION

9.13.5 Bull Beta

Calculate the bull beta of Apple relative to the S&P 500.

10

Portfolios

So far, we have mostly investigated individual investments. But to an investor, what matters is the overall profit and risk they make over multiple investments. A **portfolio** is just a group of investments considered together for the purpose of minimizing risk and maximizing overall profit.

The value of a portfolio is simply the sum of the values of all its n holdings: if each position i has a price of P_i, then the portfolio value is

$$P_{\text{Portfolio}} = \sum_i^n P_i.$$

The **weight** w_i of a portfolio holding equals its value in dollars divided by the value of the portfolio:

$$w_i = \frac{P_i}{\sum_i^n P_i}.$$

The return of a portfolio is the weighted average of returns R_i of its holdings:

$$R_{\text{Portfolio}} = \sum_i^n w_i R_i. \tag{10.1}$$

Note that the Equation (10.1) above has the form of a **sum-product**.

This sounds simple enough; but prices of financial assets change all the time, so a portfolio's value changes all the time and returns change all the time, and so even the weights of a portfolio's holdings change all the time. This makes calculations tedious, but fortunately two packages are here to help us: **`tidyquant`** and **`PerformanceAnalytics`**.

10.1 Building Portfolios Using Tidyquant

Pulling in one shot historical prices for a basket of stocks can be done as before using **`tq_get()`** but passing a vector of tickers instead of just one:

DOI: 10.1201/9781003320555-10

```
basket = tq_get(c("AAPL", "GOOG", "NFLX"),
                from = "2020-04-01",
                to   = "2021-04-01")
head(basket, 2)
```

```
# A tibble: 2 x 8
  symbol date        open  high   low close    volume
  <chr>  <date>     <dbl> <dbl> <dbl> <dbl>     <dbl>
1 AAPL   2020-04-01  61.6  62.2  59.8  60.2 176218400
2 AAPL   2020-04-02  60.1  61.3  59.2  61.2 165934000
# ... with 1 more variable: adjusted <dbl>
```

As before, the historical price levels are to be converted to returns, but only after we group the return histories by stock. From the output above, we know the relevant column is named **symbol**:

```
(basket.returns = basket %>%
                  group_by(symbol) %>%
                  tq_transmute(select     = adjusted,
                               mutate_fun = periodReturn,
                               period     = "monthly"))
```

```
# A tibble: 36 x 3
# Groups:   symbol [3]
   symbol date       monthly.returns
   <chr>  <date>               <dbl>
 1 AAPL   2020-04-30          0.220
 2 AAPL   2020-05-29          0.0851
 3 AAPL   2020-06-30          0.147
 4 AAPL   2020-07-31          0.165
 5 AAPL   2020-08-31          0.217
 6 AAPL   2020-09-30         -0.103
 7 AAPL   2020-10-30         -0.0600
 8 AAPL   2020-11-30          0.0955
 9 AAPL   2020-12-31          0.115
10 AAPL   2021-01-29         -0.00550
# ... with 26 more rows
```

Now that we have the historical returns of each stock, we can seek the performance of a blend of these stocks. A portfolio either maintains fixed weights to each of the stocks, or lets them drift. In the first case, since the value of each holding changes constantly, weights change all the time as well. Thus the holdings' weights have to be brought in line periodically in a process call **rebalance**: At the end of each period, stocks that appreciated more than the others are partially sold and the proceeds from this sale is used to increase the investment in the under-performers. The second situation is simply called a **buy-and-hold** portfolio.

We will give an example of a rebalanced portfolio of the three stocks above with fixed weights of 30%, 20% and 50% respectively. These weights are stored in a vector:

```
weights = c(0.3, 0.2, 0.5)
```

The stocks' returns can then be combined into one return series for the portfolio by calculating Equation (10.1) every month. The function to do that is **tq_portfolio()**. The first named argument, **assets_col**, indicates the name of the column containing the stocks' tickers; looking at the earlier result, we see that the name of that column in **basket.returns** is **symbol**. Then, named parameter **returns_col** points to the name of the column that contains the returns, and that name is **monthly.returns**. Finally, **tq_portfolio()** expects a value for its argument named **weights**, and that value is contained in the variable we had also called, coincidentally, **weights**.

```
(portfolio_returns =
   basket.returns %>%
   tq_portfolio(assets_col  = symbol,
                returns_col = monthly.returns,
                weights     = weights))
```

```
# A tibble: 12 x 2
   date       portfolio.returns
   <date>                 <dbl>
 1 2020-04-30         0.186
 2 2020-05-29         0.0383
 3 2020-06-30         0.0846
 4 2020-07-31         0.100
 5 2020-08-31         0.135
 6 2020-09-30        -0.0818
 7 2020-10-30        -0.0266
 8 2020-11-30         0.0656
 9 2020-12-31         0.0853
10 2021-01-29         0.0000332
11 2021-02-26        -0.00433
12 2021-03-31        -0.00780
```

We can mix **ggplot** and **tidyquant** to plot an elegant bar chart of monthly portfolio returns. We first construct a character string that describes the portfolio constituents:

```
subtitle.with.weights = paste(format(weights[1]*100, digits=2),
                              "% AAPL, ",
                              format(weights[2]*100, digits=2),
```

```
                                  "% GOOG, ",
                                  format(weights[3]*100, digits=2),
                                  "% NFLX")
```

We then use **geom_bar** with **scale_y_continuous** of **ggplot** combined with **theme_tq** and **scale_color_tq** of **tidyquant**. This results in the plot in Figure 10.1.

```
portfolio_returns %>%
  ggplot(aes(x = date, y = portfolio.returns)) +
  geom_bar(stat = "identity", fill = palette_light()[[1]]) +
  labs(title = "Monthly Portfolio Returns",
       subtitle = subtitle.with.weights,
       caption = "12 months ending April 2021",
       x = "",
       y = "Monthly Returns") +
  geom_smooth(method = "lm") +
  theme_tq() +
  scale_color_tq() +
  scale_y_continuous(labels = scales::percent)
```

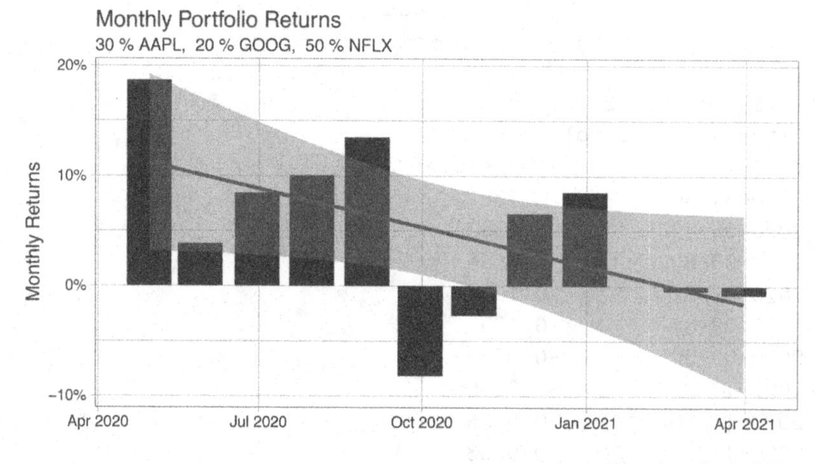

FIGURE 10.1 Using ggplot and tidyquant's color scheme

10.2 Building Portfolios Using PerformanceAnalytics

Another approach to building portfolio returns relies on the **PerformanceAnalytics** package. This time we will take three other stocks, Apple, Johnson & Johnson and JP Morgan, and we will pull return data using the **periodReturn()** function of the **quandmod** package:

```
getSymbols("AAPL", from="2018-01-01", to="2021-12-06")
```

```
[1] "AAPL"
```

```
getSymbols("JNJ",  from="2018-01-01", to="2021-12-06")
```

```
[1] "JNJ"
```

```
getSymbols("JPM",  from="2018-01-01", to="2021-12-06")
```

```
[1] "JPM"
```

```
aapl.daily = periodReturn(AAPL, period="daily")
jnj.daily  = periodReturn(JNJ,  period="daily")
jpm.daily  = periodReturn(JPM,  period="daily")
```

In preparation for portfolio calculations, we put these return time series side by side, instead of one after the other as in the previous example:

```
holdings.returns = cbind(aapl.daily, jnj.daily, jpm.daily)
```

In the previous example, the portfolio had constant weights and was rebalanced. Here, we will compare a rebalanced portfolio with a buy-and-hold one.

Let's first construct the rebalanced portfolio: To simplify matters, it will be an **equal-weighted** portfolio in which we evenly divide our money between the three stocks:

```
equal.weights = c(0.333, 0.333, 0.334)
```

Now comes the **PerformanceAnalytics** package into play. We first load it:

```
library(PerformanceAnalytics)
```

Then we calculate the returns over time for that equal-weighted portfolio using **Return.portfolio()**:

```
port.buy.n.hold =
  Return.portfolio(R = holdings.returns,
                   weights = equal.weights,
                   verbose = TRUE)
```

The above call to **Return.portfolio()** assumed the weights were set when the investments were made, and no trading would take place. This is called a **buy and hold** strategy, and the weights of each stock in the portfolio drift over time according to their relative performance. In contrast, **rebalancing** involves the periodic selling of stocks that are above their target weight (here, one third) and using these funds to buy more of stocks whose weights are now less than their targets. This is an intricate process that can be done for us by **Return.portfolio()** when setting the **rebalance_on=** parameter, as in the example below:

```
port.rebalanced =
  Return.portfolio(R = holdings.returns,
                   weights = equal.weights,
                   rebalance_on = "months",
                   verbose = TRUE)
```

Just plotting these returns would not help us appreciate the difference. A better way is to compound returns as we saw on page 160. Note that we first have to see the structure of the object returned by **Return.portfolio()**, or at least the names of its different fields:

```
names(port.rebalanced)
```

```
[1] "returns"      "contribution" "BOP.Weight"
[4] "EOP.Weight"   "BOP.Value"    "EOP.Value"
```

Now we can proceed, using the vector of returns:

```
compounded.bnh        = cumprod(1+port.buy.n.hold$returns)
compounded.rebalanced = cumprod(1+port.rebalanced$returns)
```

We plot both of them on the same plot by passing to **plot()** the two vectors of returns, bound together using **cbind()**. The plot appears in Figure 10.2.

```
buy.n.hold.VS.rebal= cbind(compounded.bnh, compounded.rebalanced)
plot(buy.n.hold.VS.rebal)
legend(1, 1, legend=c("B&H", "Monthly Rebalance"),
       col=c("black","red"))
```

In this example, the buy-and-hold portfolio outperformed the one rebalanced monthly.

By the way, we noticed thanks to **names()** that the weights at each beginning and end of periods (BOP and EOP, respectively) were also provided. Plotting the beginning-of-period weights confirms that the weight of each stock was maintained around 0.33, as shown in Figure 10.3.

```
plot(port.rebalanced$BOP.Weight)
```

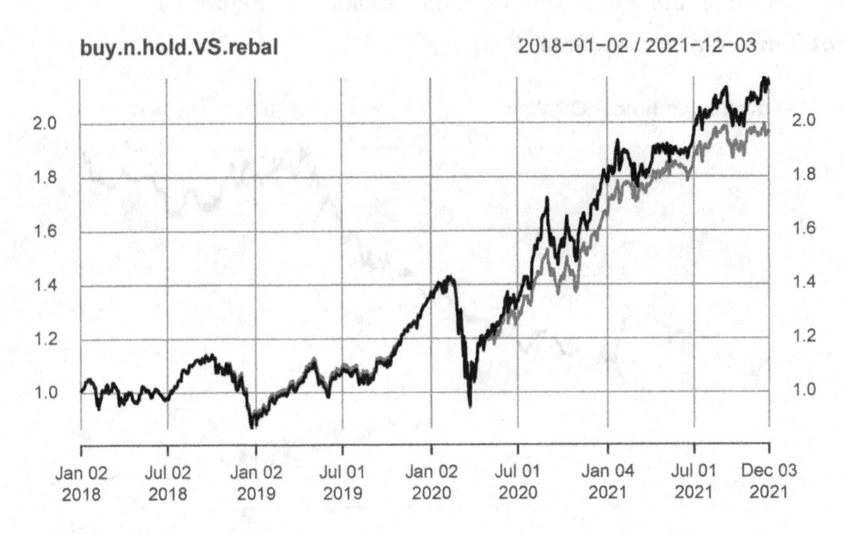

FIGURE 10.2 Cumulative value of a buy and hold portfolio versus one rebalanced monthly

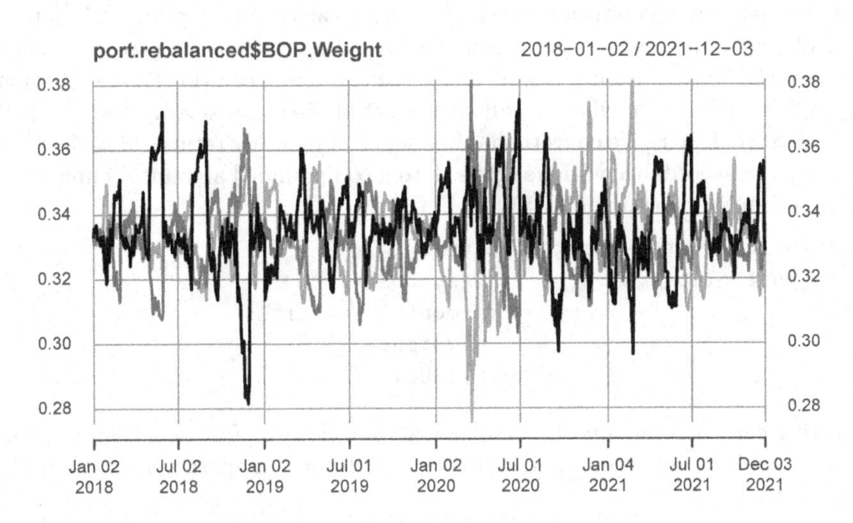

FIGURE 10.3 Weights over time of the rebalanced portfolio

In contrast, the buy-and-hold portfolio let Apple run away and become an ever increasing portion of the portfolio, as shown in Figure 10.4.

```
plot(port.buy.n.hold$EOP.Weight)
```

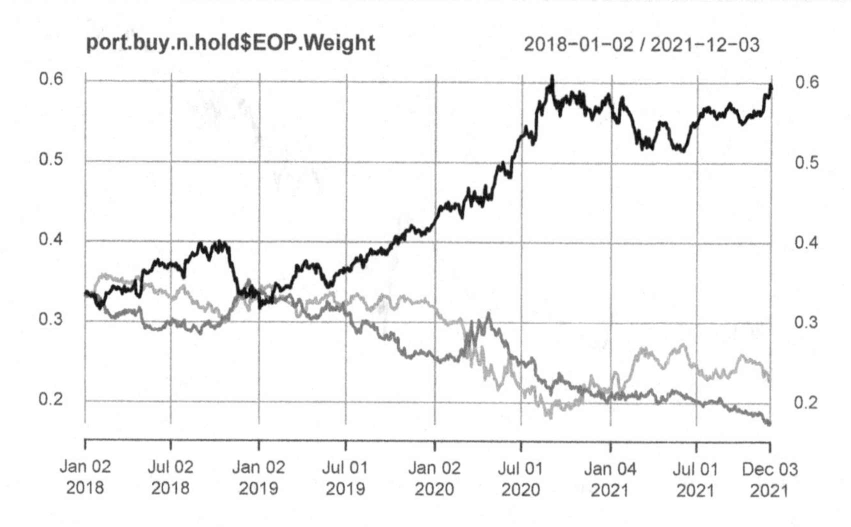

FIGURE 10.4 Weights over time of the buy-and-hold portfolio

Portfolio Growth

The previous calculations allowed us to look at each month's profit in isolation. But of course, investors benefit from compounding and want to see how much their initial contribution has grown. **tq_portfolio()** can do that calculation as well if we request what it calls a **wealth.index**. By setting that named argument to TRUE, we convert the time series of monthly returns of each stock, weighted in accordance with **weights**, into a compounded amount accumulated from an initial investment of $1:

```
portfolio.growth = basket.returns %>%
  tq_portfolio(assets_col    = symbol,
               returns_col   = monthly.returns,
               weights       = weights,
               wealth.index = TRUE)
```

Plotting can then rely on our well-known **geom_line()** function from **ggplot**, plus possibly some themes and color schemes from **tidyquant**. This results in Figure 10.5.

```
portfolio.growth %>%
  ggplot(aes(x = date, y = portfolio.wealthindex)) +
  geom_line(size = 2, color = palette_light()[[1]]) +
```

```
  labs(title = "Portfolio Growth",
       subtitle = subtitle.with.weights,
       caption = "Growth of a $1 investment in this portfolio",
       x = "",
       y = "Portfolio Value") +
  theme_tq() +
  scale_color_tq() +
  scale_y_continuous(labels = scales::dollar)
```

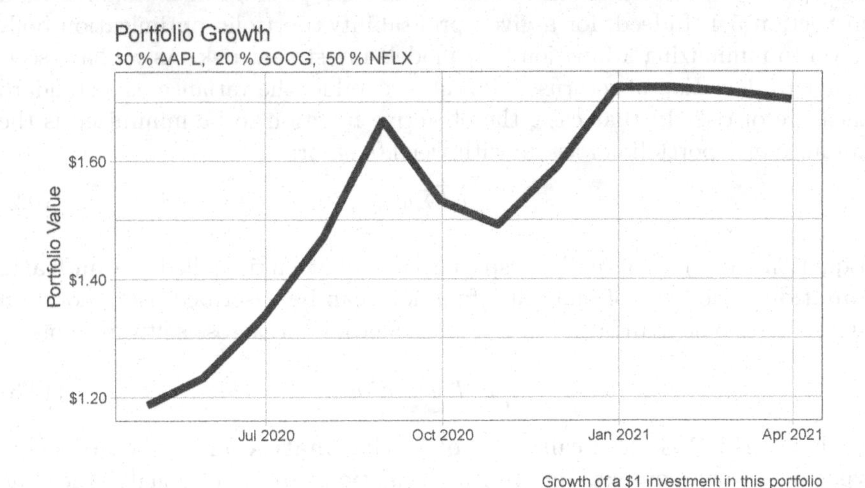

FIGURE 10.5 Another use of ggplot2 and a color scheme from tidyquant

Note the dollar scale for the y-axis using **scales::dollar** (or, alternatively, **scales::dollar_format()**).

10.3 Portfolio Optimization

We've seen how to pull the individual returns of n stocks and calculate the history of returns of a portfolio made of these n stocks, assuming we know the weights each of these stocks. We further assumed that these weights are constant, i.e., that they are reset (rebalanced) to the same initial value at the beginning of each new calculation period. What we haven't seen, however, is how to determine these weights. How can we determine the "best" weights?

This is a fundamental question in finance because higher returns come with higher risk: if two investments had the same profitability but one presented significantly less risk, then investors would flock to the safer one. Inversely, higher risk should come with a higher potential for returns – otherwise investors would divert their money to other options. There is thus a balance to be found between risk and return, and any option that offers too much risk for its return (or too low a potential return for its risk) should be excluded. The process of finding the asset weights that offer the best balance is called **portfolio optimization**.

Portfolio Optimization Using Quadratic Programming

This process is related to the mathematical concept of optimization discussed in Section 5.4. Indeed, for a given profitability, portfolio optimization boils down to minimizing a function that models investment risk. As we have seen, standard deviation of returns or its squared value, the variance, is a standard measure of risk. In that case, the objective function to be minimized is the variance of a portfolio of assets with weights w, or:

$$w^T \Sigma w \tag{10.2}$$

Equation (10.2) is actually a special case of what is called a **quadratic function**. The form of quadratic functions can be described using notation that has become standard, both in mathematics and across software tools:

$$\frac{1}{2} w^T D w - d^T w \tag{10.3}$$

Here, matrix D is often called the **objective matrix** and d the **objective vector**. In our context, referring to Equation (10.2), D equals twice the covariance matrix Σ and d is the null vector.

Now, finding the least-risky portfolio for a given return means searching the value of w so that Equation (10.3) is minimized. However, we operate under two constraints:

- First, the sum of the weights w has to equal 1 (or 100%). This can be written as $w^T \mathbf{1} = 1$, because the sumproduct of the vector of weights and the unit vector $\mathbf{1}$ equals the sum of the weights.
- Second, the mix of stocks given by the weights w provides some target portfolio return μ_P. The portfolio return μ_P is the sumproduct of the weights and the asset returns $\mu_1, ..., \mu_n$. That portfolio return is a parameter to the optimization.

Stacking the two constraints on top of each other, we get:

$$\begin{pmatrix} 1 & 1 & .. & 1 \\ \mu_1 & \mu_2 & ... & \mu_n \end{pmatrix} \begin{pmatrix} w_1 \\ w_2 \\ ... \\ w_n \end{pmatrix} = \begin{pmatrix} 1 \\ \mu_P \end{pmatrix}$$

This expression can be denoted $Aw = b$, where A is here a $2 \times N$ matrix and b is a vector of (here, 2) values that our two constraints must be equal to.

We said "equal to" because the two constraints are equations. But more generally, constraints can also be inequalities, and as long as these inequalities are linear combinations of the weights w, they can be expressed using as a sumproduct and a constant. Hence, Aw and b can capture both equations and inequalities: the first few rows of Aw and b would be understood as "equal," and the remaining rows should be understood as "less than."

Now let's do this in R. We read the data and compute the number of stocks using **ncol()**, which we called n above and call **num.stocks** in the code below:

```
library(quadprog)
library(tidyverse)
stocks = read.csv(file = 'Datasets/stocks5.csv')
```

```
num.stocks = ncol(stocks)
```

We then calculate the $\mu_1, ..., \mu_n$, which are the means of the returns listed in the respective columns of each stock. We thus use **colMeans()** to calculate these column-wise averages and place them in a vector called **mean.returns**. We also need to decide on a desired portfolio return – let's say 2% (per month, because the input data are monthly returns).

```
mean.returns = colMeans(stocks)
target.return = 0.02
```

The A matrix is made of a series of 1's and of the vector called **mean.returns**:

```
(A = cbind(rep(1,num.stocks), mean.returns))
```

```
       mean.returns
GM   1      0.01097
F    1      0.00129
JPM  1      0.01323
BAML 1      0.01419
GOOG 1      0.01613
AMZN 1      0.03323
AAPL 1      0.02484
CRM  1      0.02839
```

We then build the constraint vector b:

```
(b = c(1, target.return))
```

```
[1] 1.00 0.02
```

We now go back to Equation (10.3) and create the objective matrix D:

```
(D = 2 * cov(stocks))
```

```
           GM         F      JPM     BAML      GOOG
GM   0.011925 7.744e-03 0.004700 0.006492 2.501e-03
```

```
F    0.007744 1.146e-02 0.004711 0.005635 7.032e-05
JPM  0.004700 4.711e-03 0.007298 0.007672 2.466e-03
BAML 0.006492 5.635e-03 0.007672 0.009544 3.674e-03
GOOG 0.002501 7.032e-05 0.002466 0.003674 6.449e-03
AMZN 0.002754 1.978e-03 0.004365 0.006019 6.032e-03
AAPL 0.003924 2.774e-03 0.002448 0.003751 3.432e-03
CRM  0.002183 1.398e-03 0.001917 0.003521 3.820e-03
         AMZN     AAPL      CRM
GM   0.002754 0.003924 0.002183
F    0.001978 0.002774 0.001398
JPM  0.004365 0.002448 0.001917
BAML 0.006019 0.003751 0.003521
GOOG 0.006032 0.003432 0.003820
AMZN 0.013912 0.006048 0.007604
AAPL 0.006048 0.014172 0.004469
CRM  0.007604 0.004469 0.007188
```

We then create the objective vector – in our case, this is just a vector of zeroes:

```
(d = rep(0, num.stocks))
```

```
[1] 0 0 0 0 0 0 0 0
```

We solve the optimization problem thanks to the **quadprod** library and the **solve.QP()** function. The parameters of the function are our objective matrix and objective vector, our constraint matrix and constraint vector, and the number of constraints that are equations (as opposed to inequalities), as discussed earlier. In this example, that last argument to **solve.QP()** is 2:

```
library(quadprog)
opt = solve.QP(D, d, A, b, 2)
round(opt$solution, digits=3)
```

```
[1]  0.123  0.004  0.985 -0.861  0.208 -0.127  0.072
[8]  0.596
```

It worked! We got the 8 weights of the 8 stocks in our portfolio so that the variance of the portfolio is minimized. This is a valid portfolio, in the sense that the weights do add up to 1:

```
sum(opt$solution)
```

```
[1] 1
```

Some of the weights are negative, which means that you would have to **short** those stocks, meaning that you would be betting that their price will go down. It may be counter-intuitive that the optimizer would want to bet that Amazon's stock would go down given the performance of the stock in the early 2020's, but the stock's own variance and its covariance with the other stocks made minimizing (10.2) produce this surprising result.

Shorting involves an investment bank that lends you the stock and lets you sell it, under the condition you give it back to the bank later – after you buy it back, presumably at a cheaper price. In practice, no share changes hands but a contract between the investor and the bank is entered into. Because of these technicalities, shorting a stock is not always an option depending on the type of investor you are and what your investment constraints are, and most investors will simply pass on a stock (giving it a weight of zero) instead of shorting it. Note that the opposite of being short is to be **long**, and a portfolio where all the investments go long is said to be **long-only**.

We are going to construct such a long-only portfolio. In terms of our optimization, this translates into enforcing that all the weights w are non-negative. In matrix notation, we would write:

$$\begin{pmatrix} 1 & 0 & .. & 0 \\ 0 & 1 & .. & 0 \\ .. & .. & .. & .. \\ 0 & 0 & .. & 1 \end{pmatrix} \begin{pmatrix} w_1 \\ w_2 \\ \dots \\ w_n \end{pmatrix} \geq \begin{pmatrix} 0 \\ 0 \\ .. \\ 0 \end{pmatrix}$$

This boils down to an additional constraint matrix, a diagonal identity matrix, and an additional constraint vector of zeroes, so that $A_{\text{ineq}} w \geq b_{\text{ineq}}$. We construct the additional constraint matrix A_{ineq} as follows:

```
(Aineq = diag(1, num.stocks, num.stocks))
```

```
     [,1] [,2] [,3] [,4] [,5] [,6] [,7] [,8]
[1,]    1    0    0    0    0    0    0    0
[2,]    0    1    0    0    0    0    0    0
[3,]    0    0    1    0    0    0    0    0
[4,]    0    0    0    1    0    0    0    0
[5,]    0    0    0    0    1    0    0    0
[6,]    0    0    0    0    0    1    0    0
[7,]    0    0    0    0    0    0    1    0
[8,]    0    0    0    0    0    0    0    1
```

We then add it to the matrix `A` we already have:

```
A = cbind(A, Aineq)
```

We then construct b_{ineq}, a vector of n zeroes:

```
(bineq = rep(0, num.stocks))
```

```
[1] 0 0 0 0 0 0 0 0
```

And we add this vector to the constraint vector `b` we already have:

```
(b = c(b, bineq))
```

```
 [1] 1.00 0.02 0.00 0.00 0.00 0.00 0.00 0.00 0.00 0.00
```

Now we call **solve.QP()** again with the augmented constraint vectors. All the constraints but the first 2 are inequalities, so that last parameter to **solve.QP()** is still 2:

```
opt = solve.QP(D, d, A, b, 2)
```

The weights of each investment in the portfolio are stored in a field called **solution**; let's display them, rounding to three digits:

```
round(opt$solution, digits=3)
```

```
[1] 0.004 0.069 0.238 0.000 0.210 0.000 0.073 0.406
```

We notice that the two stocks that were shorted in the first optimization now receive a weight of zero, which sounds reasonable. As a matter of course, we verify that the weights make up a valid portfolio in the sense that the weights add up to 1:

```
sum(opt$solution)
```

```
[1] 1
```

We can go one step farther. Instead of setting the target return to 2%, we can construct all the portfolios that, for a specific portfolio return, minimize variance. This will build the **efficient frontier**, which is an arc of a curve when plotting returns against standard deviations. We first decide the number of points we want on the curve, then the different target returns – equally spaced returns within a realistic range:

```
points.on.curve = 20
possible.returns.muP = seq(0.01, 0.03, length=points.on.curve)
```

The volatilities for each of these possible portfolios will be calculated by the optimizer. We could add them one at a time to an increasing list, but since we know already how many data points we will need, we might as well allocate storage for a vector of the desired length:

```
portfolio.volatilities = rep(0, points.on.curve)
```

For the same reason, we pre-allocate storage for the portfolio weights. This requires storage for a matrix, not a vector, that will contain as many rows as there are points on the efficient frontiers and as many weights as there are stocks:

```
portfolio.weights =
  matrix(0, nrow=points.on.curve, ncol=num.stocks)
```

```
for (i in 1:length(possible.returns.muP)) {
  b = c(1, possible.returns.muP[i])
  b = c(b, bineq)
  result = solve.QP(D, d, A, b, 2)
  portfolio.volatilities[i] = sqrt(result$value)
  portfolio.weights[i,] = result$solution
}
```

We can then simply plot the vectors of specified returns against the vector of minimum volatilities reachable for these respective returns. The result of the command below appears in Figure 10.6.

```
plot(portfolio.volatilities, possible.returns.muP)
```

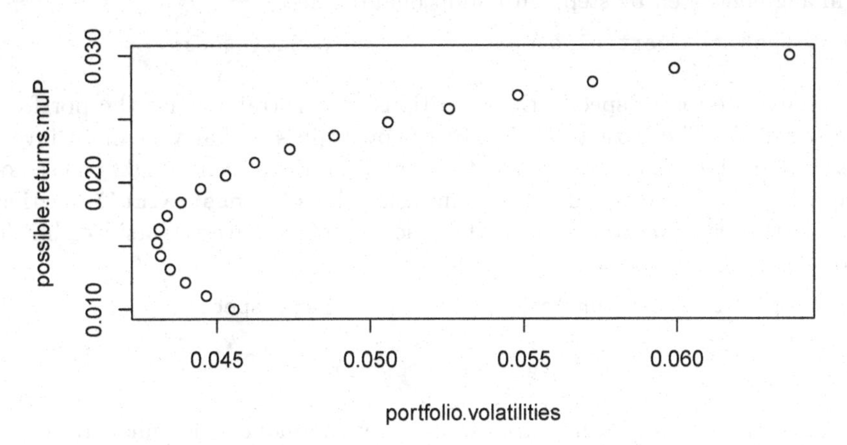

FIGURE 10.6 Plot of an efficient frontier, i.e., minimum volatilities (x-axis) for each possible return (on the y-axis)

The top right point on this plot (the efficient portfolio with the highest return) corresponds to the following weights:

```
round(portfolio.weights[points.on.curve, ], 2)
```

```
[1] 0.00 0.00 0.00 0.00 0.00 0.36 0.03 0.61
```

This corresponds to 36% in Amazon, 3% in Apple and 61% in Salesforce – hardly a well-diversified portfolio. We also note that Amazon here receives a significant allocation even though it was shorted to minimize the variance of a portfolio with a lower target return.

Be careful, however, when using **solve.QP()** this way: if the target return (in our case, any of the values in **possible.returns.muP**) is not feasible (cannot be obtained given the returns of the portfolio assets and the constraints on their weights), then **solve.QP()** will fail with the following error message:

```
Error in solve.QP(D, d, A, b, 2) :
  constraints are inconsistent, no solution!
```

Portfolio Optimization Using PortfolioAnalytics

Another way to find efficient portfolios is to use the **PortfolioAnalytics** package:

```
library(PortfolioAnalytics)
```

The first step in constructing a portfolio is to define the assets it is made of (at this point, only their names) using the **portfolio.spec()** function. The function will create a portfolio specification we call **port.specif**, which we will augment, step by step, with additional details.

```
port.specif = portfolio.spec(assets=colnames(stocks))
```

The next step is to specify the conditions or constraints that the portfolio must satisfy. The most typical of these constraints is that we cannot invest more than the money we have, so the sum of all investment weights must not exceed 100%. One can add a constraint using the **add.constraint()** function, and in this case the constraint is that the sum of the investments' weights do not exceed 1:

```
port.specif = add.constraint(portfolio=port.specif,
                             type="weight_sum",
                             max_sum=1)
```

In most cases however, investors want all of the money they allocate to the portfolio to be invested, meaning the sum of the weights should not be less than 100% either. This can be done by adding a **min_sum** constraint in addition to **max_sum**, in one shot:

```
port.specif = add.constraint(portfolio=port.specif,
                             type="weight_sum",
                             max_sum=1,
                             min_sum=1)
```

It is simpler however, and probably more elegant to state that we need the portfolio to be **fully invested**:

```
port.specif = add.constraint(portfolio = port.specif,
                             type = "full_investment")
```

Another typical constraint is that all the weights should be greater than or equal to 0, i.e. that we want a long-only portfolio without shorts. That trait can be added as a further constraint to our portfolio descriptor:

```
port.specif = add.constraint(portfolio = port.specif,
                             type = "long_only")
```

Yet another constraint placed in practice on actual portfolio is a minimum and a maximum on each asset weight. These floors and ceilings can be the same for all constituents of the portfolio, in which case a single value (a scalar) can be specified, and will be applied to all assets. That type of constraint is called “box” in **PortfolioAnalytics**:

```
port.specif = add.constraint(portfolio=port.specif,
                             type="box",
                             min=0.03,
                             max=0.4)
```

Alternatively, different minima and maxima can be enforced on each assets. Remember that the stocks in our portfolio are:

```
names(stocks)
```

```
[1] "GM"   "F"    "JPM"  "BAML" "GOOG" "AMZN" "AAPL"
[8] "CRM"
```

Let’s assume we are less bullish on General Motors and Ford, and want to cap their weight at 10%, but more sanguine on Google and Salesforce, and want to allocate at least 10% to these companies. All the other assets will continue receiving a minimum allocation of 3% with a cap of 40%. The box constraints would then be expressed as two vectors of 8 values, one per stock:

```
port.specif =
  add.constraint(portfolio = port.specif,
        type="box",
        min=c(0.03, 0.03, 0.03, 0.03, 0.1, 0.03, 0.03, 0.1),
        max=c(0.1, 0.1, 0.4, 0.4, 0.4, 0.4, 0.4, 0.4))
```

More constraint types are offered by **PortfolioAnalytics**, but they are less used in practice. One such constraint targets a certain diversification level, where diversification is quantified as the sum of the squares of asset weights. That constraint type is specified by **type="diversification"** and the target for the diversification metric is provided using the **div_target=** optional parameter.

Now, the goal of all this is to build a portfolio that has good performance or low risk, or both. Depending on the investor’s preference, a performance target can be set as a constraint and the **objective** can be to minimize risk for that performance; or, a risk level can be targetted as a constraint and the portfolio with the highest expected performance at that level of risk is desired; or finally, you can optimize two contradicting objectives, maximize return while minimizing risk. This is done (under the hood) by creating a so-called **utility function** equal to the portfolio return less a penalty for the risk taken. That penalty equals a measure of risk, typically volatility, *times* a **risk-aversion factor**.

In the example below, we are going to request a return target of 2.5% and, as the objective, will request to minimize the standard deviation of returns. This should make sense at this point:

```
port.specif = add.constraint(portfolio=port.specif,
                              type="return",
                              return_target=0.025)
port.specif = add.objective(portfolio = port.specif,
                             type = "risk",
                             name = "StdDev")
```

The third `add.objective()` below is certainly more surprising:

```
port.specif = add.objective(portfolio = port.specif,
                             type = "return",
                             name = "mean")
```

Why would we add maximizing return as an additional objective since we constrained the target return to be 0.025? That's simply a quirk of the `chart.RiskReward` function from the `PortfolioAnalytics` library that we are going to use shortly to graph our results! It should not, otherwise, be needed.

We now have specified our optimization problem and can now construct the portfolio that optimizes these objectives under the given constraints. This is done using the `optimize.portfolio()` function. The first parameter to that function is a time series[1], followed by the portfolio specification, then the optimization method of choice.

```
returns.as.ts = ts(stocks, start = c(2015,1), frequency = 12)
opt = optimize.portfolio(returns.as.ts,
                          portfolio = port.specif,
                          optimize_method = "random",
                          trace = TRUE)
```

Note that no optimizing algorithm was really used but instead that the method use to optimize was to try at `random` a large number of weight combinations. That number is 20,000 by default but can be specified using the `search_size=` optional parameter. Trying such a large number of portfolio takes quite some time, so the `optimize.portfolio()` function can be slow when using the `random` method.

[1]The documentation indicates that data frames are accepted, but this is not our experience.

And what was the result of that portfolio construction? The best portfolio that was found can be retrieved from the **weights** field of the **opt** object:

```
opt$weights
```

```
   GM     F   JPM  BAML  GOOG  AMZN  AAPL   CRM
0.032 0.030 0.100 0.052 0.116 0.392 0.072 0.206
```

We see that GM and Ford receive weights close to their lower bound of 3%. Such low weights suggest that the tool should have given minimal weights to these two stocks but that this combination simply wasn't tried out by the random process. Likewise, Google is close to its minimum, indicating better portfolios could be found if we lowered the floor on Google's weight. Reciprocally, Amazon almost maxed out its acceptable allocation.

How does that portfolio compare to other possible combinations? One way to gauge our creation is to plot that portfolio among the other assets using the **chart.RiskReward** function.

```
chart.RiskReward(opt,
                 risk.col = "StdDev",
                 return.col = "mean",
                 chart.assets = TRUE)
```

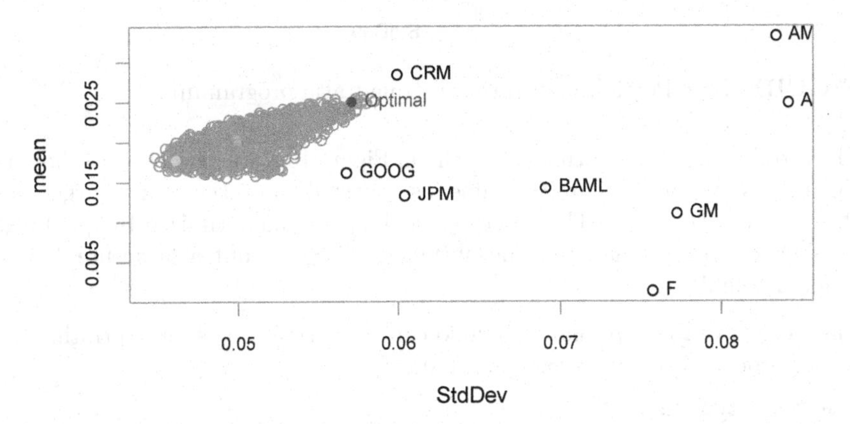

FIGURE 10.7 Volatility and return of the optimized portfolio compared to other combinations

The plot in Figure 10.7 shows that the optimal portfolio indeed offers one of the best return-over-volatility ratios and lies on the efficient frontier.

If we want to use an actual optimizer, then the **quadprog** option (which stands for quadratic programming of course) can be used:

```
opt = optimize.portfolio(returns.as.ts,
                         portfolio = port.specif,
```

```
                optimize_method = "quadprog",
                trace = TRUE)
```

We can plot that optimal portfolio in risk-return space using the **chart.RiskReward()** function again, resulting in Figure 10.8:

```
chart.RiskReward(opt,
                 risk.col = "StdDev",
                 return.col = "mean",
                 chart.assets = TRUE)
```

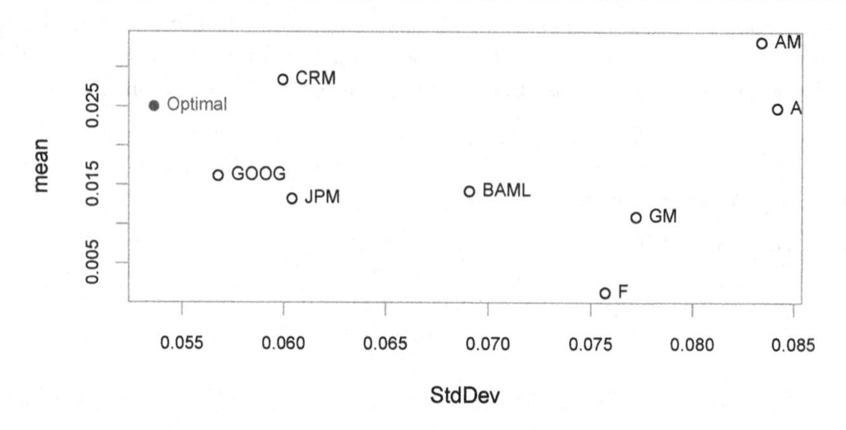

FIGURE 10.8 Portfolio optimized by quadratic programming

The coordinates of the optimal portfolio in Figure 10.8 are very close to those in Figure 10.7: the returns are the same, as expected, since the return target was the same in both cases. The volatility of the portfolio optimized by quadratic seems a bit better than the one obtained by trying out a large number of random combinations.

The weights of the "optimal" portfolio can be retrieved by simply printing the object produced by **optimize.portfolio**:

```
round(opt$weights, 3)
```

```
   GM     F   JPM  BAML  GOOG  AMZN  AAPL   CRM
0.030 0.030 0.056 0.030 0.100 0.203 0.151 0.400
```

This time, the optimizer did explore solutions at the boundaries of the allowed ranges. (If you think about it, the weight constraints define a convex polyhedron in 8-dimensional space, and a good optimizer explores the vertices of that polyhedron.) GM and Ford get as low a weight as possible, and so does Google.

10.4 Exercises

10.4.1 Correlation Matrix

Produce the correlation matrix for monthly returns of Apple, S&P 500 and oil prices.

■ SOLUTION

On page 171, we had calculated the monthly returns of Apple and the benchmark. We first merge them by date:

```
stock.and.benchmark = merge(aapl.monthly, benchmark, by="date")
```

Then merge to this set the monthly returns on oil computed in Exercise 9.13.2:

```
appl.sp.oil = merge(stock.and.benchmark, oil.monthly, by="date")
```

Change the column names into something more meaningful:

```
colnames(appl.sp.oil) = c("Date", "Apple", "SP 500", "Oil")
```

Calculate the correlations across the 3 columns of returns (drop Column 1, which contains the dates):

```
appl.sp.oil.correlations = cor(appl.sp.oil[2:4])
```

Finally, load the **corrplot** package and call **corrplot()**:

```
library(corrplot)
corrplot(appl.sp.oil.correlations)
```

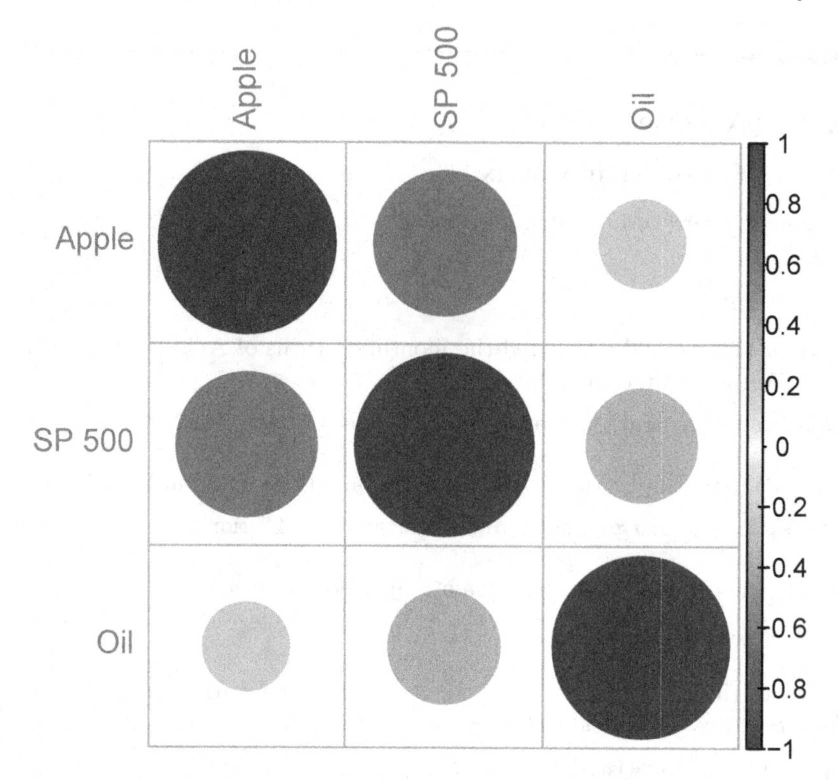

10.4.2 Improving the Portfolio Growth Graph

You may have noticed that the graph of Figure 10.5 plotting the growth of our 3-stock portfolio may not look best. First, the first graphed value is \$1.18 because that's the portfolio's value after the first month. The different values are based on a starting point of \$1, but typical graphs of investment growth start with a more meaningful value, like \$10,000. This is of course entirely a question of taste, but we can try to improve that. Secondly and more importantly, the starting amount of the investment, whatever its value, is not shown.

Adjust the graph so that it is based on a \$10,000 investment, and shows that initial investment amount.

■ SOLUTION

Looking at the top of the data frame is always a good starting point:

```
head(portfolio.growth)
```

```
# A tibble: 6 x 2
  date        portfolio.wealthindex
  <date>                      <dbl>
```

```
1 2020-04-30                    1.19
2 2020-05-29                    1.23
3 2020-06-30                    1.34
4 2020-07-31                    1.47
5 2020-08-31                    1.67
6 2020-09-30                    1.53
```

We get a confirmation of what could be suspected from the graph: that the initial value of $1 (to be $10,000) is not recorded, nor its date. We can specify that date, or calculate it backward from the end of the first month. That end-date of the first month can be extracted with:

```
portfolio.growth[1,1]
```

```
# A tibble: 1 x 1
  date
  <date>
1 2020-04-30
```

The **EOMONTH()** function provides the last day of the month relative to the specified date, expressed in a numeric format. We thus calculate the inception date of the portfolio with these two commands:

```
as.numeric(portfolio.growth[1,1])
```

```
[1] 18382
```

```
(portfolio.inception =
  EOMONTH(as.numeric(portfolio.growth[1,1]), -1))
```

```
[1] "2020-03-31"
```

We can now start creating the starting-point row that we'll add to our historical dataframe. The solution is to create a one-row data frame using **data.frame()**:

```
(portfolio.start = data.frame(portfolio.inception, 1))
```

```
  portfolio.inception X1
1          2020-03-31  1
```

So that small data frame can now be added at the very beginning of the larger **portfolio.growth** data frame thanks to the **rbind()** function. We **rename()** the columns of **portfolio.start** with the names seen in **portfolio.growth**:

```
portfolio.start = portfolio.start %>%
  rename(date = portfolio.inception,
         portfolio.wealthindex = X1)
```

Note that the same result could have been achieved with the **colnames()** function:

```
colnames(portfolio.start) = colnames(portfolio.growth)
```

We can then use **rbind()** to place **portfolio.start** "on top of" **portfolio.growth**:

```
portfolio.growth = rbind(portfolio.start, portfolio.growth)
head(portfolio.growth)
```

```
        date portfolio.wealthindex
1 2020-03-31                 1.000
2 2020-04-30                 1.186
3 2020-05-29                 1.232
4 2020-06-30                 1.336
5 2020-07-31                 1.470
6 2020-08-31                 1.669
```

Using **head()** is again a good sanity check that everything worked as planned. We can now multiply the amounts by $10,000 if we wish:

```
portfolio.growth$portfolio.wealthindex =
  portfolio.growth$portfolio.wealthindex * 10000
```

And then plot the growth of our portfolio, resulting in Figure 10.9:

```
portfolio.growth %>%
  ggplot(aes(x = date, y = portfolio.wealthindex)) +
  geom_line(size = 2, color = palette_light()[[1]]) +
  labs(title = "Portfolio Growth",
       subtitle = subtitle.with.weights,
       caption =
         "Growth of a $10,000 Investment in this Portfolio",
       x = "",
       y = "Portfolio Value") +
  theme_tq() +
  scale_color_tq() +
  scale_y_continuous(labels = scales::dollar)
```

■ END OF SOLUTION

10.4.3 Portfolio of Hedge Funds

We saw on page 14 that the **PerformanceAnalytics** package (not to be confused with **PortfolioAnalystics**) comes with a data set called **edhec** providing the monthly returns of different hedge fund strategies. Optimize a portfolio of such strategies.

■ SOLUTION

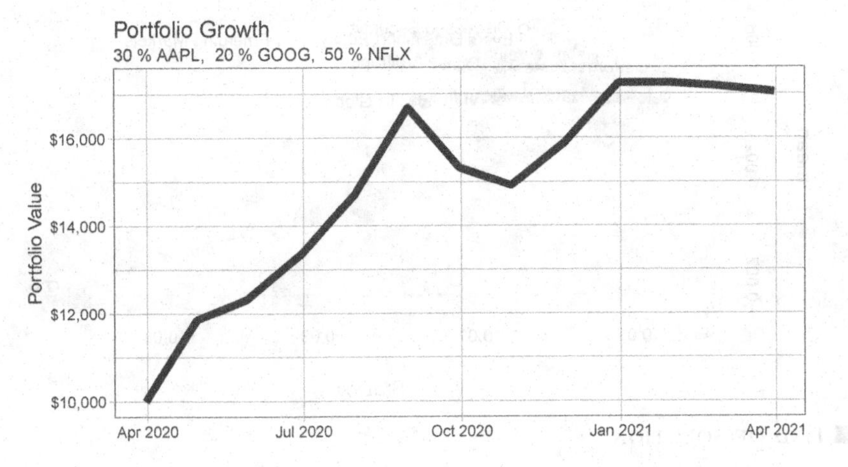

FIGURE 10.9 Growth of Portfolio Value

```
data(edhec)
```

```
port.specif = portfolio.spec(colnames(edhec))
port.specif = add.constraint(portfolio = port.specif,
 type = "full_investment")
port.specif = add.constraint(portfolio = port.specif,
 type = "long_only")
port.specif = add.objective(portfolio = port.specif,
 type = "return",
 name = "mean")
port.specif = add.objective(portfolio = port.specif,
 type = "risk",
 name = "StdDev")
```

```
opt = optimize.portfolio(edhec, portfolio = port.specif,
 optimize_method = "random",
 trace = TRUE)
chart.RiskReward(opt, risk.col = "StdDev", return.col = "mean",
 chart.assets = TRUE)
```

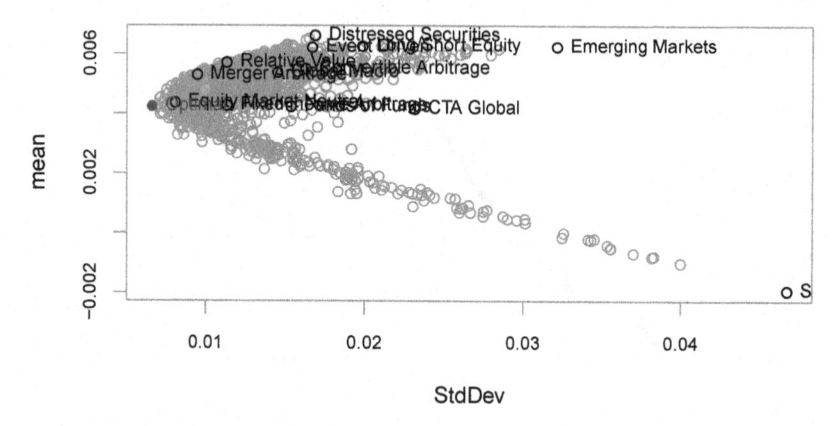

■ End of solution

10.4.4 Larger Search Space

Increase the search space of the random optimization method, from default 20,000 to 40,000, using the **search_size=** optional parameter. How do the assets weights compare to those obtained by quadratic programming?

```
opt = optimize.portfolio(returns.as.ts,
                         portfolio = port.specif,
                         optimize_method = "random",
                         search_size = 40000,
                         trace = TRUE)
chart.RiskReward(opt,
                 risk.col = "StdDev",
                 return.col = "mean",
                 chart.assets = TRUE)
```

11

Modeling Returns & Simulations

In Chapter 9, we saw how to calculate the total return on assets, including price appreciation, cash payments such as dividends, corporate events like mergers, and gimmicks[1] like stock splits. But this is all backward-looking.

As you can imagine, especially in the world of finance, people are interesting in making predictions. One can seek to predict the most probable future value for a variable given a model for its evolution, or one can seek to make prediction on the future value of that variable based on the values of other variables. This chapter and Chapter 17 on time series analysis fall in the former category, whereas Chapter 12 on linear regression and Chapter 18 on machine learning fall in the latter.

The focus of the present chapter is on modeling returns with the goal of assessing the probability of future returns, typically through simulation. The models are a bit mathematical, so we won't be able to avoid symbols and equations. Let's start with the basic assumptions.

11.1 Normal and Log-normal Models

At time $t-1$, price P_t and return R_t are not only unknown but we also do not know their probability distributions. A simple model, the **normal model**, assumes that the returns are independent, are identically distributed, and follow the same normal distribution. That's intuitive, but there are two problems with this model. First, saying that returns have a normal distribution means that returns can be, in theory, arbitrarily negative. This would imply the possibility of unlimited losses on our investments. However, an investor's liability is usually limited since you cannot lose more than your investment – hence we would prefer a model that captures that fact that $R_t \geq -1$.

[1]Stock splits are just a gimmick in that they do not change the economic value of the company.

DOI: 10.1201/9781003320555-11

Another problem is that the geometric compounding of returns, which we had defined as:

$$1 + R_t(k) = \prod_{i=0}^{k-1}(1 + R_{t-i})$$

does not follow a normal distribution. Indeed, *sums* of independent normals are normal, but not their products.

To solve these issues, the **log-normal model** assumes that the logs of the **ratio returns**, $r_t = \log(1 + R_t)$, are **i.i.d.** (independent and identically distributed) and follow a normal distribution, which we write $\log(1 + R_t) \sim N(\mu, \sigma^2)$. The **log return** is defined as $\log(1+R_t)$, and as introduced on page 84, if the price of a security changed from P_0 to P_1, then the log return can be written as $\log\left(\frac{P_1}{P_0}\right)$, which equals $\log(P_1) - \log(P_0)$. In R, we can thus calculate the log returns of a vector **close.prices** of close prices using the following expression:

```
logreturns = diff(log(close.prices))
```

As discussed on page 81, a variable has a log-normal distribution if its logarithm is normally distributed. Also, $r_t = \log(1+R_t)$ can be rewritten as $1+R_t = e^{r_t}$, then as $R_t = e^{r_t} - 1$, which is always greater than or equal to -1, as desired.

How do we use that property? As an example, suppose a ratio return $(1 + R)$ is log-normal(0, 0.01), that is: $\log(1 + R) \sim N(\mu = 0, \sigma^2 = 0.01)$. How can we find the probability that the return R is less than 5%, i.e., how can we calculate $P(R \leq 0.05)$?

The answer is to convert the return to a ratio return and then to take its log, because then we know properties on the object we just created:

$$\begin{aligned} P(R \leq 0.05) &= P(1 + R \leq 1.05) \\ &= P\left(\log(1 + R) \leq \log(1.05)\right) \\ &= P\left(N(0, 0.01) \leq \log(1.05)\right) \\ &= \Phi\left((\log(1.05) - 0)/\sqrt{0.01}\right) \end{aligned}$$

where Φ is the **CDF** of the standard normal distribution. Using standard CDF tables in all statistics textbook, we can find that:

$$\Phi\left((\log(1.05) - 0)/\sqrt{0.01}\right) = \Phi(0.4879) = 0.68719$$

We can of course do this calculation in R using **pnorm()**:

```
pnorm(log(1.05)/0.1)
```

```
[1] 0.6872
```

Note that the value passed to Φ above is a z-score, so the reference distribution is $N(0,1)$, so we do not need to specify the mean and standard deviation when calling `pnorm()`, since by default these parameters are 0 and 1, respectively. In other words, the above command is equivalent to `pnorm(log(1.05)/0.1, mean=0, sd=1)`. Note also that, in R, `log()` by default denotes the natural logarithm.

11.2 Log-normal Model – Multi-period Return

Properties for a k-periods return follow easily. That return is $1 + R_t(k) = (1 + R_t)(1 + R_{t-1}) \dots (1 + R_{t-k})$. Each of the 1-period returns are log-normal, i.e. $\log(1 + R_i) \sim N(\mu, \sigma^2)$ for all the i's, and the R_i's are independent. With these assumptions, the multi-period log-return is normally distributed with mean $k\mu$ and variance $k\sigma^2$, which we denote as $\log(1 + R_t(k)) \sim N(k\mu, k\sigma^2)$. Note that:

- The mean and the variance are multiplied by k (not squared).
- Hence standard deviation grows with $\sqrt{k}$.

Since $\log(1+R_t(k)) \sim N(k\mu, k\sigma^2)$, we can find the probabilities of multi-period returns being less than a specific value using the same trick as earlier:

$$P(R_t(k) < x) = \Phi\left(\frac{\log(1+x) - k\mu}{\sqrt{k}\sigma}\right).$$

Let's build on our earlier example. What is the probability that the two-period return R is less than 5%? First, the *two-period* ratio return is log-normal(0, 0.02) (note that the variance has doubled). Therefore

$$\begin{aligned} P(R \leq 0.05) &= P(1 + R \leq 1.05) \\ &= P\left(\log(1 + R) \leq \log(1.05)\right) \\ &= P\left(N(0, 0.02) \leq \log(1.05)\right) \\ &= \Phi\left((\log(1.05) - 0)/\sqrt{0.02}\right) = 0.6349 \end{aligned}$$

And in R:

```
pnorm(log(1.05)/sqrt(0.02))
```

```
[1] 0.635
```

Note that this probability is lower than in the one-step case: without direction drift ($\mu = 0$), the dispersion of possible values just widened over time.

11.3 Random Walk

A natural desire in finance is to forecast what returns are possible in the future. If we can say something about the distribution of one-period returns, we are able to derive probabilities on the final distribution of multi-period returns as we saw. But we should also be able to say how we get there – for example, how much can we lose and for how long before the investment reaches the return forecasted by the model?

This intuitively calls for a step by step construction of each possible future outcome. Because we assume each step, going forward, is randomly picked from the one-period distribution, these constructions are called random walks.

The random walk hypothesis states that the single-period log returns are independent, and the log returns are assumed normally distributed. Let $Z_1, Z_2, \ldots, Z_n$ be i.i.d. with mean μ and standard deviation σ. Let P_0 be the price of an asset at time 0 and, for $t \geq 1$, define:

$$P_t = P_0 + Z_1 + \ldots + Z_t \tag{11.1}$$

.

If the Z_i are normally distributed, the sequence $P_0, P_1, \ldots, P_t$ is called a **normal random walk**. Note that, because sums of normal variables are themselves normal, the normality of single-period Z_t's implies the normality of multi-period sum P_t.

A normal random walk has many properties. First, the expected value at time t equals $E(P_t) = P_0 + t\mu$. The variance of the P's, as of time t, is $Var(P_t) = \sigma^2 t$ so their standard deviation is $\text{StdDev}(P_t) = \sigma\sqrt{t}$.

The mean of the distribution, μ, is also called the **drift** because it determines the general direction of the random walk. If $\mu > 0$, there is an upward trend, and a downward trend if μ is negative. Note, however, that the mean of P grows with time t! Also, observes that the standard deviation grows with the square root of time: That's why you multiply by $\sqrt{12}$ when converting from monthly volatility to annualized volatility, for example.

At any time, we know far less about where the random walk will be in the distant future compared to where it will be in the immediate future. The uncertainty grows with how far in the future we try to project. More precisely, the width of this uncertainty grows proportionally to $\sqrt{t}$.

Let's now turn to R, and on how to construct a random walk. This is actually easy: we decide the number of time steps we want to simulate, draw as many random normal samples and, because of Equation (11.1), do a cumulative sum starting from the assumed starting point. Below, the initial value (at time step

0) is set to be 0, and we construct and plot (in Figure 11.1) a random walk of drift 0.1 and standard deviation 2:

```
num.timesteps = 100
t = 0:num.timesteps
values = rnorm(num.timesteps, mean=0.1, sd=2)
cumulative.values = cumsum(values)
values = c(0, cumulative.values)
plot(t, values, type="l")
```

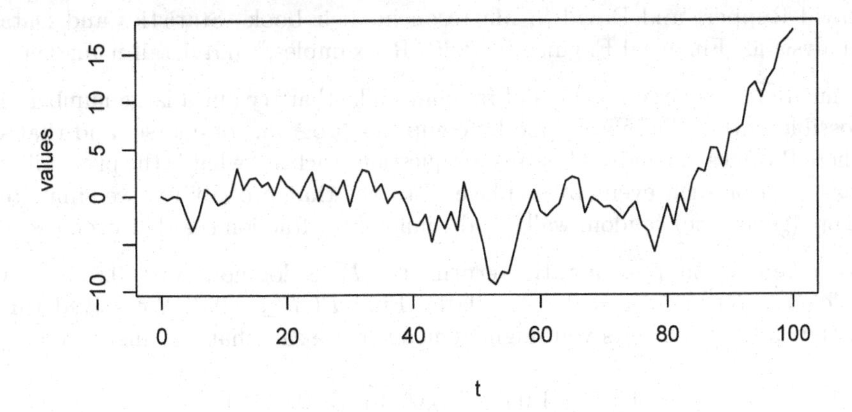

FIGURE 11.1 Our first simple random walk

With 100 time steps as selected above, we typically see the upward drift caused by the positive μ – but there is no guarantee of such trend in any given randomly selected path.

11.4 Geometric Random Walk

A **geometric random walk** is a special case of random walks. Recall that:

$$\begin{aligned}\log(1+R_t(k)) &= \log(1+R_t)+\log(1+R_{t-1})+\ldots+\log(1+R_{t-k+1}) \\ &= r_t+r_{t-1}+\ldots+r_{t-k+1}\end{aligned}$$

This is very similar to Equation (11.1)! The $r_1, r_2, \ldots, r_t$ are independent and follow distribution $N(\mu, \sigma^2)$. And since sums of normal variables are themselves normal, the normality of single-period log returns implies the normality of multi-period log returns. We can thus say that $\log(1+R_t(k)) = r_1+r_2+\ldots+r_{t-k+1}$ is a random walk and that $1+R_t(k) = \exp(r_t+r_{t-1}+\ldots+r_1)$ is log-normal. Moreover, the series of prices at time t, P_t, is equal to $P_0\exp(r_t+r_{t-1}+\ldots+r_1)$

and is the exponential of a random walk. We say the P_t's follow a **geometric random walk**. As before, if μ is positive, there is an upward drift.

11.5 Toward Simulations

We are going to investigate a case study based on an example provided by David Ruppert and David S. Matteson in their book "Statistics and Data Analysis for Financial Engineering with R Examples," and detail it further.

Simulations are large numbers of random walks that try out a large number of possible paths. There are done by computer programs of course, and that's where R comes in handy. They answer questions such as "what is the probability that such or such event takes place?" To get that probability, we simulate many (geometric) random walks and count which fraction see that even occur.

For example, suppose a ratio return $(1 + R)$ is log-normal(0, 0.01), that is $\log(1 + R) \sim N(\mu = 0, \sigma^2 = 0.01)$. Find $P(R \leq 0.05)$. We solved this analytically already! As you might remember, we saw that the answer was:

$$\Phi\left((\log(1.05) - 0)/\sqrt{0.01}\right) = 0.68719$$

Let's build a simple simulation that calculates this probability. In this first step toward simulation, we try out 10,000 scenarios for 1 time step. We create a vector called **below** that will store, for each of these scenarios, whether it resulted in a return below 5%. Each entry will be either 1 or 0, respectively, to the probability that $R \leq 0.05$ is the count of 1's in **below** divided by 10,000, i.e. the mean of the vector's values.

```
num.scenarios = 10000
below = rep(0, num.scenarios)
for(i in 1:num.scenarios) {
  random.number = rnorm(1, mean=0, sd=0.1)
  log.return = random.number
  below[i] = log.return <= log(1.05)
}
mean(below)
```

```
[1] 0.6804
```

This is very close to the exact number we saw earlier. As an aside, a mathematically equivalent but more legible code would be:

```
num.scenarios = 10000
below = rep(0, num.scenarios)
for(i in 1:num.scenarios) {
```

```
  random.number = rnorm(1, mean=0, sd=0.1)
  ratio.return = exp(random.number)
  below[i] = ratio.return <= 1.05
}
mean(below)
```

```
[1] 0.6853
```

The small difference in results comes of course from the random generation of the scenarios' returns.

Now, assume again that $(1 + R)$ is log-normal(0, 0.01), and let's find the probability that a ratio two-period return is less than 5%. We saw that the two period ratio return is log-normal(0, 0.02), so:

$$P(R \leq 0.05) = \Phi\left((\log(1.05) - 0)/\sqrt{0.02}\right) = 0.6349$$

We can then easily extend our code to simulations of 2 time steps. All that needs to change is the number of returns that **rnorm()** picks, that we have to sum these returns, and that the last **log.return** is the final one that we compare to log(1.05).

```
num.scenarios = 10000
num.timesteps  = 2
below = rep(0, num.scenarios)
for(i in 1:num.scenarios) {
  random.numbers = rnorm(num.timesteps, mean=0, sd=0.1)
  log.return = cumsum(random.numbers)
  final.log.return = log.return[num.timesteps]
  below[i] = final.log.return <= log(1.05)
}
mean(below)
```

```
[1] 0.6361
```

Here again, the result of the simulation is quite close to the exact answer, so we can be confident in our simulation framework.

11.6 The Multiple Questions Simulations Can Answer

The strongest benefit of simulations is their capacity to answer questions that have no analytic solutions. In other words, financial instruments can be so complex, and contractual terms so specific, that no formula can be derived to answer all specific questions.

One example, yet a stylistic one, is suggested in Ruppert's book[2]. This example supposed that a hedge fund invests a million dollar in stocks, but borrows most of it ($950,000 to be exact) from a bank – and thus has only $50,000 of its own money at risk. But if the value of the stock portfolio plunges below $950,000 at the end of any trading day, then the hedge fund is obligated to liquidate the entire portfolio and pay back the loan. The first question asked in this example is the probability that the value of the portfolio will fall below $950,000 at the close of any of the first 100 days.

In this example, we assume that the daily log return of the equity portfolio has a mean of 5%, annualized, with an annualized volatility of 23%. Given that, the first step will be to generate random paths of 100 steps decided by 100 random samples out of a normal distribution of mean $\frac{0.05}{253}$ and standard deviation $\frac{0.23}{\sqrt{253}}$.

We recall that:

$$P_t = P_0 \exp(r_t + r_{t-1} + \ldots + r_1)$$

So the code below sums up the 100 r_t's picked from a normal distribution of appropriate characteristics, then sums them cumulatively, and finally raises e to these successive powers to get the successive prices, starting with an initial value P_0 of $1,000,000.

```
num.scenarios = 200
num.timesteps = 100
below = rep(0, num.scenarios)
for(i in 1:num.scenarios) {
  random.numbers = rnorm(num.timesteps,
                         mean=0.05/253,
                         sd=0.23/sqrt(253))
  price = 1e6 * exp(cumsum(random.numbers))
```

This is similar to what we did before, but we add a line of code that calculates the smaller value that prices reached. That way, we can test whether any of these prices went below $950,000 and store the results into **below**:

```
  min.price = min(price)
  below[i] = min.price < 950000
}
mean(below)
```

```
[1] 0.655
```

So this simple code already answers a complex question: The hedge fund should expect to have to liquidate its investment 66% of the time it deploys that

[2]David Ruppert and David S. Matteson, "Statistics and Data Analysis for Financial Engineering with R Examples."

strategy. However, this code is not enough to answer more complex questions, such as:

- How much did we lose when the investment had to be liquidated?
- How often is the strategy profitable?
- What is the distribution of profits and losses?

We are assuming here that we hold our investment until the end of the 100-day period, unless of course we hit the trigger that forces to sell our position.

```
num.scenarios = 400
profit.or.loss = rep(0, num.scenarios)
min.price = rep(0, num.scenarios)
final.price = rep(0, num.scenarios)
for(i in 1:num.scenarios) {
```

We then do a geometric random walk of 100 steps:

```
  random.numbers = rnorm(100, mean=0.05/253, sd=0.23/sqrt(253))
  growth.factors = exp(cumsum(random.numbers))
  prices = 1e6 * growth.factors
```

The lowest price level reached on this price path i is recorded, together with the final price at the end of that path:

```
  min.price[i] = min(prices)
  final.price[i] = prices[100]
```

We then test if the price went below the liquidation threshold at one point:

```
  if(min.price[i] < 950000) {
```

If the simulation enters this point of the code, then the price did go below the liquidation trigger. But what was that price? To get that information, we need to collect all prices below that level, then pick the first one in that vector. The P&L is then the price at the time of liquidation less the one million dollars initially invested.

```
    prices.below.floor = prices[ prices<950000 ]
    price.at.time.of.event = prices.below.floor[1]
    profit.or.loss[i] = price.at.time.of.event-1e6
```

If the simulation did not satisfy the condition of the **if()**, then the execution goes into the **else** clause below, meaning that the investment value never went below $950,000. The P&L is the final price less the amount invested:

```
  } else {
    profit.or.loss[i] = final.price[i]-1e6
  }
}
```

In the code above, we collected the minimum prices reached in each iteration (i.e., in each scenario). Let's look at the distribution of these returns. To do that, we need to first define the break points of the histogram. To make the histogram more pleasing aesthetically, we can round down the minimum price to the next 10 thousand, round up the maximum price to the next 10 thousand, and use a round number for the bin width, such as, here again, 10,000.

```
breaks = seq(floor(min(min.price)/10000)*10000,
          ceiling(max(min.price)/10000)*10000,
          by=10000)
```

We are now ready to plot the histogram, and the result appears in Figure 11.2.

```
hist(min.price, breaks = breaks)
```

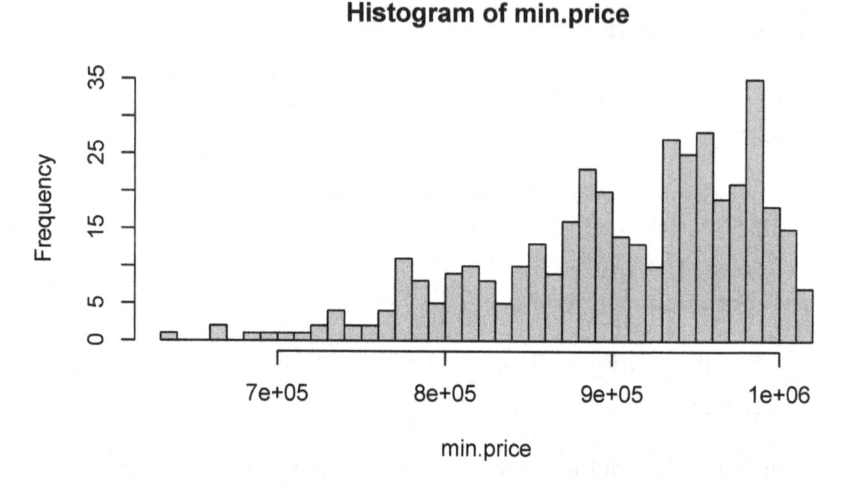

FIGURE 11.2 Distribution of minimum prices reached in our simulation

We learn from Figure 11.2 that the value of the investment often reached very low levels: the distribution of minimal valuations has a long left tail, meaning that, even if the trade gets unwound because the liquidation threshold is triggered, the loss to the investor can be deep.

Likewise, we can plot the distribution of final prices, using the same breaks. This results in Figure 11.3.

```
breaks = seq(floor(min(final.price)/10000)*10000,
          ceiling(max(final.price)/10000)*10000,
          by=10000)
hist(final.price, breaks = breaks)
```

And finally, we can plot the distribution of P&L of the strategy. That distribution appears in Figure 11.4.

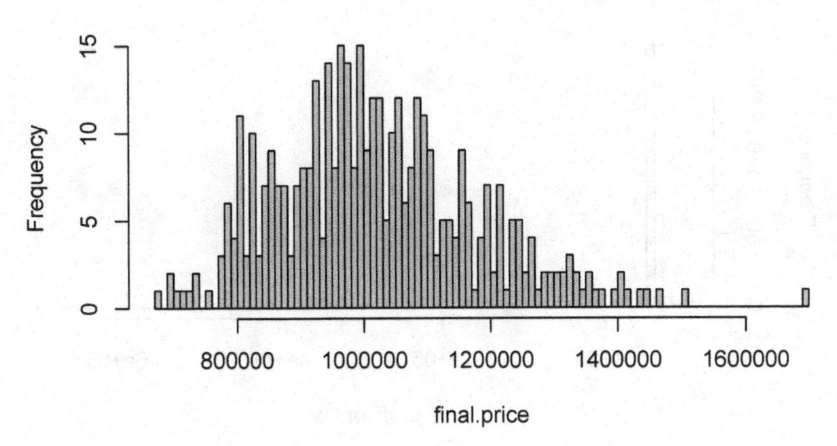

FIGURE 11.3 Distribution of final prices

```
breaks = seq(floor(min(profit.or.loss)/10000)*10000,
             ceiling(max(profit.or.loss)/10000)*10000,
             by=10000)
hist(profit.or.loss, breaks=breaks)
```

The expected average profit of this trading strategy is then:

```
mean(profit.or.loss)
```

```
[1] 19311
```

Despite its left tail, it has a positive expected profit.

We can also calculate the probability that the profit exceeds $100,000. To do that, we could count the number of times the P&L was above that value and divide it by the number of scenarios:

```
(num.profitable.scenarios =
   length(profit.or.loss[ profit.or.loss>100000 ]))
```

```
[1] 89
```

```
num.profitable.scenarios / num.scenarios
```

```
[1] 0.2225
```

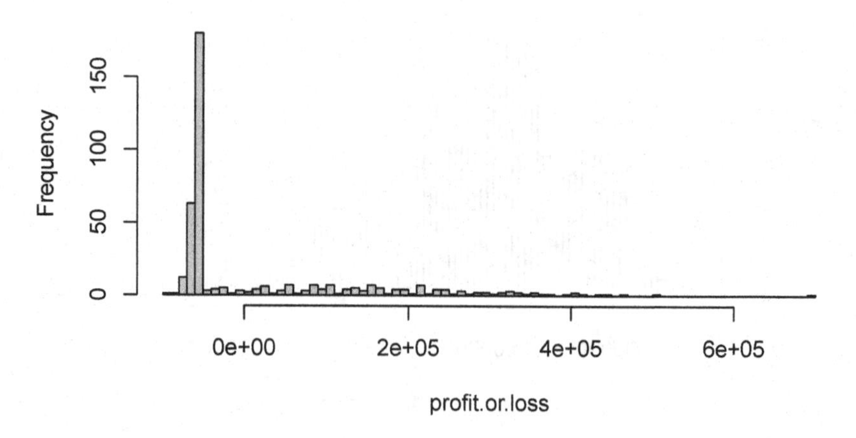

FIGURE 11.4 Profit and Loss distribution of our example hedge fund simulations

11.7 Exercises

11.7.1 Probability of a Loss

On page 237, we calculated the probability that the profit exceeds $100,000. Extend the code to calculate the probability of a loss.

■ Solution

```
(num.loss.scenarios =
   length(profit.or.loss[ profit.or.loss < 0 ]))
```

```
[1] 273
```

```
num.loss.scenarios / num.scenarios
```

```
[1] 0.6825
```

■ End of solution

12

Linear and Polynomial Regression

The simulations discussed in the previous chapter were tools to determine what could potentially happen, in the future, to a variable's value based on a statistical model of that variable. Another way to predict future values is to craft a model between our hard-to-predict variable and another, hopefully easier to predict, variable. **Regression** is one method to establish such a model, and a regression is linear or polynomial depending on the form of that model.

In R, the function that builds a linear model (i.e., performs a linear regression) is `lm()`. The `lm()` function is easy to use but its syntax can be a bit intimidating: a call to `lm()` looks like **`lm(dependentVariable ~ independentVariable)`** with a tilde ~ symbol following the dependent variable.

Let's put it to use and see if a simple vector **`ys`** of four values can be modeled as a linear function of a vector **`xs`**:

```
xs = c(1, 3, 4, 8)
ys = c(-2, 5, 7, 17)
lm(ys ~ xs)
```

```
Call:
lm(formula = ys ~ xs)

Coefficients:
(Intercept)           xs
      -3.87         2.65
```

We learn that a linear model of **`ys`** as a function of **`xs`** is:

$$ys = -3.865 + 2.654 \times xs + \epsilon,$$

where ϵ represents the model error, or **residual**. In other words, the relationship is a line whose slope is 2.654 and whose intercept with the y-axis is -3.865. The slope is the coefficient of the independent variable xs, which is sometimes called the variable's **loading**.

The output of `lm()` is much more than what gets printed by default: the output of `lm()` is actually an object that can be manipulated and passed to

DOI: 10.1201/9781003320555-12

other functions. For example, we can pass it to **abline()** to draw the line of best fit calculated by **lm()**, producing the plot in Figure 12.1:

```
plot(xs, ys)
linearRegr = lm(ys ~ xs)
abline(linearRegr)
```

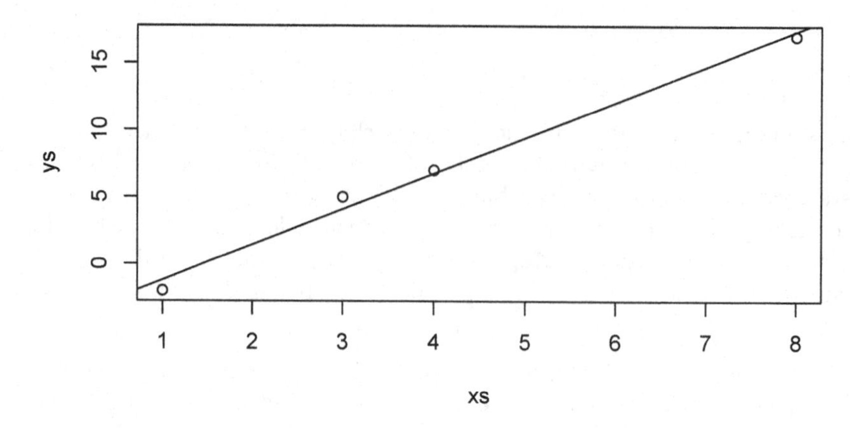

FIGURE 12.1 Our first line of best fit

The **summary()** function helps us extract additional details from the linear model:

```
summary(linearRegr)
```

```
Coefficients:
            Estimate Std. Error t value Pr(>|t|)
(Intercept)  -3.8654     0.8410  -4.596  0.04422 *
xs            2.6538     0.1773  14.968  0.00443 **
---
Signif. codes:  0 '***' 0.001 '**' 0.01 '*' 0.05 '.' 0.1 ' ' 1

Residual standard error: 0.9041 on 2 degrees of freedom
Multiple R-squared:  0.9912,    Adjusted R-squared:  0.9867
F-statistic:   224 on 1 and 2 DF,  p-value: 0.004434
```

We can find the p-value for the entire model on the very last line and, for each coefficient (intercept and **xs**), under **Pr(>|t|)**. Appreciating their significance is made easy by the stars ** on the different rows. For **xs** for instance, and as explained in the caption that immediately follows, the ** code implies significance at the 100% - 1% = 99% confidence level.

Statistics Note: The low p-value of 0.0044 implies rejecting $H_0 : \beta = 0$ vs $H_1 : \beta \neq 0$, where β is the coefficient of the independent variable.

The **R-squared** and **adjusted R-squared** are also two key metrics. The latter penalizes (reduces) the former depending on the number of variables used in the model. Indeed, a model using too many variables can cause **overfit**. The adjusted R-squared favors parsimonious models.

An even more comprehensive view of the details contained in the **linearRegr** object is provided by the following command:

```
str(summary(linearRegr))
```

An excerpt of the output is:

```
 $ coefficients : num [1:2, 1:4] -3.865 2.654 0.841 0.177 -4.596 ...
  ..- attr(*, "dimnames")=List of 2
  .. ..$ : chr [1:2] "(Intercept)" "xs"
  .. ..$ : chr [1:4] "Estimate" "Std. Error" "t value" "Pr(>|t|)"
 $ aliased      : Named logi [1:2] FALSE FALSE
  ..- attr(*, "names")= chr [1:2] "(Intercept)" "xs"
 $ sigma        : num 0.904
 $ df           : int [1:3] 2 2 2
 $ r.squared    : num 0.991
 $ adj.r.squared: num 0.987
```

In particular, we find that the object produced by the regression contains a field storing the **R-squared**, **r.squared**, and another containing the **adjusted R-squared**, **adj.r.squared**. So we can write R code referencing these values, such as:

```
summary(linearRegr)$adj.r.squared
```

```
[1] 0.9867
```

We can also observe a field named **coefficients** containing a vector whose first two entries are the loadings of the intercept and the independent variable, respectively.

12.1 The House Price Dataset

For a more realistic example coming from finance, we turn to one of the numerous house price datasets available on the internet[1]:

```
house = read.csv("Datasets/HousePrices.csv")
```

[1]The one we are using comes from https://www.kaggle.com/datasets/shubhendra7/houseprices-dataset

In this example, we plot prices against lot sizes in the house dataset, resulting in Figure 12.2:

```
plot(house$lotsize, house$price,
    xlab = "Lot Size", ylab = "House Price")
```

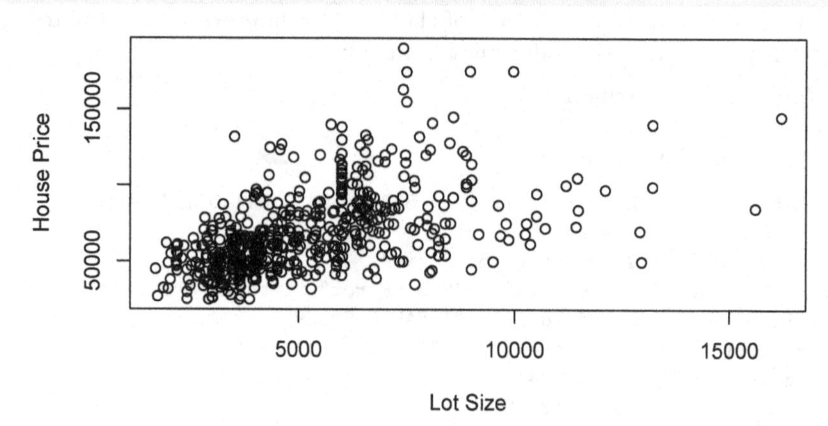

FIGURE 12.2 Prices vs Lot Sizes in the HousePrices data set

There clearly is a relationship between lot size and house value, but how linear is that relationship?

Calculating The Linear Regression

Since we are now dealing with data stored in a dataframe, we have to use the **$** operator to refer to the columns containing the dependent and independent variables:

```
linearRegr = lm(house$price ~ house$lotsize)
```

R offers an alternate syntax to produce a similar result:

```
linearRegr = lm(price ~ lotsize, data=house)
```

Note that the new parameter, **data**, specifies which dataframe the names **price** and **lotsize** implicitly refer to. Without it, **lm()** would not recognize **price** or **lotsize** since there are no variables with those names outside of **house**.

We can then plot the line of best fit, and while at it, add the adjusted R-squared as a caption in the **topright** corner. The output appears in Figure 12.3:

```
plot(house$lotsize,house$price)
abline(linearRegr)
legend("topright",
      legend=paste("R-squared = ",
```

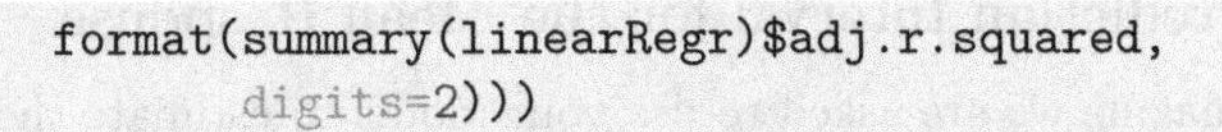

```
          format(summary(linearRegr)$adj.r.squared,
                 digits=2)))
```

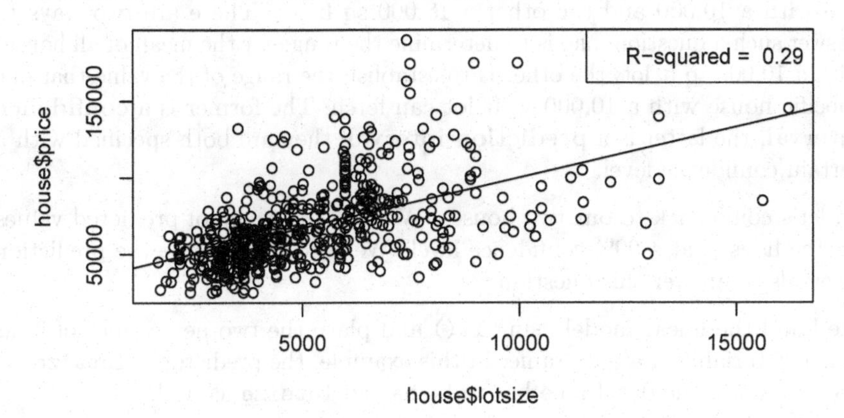

FIGURE 12.3 Linear regression and R-squared of price vs lot size

We extract the coefficient of the independent variable, plus other data, as follows:

```
summary(linearRegr)$coefficients
```

```
              Estimate Std. Error t value  Pr(>|t|)
(Intercept) 34136.192  2491.0636    13.7 6.275e-37
lotsize         6.599     0.4458    14.8 6.770e-42
```

The low p-value of 6.7699×10^{-42} implies rejecting the null hypothesis that the loading β of **`lotsize`** is zero, $H_0 : \beta = 0$, and we favor the alternative, $H_1 : \beta \neq 0$.

Confidence intervals on the estimate of a loading are easy to get in R. Here is an example for a 99% confidence interval on the coefficient of **`lotsize`**. Note that the "remaining" 1% is spread between both sides:

```
confint(linearRegr, level = 0.99)
```

```
                0.5 %    99.5 %
(Intercept) 27697.049 40575.334
lotsize         5.446     7.751
```

So we can be "99% confident" that the "true" loading is between 5.446 and 7.751. This confidence interval does not contain the hypothesized value of 0. Since it does not contain 0, it is equivalent to rejecting the test of $H_0 : \beta = 0$ vs $H_1 : \beta \neq 0$ at $\alpha = 0.01$, which we had seen previously.

Prediction Interval for the Mean Response

Imagine we are asked to use your model to estimate the prices of two houses, one with a 10,000 and the other a 18,000 sq ft lot. There are two ways to answer such a question: one is to determine the range for the mean of all houses with a 10,000 sq ft lot; the other is to establish the range of the value that one specific house with a 10,000 sq ft lot can fetch. The former is a **confidence interval**, the latter is a **prediction interval**; they are both specified with a certain confidence level.

So let's come back to our two houses. What is the range of predicted values for the houses, at a 99% confidence level? We need to calculate the prediction intervals to answer this question.

We build the linear model using `lm()` and place the two new values of independent variable in a dataframe. In this example, the predictor is `lotsize`, so the column in the dataframe has to be named `lotsize` as well:

```
linearRegr = lm(price ~ lotsize, data = house)
new_houses = data.frame(lotsize = c(10000, 18000))
```

We then use the generic `predict()` function offered by `stats` using the appropriate value for `interval`:

```
predict(linearRegr,
        newdata = new_houses,
        interval = c("prediction"),
        level = 0.99)
```

```
     fit   lwr    upr
1 100124 41470 158778
2 152914 92678 213150
```

So the linear regression predicts that, all else being equal, the value of a house with a lot of 10,000 square feet is $100,124 and that the true value, at a 99% confidence level, lies between $41,470 and $158,778, and between $92,678 and $213,150 for a lot size of 18,000 square feet.

If we want to know the average price of all houses with a 10,000 square feet lot, then we compute a confidence interval. We use the `predict()` function again:

```
predict(linearRegr,
        newdata = new_houses,
        interval = c("confidence"),
        level = 0.99)
```

```
     fit    lwr    upr
1 100124  94003 106245
2 152914 137896 167932
```

So the linear regression predicts that, all else being equal, the average value of all houses lies, at a 99% confidence level, between $94,003 and $106,245, and between $137,896 and $167,932 for the average of all houses with a lot size of 18,000 square feet.

Statistics Note:
Note that the confidence interval for the mean response

$$\hat{y}(x) \pm t_{\alpha/2,n-2} \cdot s_e \sqrt{\frac{1}{n} + \frac{(x-\bar{x})^2}{S_{xx}}}$$

This is interesting mostly in comparison to the prediction interval: Prediction intervals must account for both the uncertainty in estimating the population mean, plus the random variation of the individual values. This **prediction interval** looks like this:

$$\hat{y}(x) \pm t_{\alpha/2,n-2} \cdot s_e \sqrt{1 + \frac{1}{n} + \frac{(x-\bar{x})^2}{S_{xx}}}.$$

Note the added "1 +" in the square root, so the prediction interval is larger than the confidence interval.

ICE BofA AAA US Corporate Index

Let's apply what we've just learned on two time series in the `fredgraph.xlsx` file introduced earlier:

```
rates = read.xlsx("fredgraph.xlsx", startRow=14)
```

Specifically we are going to regress the monthly changes in the yield of the ICE BofA AAA US Corporate Index against the monthly changes in 10-Year Treasury Constant Maturity rates. Note that we are using **diff()** to calculate differences with a lag of 1:

```
rates = rates %>%
  rename(baml.AAA = BAMLC0A1CAAAEY, tsy.10yr=DGS10)
AAA.bonds = diff(rates$baml.AAA)
tsy.10yr = diff(rates$tsy.10yr)
```

We can then construct a linear model, whose characteristics are as follows:

```
summary(lm(AAA.bonds ~ tsy.10yr))

Call:
lm(formula = AAA.bonds ~ tsy.10yr)

Residuals:
```

```
    Min       1Q   Median       3Q      Max
-0.2989 -0.0390  0.0059  0.0286  0.4639

Coefficients:
            Estimate Std. Error t value Pr(>|t|)
(Intercept)  -0.0065     0.0124   -0.53      0.6
tsy.10yr      0.5665     0.0717    7.90  8.9e-11 ***
---
Signif. codes:
0 '***' 0.001 '**' 0.01 '*' 0.05 '.' 0.1 ' ' 1

Residual standard error: 0.0958 on 58 degrees of freedom
Multiple R-squared:  0.518, Adjusted R-squared:  0.51
F-statistic: 62.4 on 1 and 58 DF,  p-value: 8.94e-11
```

Note again the little stars that R provides as helpers: the `***` next to the p-value for `tsy.10yr` tell us that the loading for that variable is significantly different from zero. The absence of stars next to the p-value for the intercept indicates that we cannot reject the null hypothesis that the intercept is zero. Well.. its estimated value is very close to zero anyway: -0.0065.

Moreover, the standard deviation of the residual (the part unexplained by the model and often denoted ϵ) is provided on the third line up from the bottom: 0.09584. From that, we can conclude that the fitted regression model is

$$\texttt{AAA.bonds} = 0.00 + 0.566 \times \texttt{tsy.10yr} + \epsilon,$$

with $\sigma_\epsilon = 0.09584$.

12.2 Multi-linear Regression

We now want to improve on the previous model by incorporating the impact of additional variables using **multi-linear regression**. The two new independent variables are the time series of monthly 3-Month Treasury Constant Maturity Rates and the monthly 3-Year Treasury Constant Maturity rates. As said, we would be interested in quantifying the combined impact of monthly changes in all 3 rates, including the 10-year rate seen earlier. We first compute the time series of differences for the other two series we didn't consider in the previous section, again using `diff()`:

```
rates = rates %>% rename(tsy.3mo=DGS3MO, tsy.3yr=DGS3)
tsy.3yr = diff(rates$tsy.3yr)
tsy.3mo = diff(rates$tsy.3mo)
```

The multi-linear regression is specified by separating the independent variables using the + operator:

```
summary(lm(AAA.bonds ~ tsy.10yr + tsy.3yr + tsy.3mo))
```

```
Call:
lm(formula = AAA.bonds ~ tsy.10yr + tsy.3yr + tsy.3mo)

Residuals:
     Min       1Q   Median       3Q      Max
-0.29663 -0.03112  0.00345  0.04511  0.18087

Coefficients:
            Estimate Std. Error t value Pr(>|t|)
(Intercept) -0.00469    0.01083   -0.43  0.66672
tsy.10yr     0.54015    0.16389    3.30  0.00171 **
tsy.3yr      0.34992    0.22865    1.53  0.13156
tsy.3mo     -0.43670    0.10618   -4.11  0.00013 ***
---
Signif. codes:
0 '***' 0.001 '**' 0.01 '*' 0.05 '.' 0.1 ' ' 1

Residual standard error: 0.0829 on 56 degrees of freedom
Multiple R-squared:  0.652, Adjusted R-squared:  0.634
F-statistic:    35 on 3 and 56 DF,  p-value: 6.97e-13
```

The order in which the independent variables are listed in a multi-linear regression does not matter: the results will be almost equal, with perhaps some rounding error. This is in contrast to ANOVA, discussed later.

The p-values for the intercept and **tsy.3yr** are high, so we do not reject the hypothesis that their values are 0. But those for the two other predictors are very small, hence we can conclude that these slopes are not 0. So, we can accept the null hypothesis that, conditional on **tsy.10yr** and **tsy.3mo**, **AAA.bonds** and **tsy.3yr** are not linearly related. This result should not be interpreted as stating that **AAA.bonds** and **tsy.3yr** are unrelated, but only that **tsy.3yr** is not useful in predicting **AAA.bonds** when **tsy.10yr** and **tsy.3mo** are included in the regression model. In fact, **AAA.bonds** and **tsy.3yr** have a significant correlation and the linear regression of **AAA.bonds** on **tsy.3yr** alone is highly significant. You can check the first statement using the **cor()** function seen on page 42:

```
cor(AAA.bonds, tsy.3yr)
```

```
[1] 0.5789
```

Your can verify the relationship significance by creating a linear model using **lm()**, extracting a summary, and printing the coefficients corresponding to the p-value for **tsy.3yr**:

```
summary(lm(AAA.bonds ~ tsy.3yr))$coefficients[2,4]
```

```
[1] 1.266e-06
```

Finally, we should always check for the reasonableness or intuition behind regression coefficients: AAA corporate bonds are issued by businesses, thus have default risk, but that risk is presumably low; instead, their sensitivity to interest rates is much more significant. And as far as sensitivity to interest rates, the maturities of corporate bonds are typically of 5 to 10 years, so 10-year Treasury should have a strong explanatory power. 3-year Treasury rates should make sense too but 3-month rates, which are very short term, probably much less. The fact that 3-month rates had a low p-value hints that there could be collinearity among the independent variables, or that there is an economic rationale (maybe, that 3-month rates are an indicator of easy financing, which does help corporations).

12.3 Collinearity

The issue of collinearity is critical, but instead of explaning the theory of it, we will illustrate its danger on an artificial but revealing example.

Assume you are given the following data and are tasked to assess whether age, in years, or experience (however, it is measured!) is a better predictor of income (in dollars):

Income	Age	Experience
20000	20	18
33000	32	30
40000	41	39
50000	50	52
60000	59	61

As usual, it is a good idea to form an intuition (but not a preconception) of what to expect from a linear regression. This simple made-up table was constructed so that an approximate increase of 10 in age or experience corresponds to an increase of $10,000 in income, so the loading or either age or experience should be around 1,000 – by design!

Regressing income against age shows a strong R-squared and a very low p-value for age (only part of the output is shown for brevity):

```
summary(lm(Income ~ Age))
```

```
             Estimate Std. Error t value Pr(>|t|)
(Intercept)   -219.20    1329.65  -0.165     0.88
Age           1010.38      31.19  32.393 6.47e-05 ***
Multiple R-squared:  0.9971,    Adjusted R-squared:  0.9962
```

We also check that the coefficient of **Age** is around 1,000, as expected.

Likewise, regressing income against experience also indicates that experience is highly significant (again, partial output displayed):

```
summary(lm(Income ~ Experience))
```

```
             Estimate Std. Error t value Pr(>|t|)
(Intercept)  4702.56     1598.83   2.941 0.060449 .
Experience    897.44       37.33  24.038 0.000158 ***
Multiple R-squared:  0.9948,    Adjusted R-squared:  0.9931
```

We note again that the beta of **Experience** is roughly 1,000.

But regressing income against both age and experience concludes that neither independent variable is significant!

```
summary(lm(Income ~ Age + Experience))
```

```
             Estimate Std. Error t value Pr(>|t|)
(Intercept)     982.9     2947.3   0.333    0.770
Age             757.1      530.3   1.428    0.290
Experience      225.7      471.6   0.479    0.679
Multiple R-squared:  0.9974,    Adjusted R-squared:  0.9949
```

The **R-squared** is very high though! So there's something fishy here, and that's due to collinearity.

Note also how dramatically different (and wrong) the loadings now are. In fact, they *add up* to about 1,000 – that amount was spread, about "randomly", between the two betas. (See Exercise 12.9.1 for more on this.)

Back to the ICE BAML AAA Series

We are now alerted on the impact of collinearity on R-squared, significance and even coefficient values. Let's see now how this can impact our analysis of the impact of rate changes on the rate of AAA bonds. But first, let's look at the cross-correlations among our data series thanks to the **corrplot()** function from the **corrplot** package. We first load the package then create a single dataframe (using **cbind()**) whose columns are the data we want to compute cross-correlations on:

```
rates.changes = cbind(AAA.bonds, tsy.3mo, tsy.3yr, tsy.10yr)
```

We then calculate the correlations using **cor()**, then plot them using **corrplot()** offered by the package of the same name. The output appears in Figure 12.4.

```
rates.cor = cor(rates.changes)
corrplot(rates.cor)
```

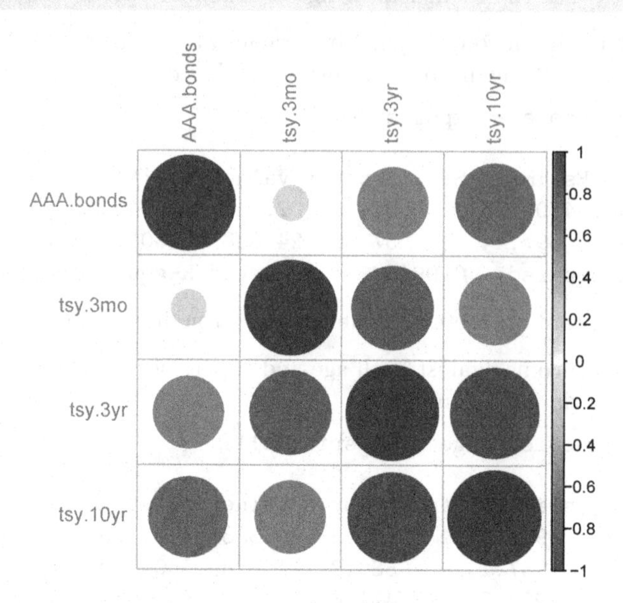

FIGURE 12.4 Correlation Plot using Corrplot

Note that **corrplot()** needs the results of **cor()**. Also, please see on page 42 for another example of visualizing a correlation matrix, and Exercise 10.4.1 for a related exercise.

We notice a low correlation between **tsy.3mo** and the independent variable, which is not necessarily a bad thing. A more serious red flag comes from the high correlation between **tsy.3mo** and **tsy.3yr**. Let's keep that in the back of our minds for now.

Individually, the (simple) linear regressions are as follows (again, we show only partial outputs for brevity). Against 3-month Treasuries:

```
summary(lm(AAA.bonds ~ tsy.3mo))
```

```
             Estimate Std. Error t value Pr(>|t|)
(Intercept) -0.007779   0.017660  -0.441    0.661
tsy.3mo   0.104596   0.097097   1.077    0.286
Multiple R-squared:  0.01961,   Adjusted R-squared:  0.002712
```

The R-squared is terrible and the beta not significant. We now regress against 3-year Treasuries:

```
summary(lm(AAA.bonds ~ tsy.3yr))
```

```
              Estimate Std. Error t value Pr(>|t|)
(Intercept) -0.002963  0.014572  -0.203     0.84
tsy.3yr  0.488718  0.090394   5.407 1.27e-06 ***
Multiple R-squared:  0.3351,    Adjusted R-squared:  0.3236
```

And finally against 10-year so-called **T-notes**:

```
summary(lm(AAA.bonds ~ tsy.10yr))
```

```
              Estimate Std. Error t value Pr(>|t|)
(Intercept)  -0.006502   0.012375  -0.525    0.601
tsy.10yr  0.566446   0.071683   7.902 8.94e-11 ***
Multiple R-squared:  0.5184,    Adjusted R-squared:  0.5101
```

That R-squared is reasonably high. Now, what is the effect of adding `tsy.3yr` to `tsy.10yr`?

```
summary(lm(AAA.bonds ~ tsy.10yr + tsy.3yr))
```

```
              Estimate Std. Error t value Pr(>|t|)
(Intercept)  -0.009269   0.012188  -0.760   0.4501
tsy.10yr  0.850737   0.164519   5.171 3.12e-06 ***
tsy.3yr  -0.337238   0.176556  -1.910   0.0612 .
Multiple R-squared:  0.5474,    Adjusted R-squared:  0.5315
```

Regressing against `tsy.10yr` alone had an adjusted R-squared of 0.51, but that metric only increases to 0.53 if we add `tsy.3yr` as a second predictor. This might suggest that `tsy.3yr` might not be related to the dependent variable (here, `AAA.bonds`) but this is not necessarily the case – even though in this example, the R-squared for `tsy.3yr` was a modest 0.32.

Also, as we could expect from the earlier discussion, the loading of `tsy.3yr` is now non-significant. Moreover, its value went from 0.49 to *negative* 0.34. Another effect of the high correlation between the predictor variables is that the regression coefficient for each variable is very sensitive to whether the other variable is in the model. The correlation of `tsy.3yr` with `tsy.10yr` is indeed high – the two variables provide redundant information; using the `cor()` function again:

```
cor(tsy.3yr, tsy.10yr)
```

```
[1] 0.9047
```

And what about adding `tsy.3mo` to `tsy.10yr`? Let's see the results:

```
summary(lm(AAA.bonds ~ tsy.3mo + tsy.10yr))
```

```
              Estimate Std. Error t value Pr(>|t|)
(Intercept)  -0.007273   0.010827  -0.672    0.504
tsy.3mo   -0.317967   0.073338  -4.336 5.98e-05 ***
tsy.10yr  0.762079   0.077252   9.865 6.18e-14 ***
Multiple R-squared:  0.6379,     Adjusted R-squared:  0.6252
```

Adding a variable, which alone offers a weak R-squared and whose loading was non-significant, offers the best boost to the adjusted R-squared, and its loading is now highly significant.

12.4 Variance Inflation Factor

The **variance inflation factor** (**VIF**) of a variable tells us how much the squared standard error, i.e., the variance of the coefficient of that variable is increased by having the other predictor variables in the model. For example, if a variable has a VIF of 4, then the variance of its beta is four times larger than it would be if the other predictors were either deleted or were not correlated with it. The standard error is increased by a factor of 2.

Suppose we have predictor variables $X_1, .., X_p$. Then the VIF of X_j is found by regressing X_j on the $p-1$ other predictors. Let R_j^2 be the R^2-value of this regression, so that R_j^2 measures how well X_j can be predicted from the other Xs. Then the VIF of X_j is

$$VIF_j = \frac{1}{1 - R_j^2}$$

.

For example, if we regress age against experience, we find a **R-squared** of:

```
age.exp = summary(lm(Age ~ Experience))
age.exp$r.squared
```

```
[1] 0.9953
```

A value of R_j^2 close to 1 implies a large VIF. I.e., the more accurately that X_j can be predicted from the other Xs, the more redundant it is and the higher its VIF. The minimum value of VIF_j is 1 and occurs when R_j^2 is 0, and there is no upper bound to VIF_j. In the case of `Age`, its VIF is an elevated 1 / (1 - 0.9953) = 215.

It is important to keep in mind that VIF tells us nothing about the relationship between the response and jth predictor. Rather, it tells us only how correlated the jth predictor is with the other predictors. In fact, the VIFs can be computed without knowing the values of the response variable.

12.5 ANOVA

ANOVA is a decomposition of the variance of the dependent variable into additive contributions from the independent variables.

The **anova()** function outputs a table that describes how much of the variation in Y is predictable if one knows $X_1, ..., X_p$. The total variation in Y can be partitioned into two parts: the variation that can be predicted by a linear function of these variables, and the amount that cannot be predicted.

That function takes a linear regression model as its input, so **lm()** is "inside" **anova()** below:

```
anova(lm(AAA.bonds ~ tsy.10yr + tsy.3mo + tsy.3yr))
```

```
Analysis of Variance Table

Response: AAA.bonds
          Df Sum Sq Mean Sq F value  Pr(>F)
tsy.10yr   1  0.574   0.574   83.53 1.1e-12 ***
tsy.3mo    1  0.132   0.132   19.24 5.1e-05 ***
tsy.3yr    1  0.016   0.016    2.34    0.13
Residuals 56  0.385   0.007
---
Signif. codes:
0 '***' 0.001 '**' 0.01 '*' 0.05 '.' 0.1 ' ' 1
```

How can we interpret this output? The regression sum of squares for the model that uses only **tsy.10yr** is in the first row of the ANOVA table and is 0.574. The entry in the second row, 0.132, is the *increase* in the regression sum of squares when **tsy.3mo** is added to the model. Similarly, 0.016 is the increase in the regression sum of squares when **tsy.3yr** is added. Thus, $0.574 + 0.132 + 0.016 = 0.722$ is the regression sum of squares with all three predictors in the model.

The p-values indicate the confidence at which we can reject the null hypothesis that the coefficient is 0. For **tsy.10yr** and **tsy.3mo**, the p-values are very small, so the null hypotheses (that either is 0) are rejected. The p-value for **tsy.3yr** is high however, so the null hypothesis that the loading of **tsy.3yr** is nil is not rejected.

Order of the ANOVA Predictors

In contrast to linear regression, the order in which independent variables are specified matters in ANOVA. Let's take the same three independent variables as in the previous example, but specified in a different order:

```
anova(lm(AAA.bonds ~  tsy.3yr + tsy.3mo + tsy.10yr ))
```

```
Analysis of Variance Table

Response: AAA.bonds
          Df Sum Sq Mean Sq F value  Pr(>F)
tsy.3yr    1  0.371   0.371    54.0 9.2e-10 ***
tsy.3mo    1  0.276   0.276    40.3 4.2e-08 ***
tsy.10yr   1  0.075   0.075    10.9  0.0017 **
Residuals 56  0.385   0.007
---
Signif. codes:
0 '***' 0.001 '**' 0.01 '*' 0.05 '.' 0.1 ' ' 1
```

Now that **tsy.3yr** is the first predictor, its sum of squares is much larger than before and its p-value is highly significant; it was non-significant in the earlier ANOVA table. The sum of squares for **tsy.3yr** is now much larger than that of **tsy.10yr**, the reverse of what we saw earlier, since **tsy.3yr** and **tsy.10yr** are highly correlated and the first of them in the list of predictors will have the larger sum of squares.

Note that the sums of squares are the same in the two models:

```
0.02170 + 0.62551 + 0.07459 + 0.38457
```

```
[1] 1.106
```

```
0.37074 + 0.27647 + 0.07459 + 0.38457
```

```
[1] 1.106
```

So which of the two models is best? Let's see how we can perform model selection.

Model Selection Using ANOVA

Suppose we have two models, I and II, and the predictor variables in model I are a subset of those in model II, so that model I is a submodel of II. A common null hypothesis is that the data are generated by model I. Equivalently, we want to test the hypothesis that, in model II, the slopes are zero for the variables that were not also in model I.

In this use case, we build the two linear models separately, then request an **anova()** analysis of one against the other:

```
model.I  = lm(AAA.bonds ~ tsy.10yr)
model.II = lm(AAA.bonds ~ tsy.10yr + tsy.3mo + tsy.3yr)
anova(model.I, model.II)
```

```
Analysis of Variance Table

Model 1: AAA.bonds ~ tsy.10yr
Model 2: AAA.bonds ~ tsy.10yr + tsy.3mo + tsy.3yr
  Res.Df   RSS Df Sum of Sq    F  Pr(>F)
1     58 0.533
2     56 0.385  2     0.148 10.8 0.00011 ***
---
Signif. codes:
0 '***' 0.001 '**' 0.01 '*' 0.05 '.' 0.1 ' ' 1
```

In the last row of the output, the entry 2 in the "Df" column is the difference between the two models in the number of parameters and 0.14821 in the "Sum of Sq" column is the difference between the residual sum of squares (RSS) for the two models. The small p-value leads us to reject the null hypothesis that the variables in Model II and not in Model I have a slope of zero. In other words, we need *at least one* of the two variables in in Model II and not in Model I, but we don't know which one. So let's compare two models that differ in only one of the two:

```
model.I =
  lm(AAA.bonds ~ tsy.10yr + tsy.3mo)
model.II =
  lm(AAA.bonds ~ tsy.10yr + tsy.3mo + tsy.3yr)
anova(model.I, model.II)
```

```
Analysis of Variance Table

Model 1: AAA.bonds ~ tsy.10yr + tsy.3mo
Model 2: AAA.bonds ~ tsy.10yr + tsy.3mo + tsy.3yr
  Res.Df   RSS Df Sum of Sq    F Pr(>F)
1     57 0.401
2     56 0.385  1    0.0161 2.34   0.13
```

The large p-value (0.13) leads us to accept the null hypothesis that the variable added in Model II, **tsy.3yr**, has a loading of zero. Notice that this is the same as the p-value for **tsy.3yr** in the first ANOVA table. This is not a coincidence: Both p-values are the same because they are testing the same hypothesis.

12.6 Response Transformation

Linear regression makes several assumptions that are often forgotten. One is that, of course, the variables have a linear relationship. Another assumption is that the residuals follow a constant distribution. When some of these assumptions are not satisfied, the data set can be adjusted by transforming the dependent variable, also called **response**, in a transformation called **response transformation**.

Note that this is not "cheating": there is a one-to-one mapping between the old response and the transformed one, so we can always go back to the original values of the dependent variable.

First, let's examine the case where the data have a strong relationship but that relationship is not linear, and how we can transform the dependent (response) variable to best use linear regression. The example here is a dataset that only consists of the number of years of employees at a firm, and their salary. (This example is adapted from `https://daviddalpiaz.github.io/appliedstats`.) Plotting the two variables in Figure 12.5 shows a relationship that's almost linear but with some concavity up, with the line of best fit undershooting a bit on the low and high ends of the independent variable. It also appears that there is a dispersion of the response (dependent variable) on the right side of the plot.

```
initech = read.csv("Datasets/initech.csv")
plot(salary ~ years,
     data = initech,
     main = "Salaries at Initech, By Seniority")
initech_fit = lm(salary ~ years, data = initech)
abline(initech_fit)
```

The linear regression of salaries as a function of years of experience shows a solid R-squared:

```
summary(initech_fit)
```

```
Call:
lm(formula = salary ~ years, data = initech)

Residuals:
   Min     1Q Median     3Q    Max 
-57225 -18104    241  15589  91332 

Coefficients:
            Estimate Std. Error t value Pr(>|t|)
```

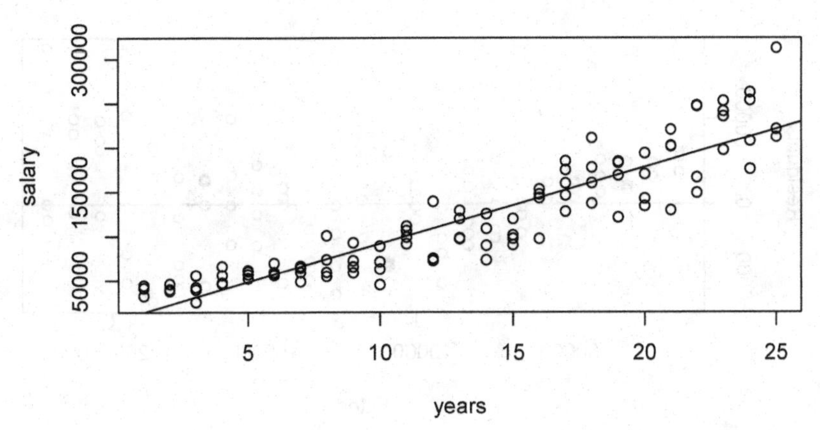

FIGURE 12.5 Salary as a function of years of experience in the Initech dataset

```
(Intercept)        5302         5750     0.92       0.36
years              8637          389    22.20     <2e-16 ***
---
Signif. codes:
0 '***' 0.001 '**' 0.01 '*' 0.05 '.' 0.1 ' ' 1

Residual standard error: 27400 on 98 degrees of freedom
Multiple R-squared:  0.834, Adjusted R-squared:  0.832
F-statistic:  493 on 1 and 98 DF,  p-value: <2e-16
```

In that output, we read in particular that the intercept equals 5,302 and the slope equals 8,637. In other words, the linear model for salaries is $5,302 + 8,637 \times$ years.

But plotting the residuals confirms that their dispersion gets higher as the fitted values (i.e. the values of the dependent variable, as predicted by the model) increase, as shown in Figure 12.6.

```
plot(fitted(initech_fit),
     resid(initech_fit),
     xlab = "Fitted",
     ylab = "Residuals",
     main = "Residuals vs Fitted Values")
abline(h = 0)
```

In other words, from the fitted versus residuals plot, it appears there is non-constant variance. This is a big problem because using the model for extrapolation can lead to very wrong results. For example, for 30 years of

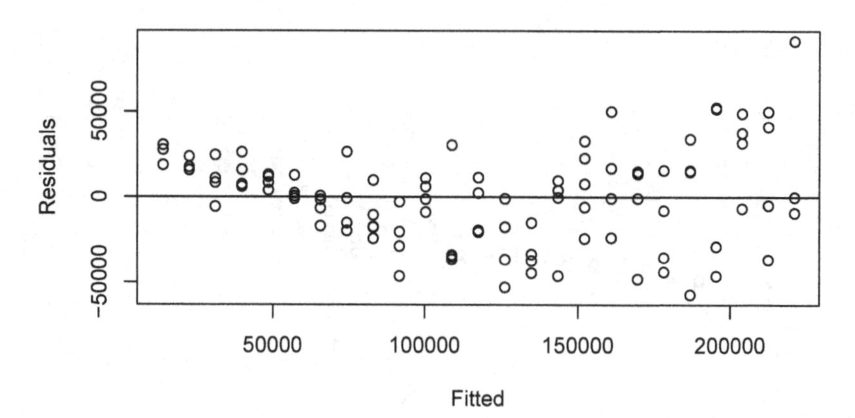

FIGURE 12.6 Plot of the residuals as a function of predicted values

experience, the model predicts $5,302 + 30 \times 8,637$ equals about \$264,000. From Figure 12.5, this seems to be too low as the cloud of points is curving up, and we will come back to this.

We can also check whether the residuals satisfy the assumption of normality that underlies linear regression. One way to check if the distribution of residuals is normal is to perform a **QQ-plot**, as explained on page 186. The QQ-plot for our example is plotted in Figure 12.7 using the `qqnorm()` function we introduced on page 186.

```
qqnorm(resid(initech_fit), main = "Normal QQ-Plot")
qqline(resid(initech_fit), col = "red")
```

The QQ-plot shows significant deviation from a normal distribution in both tails, here again suggesting to transform one of the variables to reduce the dispersion in the regression residuals.

Variance Stabilizing Transformations

A transformation is called **variance stabilizing** if it removes a dependence between the conditional variance and the conditional mean of a variable. One such transformation is to take the log of the dependent variable:

```
initech_fit_log = lm(log(salary) ~ years, data = initech)
summary(initech_fit_log)
```

```
Call:
lm(formula = log(salary) ~ years, data = initech)
```

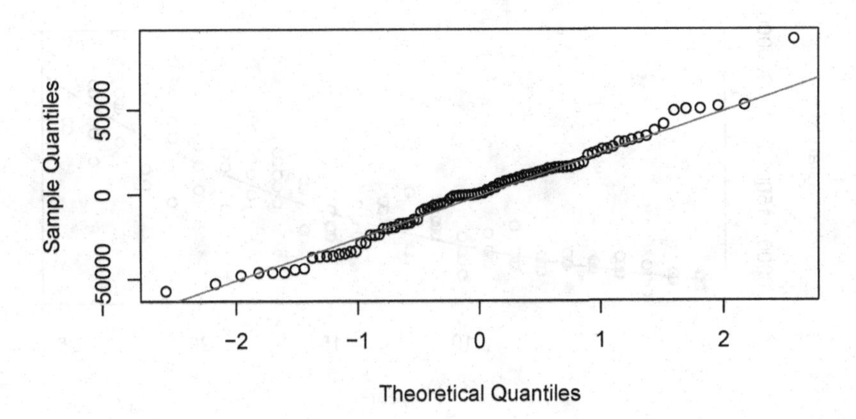

FIGURE 12.7 QQ-plot of the residuals in the Initech linear regression

```
Residuals:
    Min      1Q  Median      3Q     Max
-0.5702 -0.1356  0.0305  0.1416  0.4137

Coefficients:
            Estimate Std. Error t value Pr(>|t|)
(Intercept) 10.48381    0.04108   255.2   <2e-16 ***
years        0.07888    0.00278    28.4   <2e-16 ***
---
Signif. codes:
0 '***' 0.001 '**' 0.01 '*' 0.05 '.' 0.1 ' ' 1

Residual standard error: 0.195 on 98 degrees of freedom
Multiple R-squared:  0.891, Adjusted R-squared:  0.89
F-statistic:  805 on 1 and 98 DF,  p-value: <2e-16
```

We can then plot, in Figure 12.8, the curve of best fit using the **curve()** function and the polynomial **coef**ficients stored in the model:

```
plot(salary ~ years,
     data = initech,
     main = "Salaries at Initech, By Seniority")
curve(exp(initech_fit_log$coef[1] + initech_fit_log$coef[2] * x),
      from = 0, to = 30, add = TRUE)
```

As we can see in Figure 12.8, using log stabilizes the variance of a variable whose conditional standard deviation is proportional to its conditional mean.

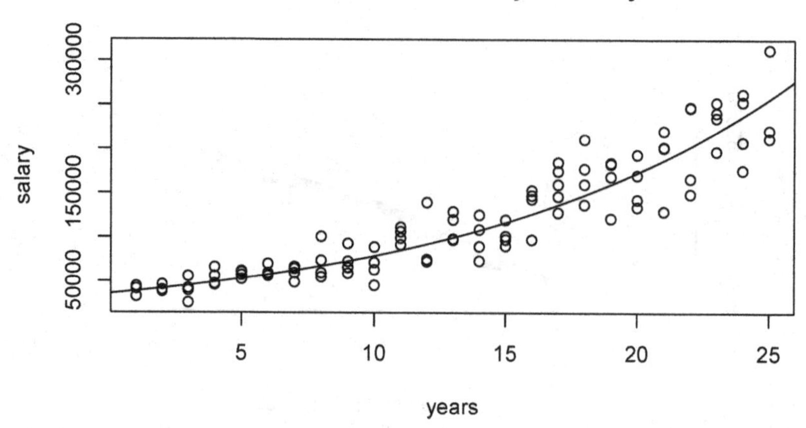

FIGURE 12.8 Example of exponential fit

Is Variance Now Constant?

We can now verify from Figure 12.9 that variance is more constant thanks to our transformation.

```
plot(fitted(initech_fit_log),
     resid(initech_fit_log),
     xlab = "Fitted values",
     ylab = "Residuals",
     main = "Fitted versus Residuals")
abline(h = 0)
```

Thanks to the code below and to Figure 12.10, we can also verify that the residuals follow a normal distribution more closely.

```
qqnorm(resid(initech_fit_log), main = "Normal Q-Q Plot")
qqline(resid(initech_fit_log))
```

The fitted-versus-residuals plot looks much better. It appears that the constant variance assumption is no longer violated.

The above should convince us that this model is better and that we can more confidently extrapolate. For 30 years of experience, this new model gives the following salary prediction:

```
exp(10.48381 + 30*0.07888)
```

```
[1] 380869
```

A salary of $380,000 sounds more reasonable given the curvature of Figure 12.5 and the exponential fit of Figure 12.8.

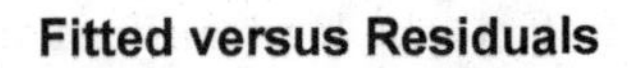

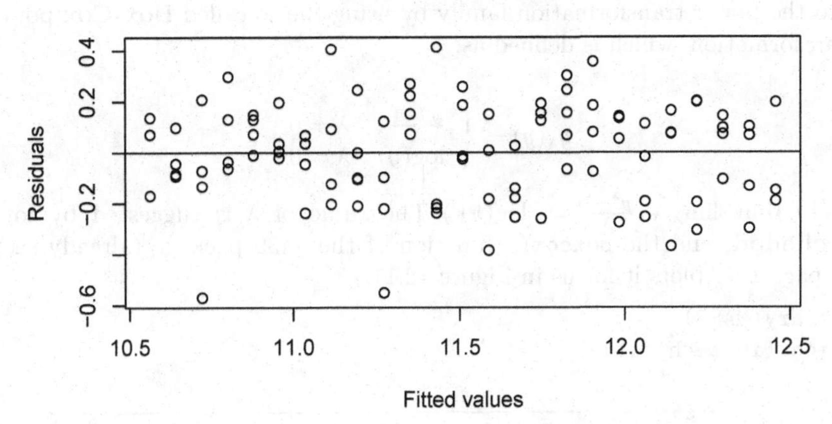

FIGURE 12.9 Residuals a.f.o. predicted values are more constant after transformation

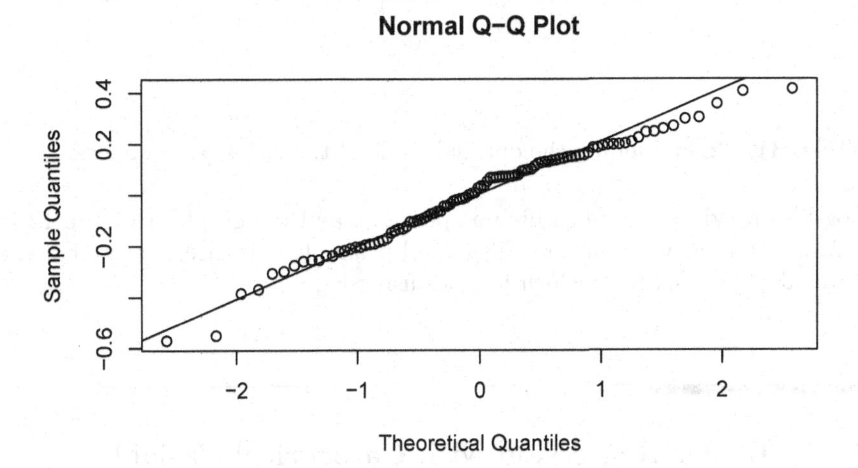

FIGURE 12.10 The residuals follow a distribution closer to normal after transformation

Box-Cox Transformations

We saw that taking the log stabilizes the variance; but taking the square root is often used as well. In fact, the log transformation is sometimes embedded into the power transformation family by using the so-called **Box-Cox** power transformation, which is defined as:

$$g_\lambda(y) = \begin{cases} \frac{y^\lambda - 1}{\lambda} & \lambda \neq 0 \\ \log(y) & \lambda = 0 \end{cases}$$

(Note that $\lim_{\lambda \to 0} \frac{y^\lambda - 1}{\lambda} = \log(y)$.) The value of λ is suggested by **log-likelihood**, and the `boxcox()` function of the `MASS` package (already seen on page 191) plots it for us in Figure 12.11.

```
library(MASS)
boxcox(initech_fit)
```

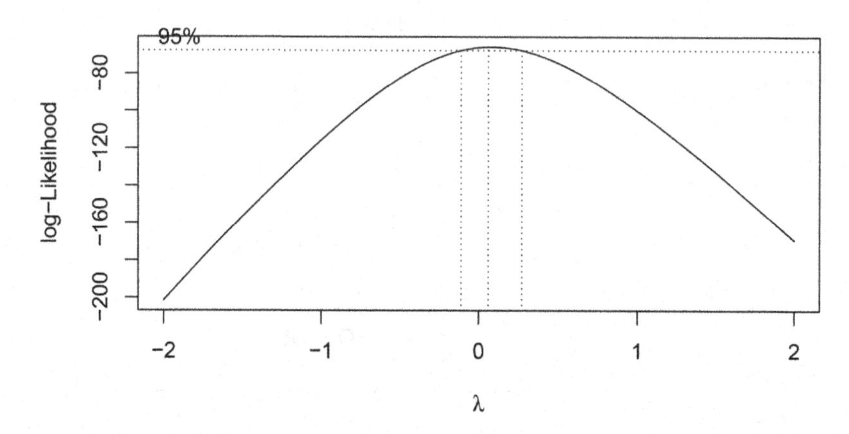

FIGURE 12.11 Finding the optimal value of the Box-Cox parameter

Log-likelihood has to be as high as possible, and the graph in Figure 12.11 tells us for what value of λ log-likelihood peaks. In this case, we see that the optimal λ is closed to 0, which justifies using log.

12.7 Linear Regression with Categorical Variables

Sometimes a single independent variable is expected to be a good predictor for a dependent variable but different situations alter the relationship. These "situations" are often described in the form of a qualitative **categorical variable**.

For instance, let's assume we are interested in year-to-date returns of stocks as a function of dividend yields across stocks in the S&P 500 index, but we suspect that the GICS sector a stock belongs to (which is a categorical variable) is also an important parameter. To prepare that analysis, let's read the data as we did before and make sure returns and dividend yields are in numeric form:

```
sp500 = read.csv("Characteristics Overview.csv")
```

```
sp500$Dividend.Yield    = as.numeric(sp500$Dividend.Yield)
sp500$Total.Return.YTD = as.numeric(sp500$Total.Return.YTD)
```

As said earlier, we suspect another piece of information could influence the relationship between return and dividend yield: the GICS sector a stock belongs to. We make it an explicit categorical variable, which R calls a **factor**:

```
sp500$GICS.Sector           = as.factor(sp500$GICS.Sector)
```

We also remove the NA's:

```
sp500 = na.omit(sp500)
```

To simplify the example further, let's assume (without loss of generality) that we are only interested in two sectors, Energy and Utilities:

```
sp500.en.ut = sp500 %>%
  filter(GICS.Sector == "Energy" | GICS.Sector == "Utilities")
```

To avoid repeating the name of the dataset, we make it the default dataframe using **attach()**:

```
attach(sp500.en.ut)
```

(Please see page 54 regarding **attach()**.)

Regressing returns against dividend yields alone would lose information that sectors can bring, and doing two separate regressions (one for each sector) would rely on data sets that would be half as large. So the idea is to add a dummy variable to encode the two options:

$$x_2 = \begin{cases} 0 & \text{Energy} \\ 1 & \text{Utilities} \end{cases}.$$

That new variable is added as an independent variable to the regression so that, mathematically, the model is $Y = \beta_0 + \beta_1 x_1 + \beta_2 x_2 + \epsilon$, where Y is the response variable (here, returns), x_1 is the continuous variable (dividend yield) and x_2 selects between Energy and Utilities, and ϵ is the error term (which we will drop for now from our model descriptions).

This formulation means that for Energy stocks, the regression model is $Y = \beta_0 + \beta_1 \times x_1 + \beta_2 \times 0$, i.e., $Y = \beta_0 + \beta_1 x_1$, and the model for Utilities is $Y = \beta_0 + \beta_1 \times x_1 + \beta_2 \times 1$, or equivalently $Y = (\beta_0 + \beta_2) + \beta_1 x_1$.

It is key to keep in mind that these two models share the same slope, β_1, but have different intercepts, differing by β_2. So the change in return for a percentage point increase in dividend yield is the same for both models, but on average returns differ by β_2 between the two types of stocks.

```
(multilin.regr = lm(Total.Return.YTD ~
                    Dividend.Yield +
                    GICS.Sector))
```

```
Call:
lm(formula = Total.Return.YTD ~ Dividend.Yield + GICS.Sector)

Coefficients:
         (Intercept)          Dividend.Yield
               97.37                   -7.97
GICS.SectorUtilities
              -62.14
```

So to recap, $\beta_0 = 97.3655$ is the estimated average return for an Energy stock with a dividend yield of 0; $\beta_1 =$ -7.9718 is the estimated change in average return for an increase of one percentage point in dividend yield, in either sector; and $\beta_2 =$ -62.1384 is the estimated *difference* in average returns for Utilities stocks as compared to Energy stocks, for any given dividend yield. Therefore, $\beta_0 + \beta_2 = 35.2271$ is the estimated average return for a Utilities stock that pays no dividend;

The code below then lets us visualize the lines of best fit in Figure 12.12.

```
ggplot(sp500.en.ut,
       aes(y = Total.Return.YTD, x = Dividend.Yield,
           color = GICS.Sector)) +
  geom_point() +
  geom_abline(slope=slope, intercept=intercept.energy, color="red") +
  geom_abline(slope=slope, intercept=intercept.utilities)
```

As we can see, the two models are lines of the same slope, but one line of best fit for Utilities is lower than that for Energy stocks.

12.8 Polynomial Regression

On page 174, we asserted without proof that the plot of Apple's returns against that of the S&P 500 showed an upward curvature. One way to assess that

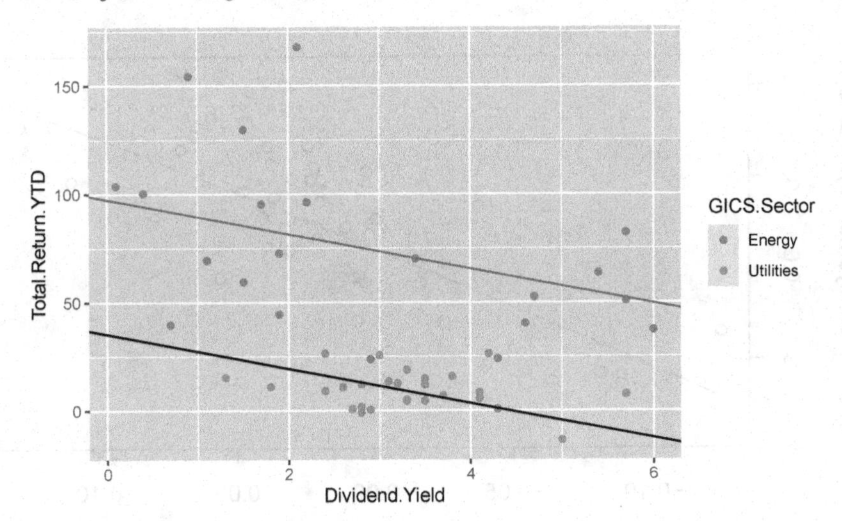

FIGURE 12.12 Linear regression with GICS sectors as a categorical variable

claim is to perform a **polynomial regression**. A **polynomial model** of degree 2 can be constructed using the **lm()** function, but with the independent variable in a polynomial of degree 2 specified using the **poly()** function. In this example, we use the **aapl.monthly** and **benchmark** data read from the Internet on page 171:

```
model = lm(aapl.monthly$monthly.returns ~
            poly(benchmark$monthly.returns, 2))
```

Plotting the scatter plot is easy; plotting the curve of best fit is a bit tricky, however. The issue comes from the fact that the **line()** function expects points to go from left to right, i.e., from the lowest x-value to the largest. We thus have to sort the returns of the S&P 500 in increasing order, and re-order the returns of Apple in that order. One way to do that is to keep the *indices* of ordered values using the **index.return=** optional argument of the **sort()** function. That way, **sort()** returns not only the sorted values but their indices in the original vector in a new column called **ix**. That column of indices must be used to index the corresponding entries in the vector of Apple returns:

```
plot(benchmark$monthly.returns, aapl.monthly$monthly.returns)
sorted.indices =
  sort(benchmark$monthly.returns, index.return=TRUE)$ix
lines(benchmark$monthly.returns[sorted.indices],
      model$fitted.values[sorted.indices])
```

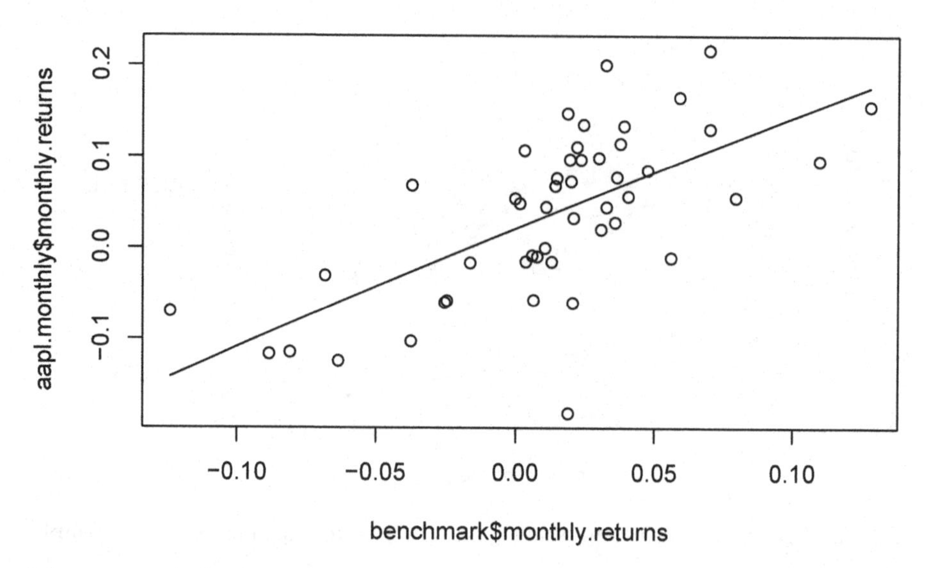

We can conclude that there is no strong curvature, in contrast to what was claimed on page 174.

12.9 Exercises

12.9.1 Collinearity

Coming back to the income, age and experience example, assume now that the data were as follows:

Income	Age	Experience
20000	20	19
33000	32	30
40000	41	39
50000	50	52
60000	59	58

Regress income against age and experience at the same time, and comment on the loadings of the two independent variables. Also, is this data set very different from the previous one? Yet, did the loadings change much?

12.9.2 Order of Independent Variables in Multi-linear Regressions

There are $3 \times 2 \times 1 = 6$ ways to do a linear regression against 3 independent variables. Try all 6 on the BAML AAA example and report on the differences you observe.

13

Fixed Income

Until now, we have been assuming that the price of the financial instruments we considered would be provided to us; from prices, we defined returns and their properties, studied the distribution of these returns and how to model them, and investigated multi-linear regression on them.

This assumption is easy to forget in the case of equities, i.e., of stocks. But it is much less true for a large area of finance, **fixed income**, which encompasses all instruments, typically debt, whose value depends on fixed recurring payments. Yes, there is a market for fixed income securities, but it's far from being as transparent as it is for equities. And more more critically, the price of many fixed income instruments can be objectively established.

The most widely known example of a fixed income security is the **bond**. Bonds are essentially receipts that a company borrowed money. The company owes you to pay back its debt and interests, nothing more. In contrast, when you own a share of a stock, you have partial ownership in that company, which means you share in both the profits and losses (but not the liabilities). Nothing is guaranteed.

But as said, when you buy a bond, you make a loan to the company. The corporation is obligated (unless it defaults...) to pay back the principal amount, called **par**, after a number of years (called **maturity**) specified by the contract, plus periodic payments called **coupon**. (Coupons are similar to dividends paid by stocks, except that coupons are specified contractually, whereas stock dividends are at the discretion of the company's management.) You will receive a fixed stream of income, and that's why bonds are called fixed-income securities.

Zero-coupon bonds pay no principal or interest until maturity. The par value (face value) is the payment made to the bond holder at maturity. A zero-coupon bond sells for less than par – assuming its yield is positive. Any bond selling for less than par is called a **discount bond**. A zero-coupon bond is also called a **pure discount bond**.

Borrowers (a.k.a. issuers) can be corporations or governments (e.g., US Treasury, and the bonds it issues are called **Treasuries**). Bonds might appear risk-free but actually this is not true: Obviously, the issuer could default on their payment. This is called **credit risk**. The interest that lenders demand on top

DOI: 10.1201/9781003320555-13

of prevailing interest rates to be compensated for that additional risk is called **credit spread.**

We stated earlier that the price of a fixed income instrument can often be objectively established. How can we give a price to payments made in the future? The concept at the core of this pricing is **present valuation.**

13.1 Present Value

The pricing of a bond (and any financial instrument, for that matter) is based on a simple principle called present valuation. For a promised payment in the future, what amount should I be willing to pay today?

One way to answer that question is to look at an alternative: Given prevailing interest rates, what amount would I need to invest today so that, after compounding, I get the same payment at the same future date?

Since x dollars today become $x(1+r)$ after one year if invested with annualized interest rate r, it follows that the value today (the **present value**) of 1,000 dollars to be received in one year is:

$$\frac{1000}{1+r} = 1000 \times (1+r)^{-1}$$

The rate r used in present valuation is called **discount rate** and the multiplicative factor, $(1+r)^{-1}$ in this example, is the **discount factor**. The discount rate is typically not constant (it depends on how long money is borrowed for), but for now we will assume it is. But even then, the discount factor is in general a function of time: If the present value of 1,000 dollar received one year from now is $1000 \times (1+r)^{-1}$, then what would be the present value of that amount if it was to be received two years from now? The present value in one year (i.e., when another year would be remaining) would be $1000 \times (1+r)^{-1}$, and that amount should be present-valued to today, resulting in $1000 \times (1+r)^{-1} \times (1+r)^{-1}$, or $1000 \times (1+r)^{-2}$.

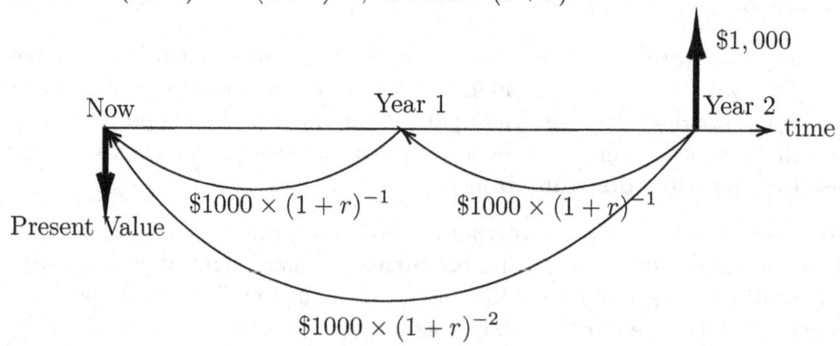

Thus, the **discount function** depends today on the remaining time T until the **maturity** of the bond (the day par value will be paid back), and we will denote it $D(T)$.

For now, we assume that the appropriate discount rate is a constant risk-free rate, or the risk-free rate plus the credit spread discussed earlier.

We just said that the value today (the **present value**) of 1,000 dollars to be received in one year is $\frac{1000}{1+r}$. So the price of a zero coupon bond is given by:

$$\text{PRICE} = \frac{\text{PAR}}{(1+r)^T} = \text{PAR}(1+r)^{-T}$$

if we assume again that T is the time to maturity in years.

For a 10-year zero coupon bond with a face value of $1000 and interest rate at r=4%, the present value is:

$$\frac{1000}{(1+0.04)^{10}}$$

if the interest is compounded annually

The price of a 20-year zero with par value of $1000 when the interest rate is $r= $6% and compounded annually is:

$$\frac{1000}{(1+r)^{20}} = \frac{1000}{1.06^{20}} = 311.80$$

All the above (and most of what follows) assumes no risk of default. But assume the same 20-year zero sells for $290. What is its credit spread c? By definition, that spread must such that the following equality holds:

$$\frac{1000}{(1+(r+c))^{20}} = 290$$

There is a lot of research (incl. math and statistics research) to link c to the **probability of default** of the issuer and the **recovery rate**.

The above showed how to calculate a "fair" price for a bond given prevailing rates. But sometimes, we have the opposite problem: we know the market price of a bond, but don't know the corresponding discount rate. The annual rate r such that the discounted PAR of a zero equals its current market price is called the **yield to maturity (YTM)** of that bond.

Semi-Annual Compounding

The price of a 20-year zero with par value of $1000 when the interest rate is $r = 6\%$ and compounded annually is:

$$\frac{1000}{(1+r)^{20}} = \frac{1000}{1.06^{20}} = 311.80$$

However, bonds are usually present-valued using semi-annual discount rates. What that means is that instead of having T remaining years, we have $2T$ remaining periods. What's the relevant discount rate then? By convention, it is half the annual rate. So if we assume the interest rate r is per year with semi-annual compounding, the price of a zero is:

$$\text{PRICE} = \text{PAR}(1 + r/2)^{-2T}$$

For example, if the interest rate is $r = 6\%$ and is compounded every six months, the present value of the 20-year bond is:

$$\frac{1000}{(1 + r/2)^{40}} = \frac{1000}{1.03^{40}} = 306.56 \tag{13.1}$$

Continuous Compounding

If we follow the same logic as for semi-annual compounding, that happens if compounding is done every month? Every day? The logical extension of semi-annual compounding to compounding n times a year is

$$\text{Price} = \text{PAR}(1 + r/n)^{-nT}$$

for a T-year zero-coupon bond.

What if we push that logic to its extreme, i.e., what if n becomes arbitrarily large? Then, compounding happens every seconds in what is called **continuous compounding**, and if you remember your calculus classes on limits, you'll appreciate that the limit of the above function is an exponential. Specifically, if interest rate $r = 6\%$ is compounded continuously, the present value of the 20-year bond is:

$$\frac{1000}{e^{20r}} = \frac{1000}{e^{20(0.06)}} = 301.19$$

Continuous compounding also applies to a **dividend yield**, which we defined on page 10. A **continuously compounded dividend yield** of 3.20% is translated in its quarterly equivalent as follows:

$$4 \times e^{(3.20\%/4)} - 1 = 0.0321283$$

or 3.2113% per year. This will be important when discussing Black-Scholes on page 287.

Sensitivity of the Price of a Zero to Rate Increases

Assume you just bought for \$306.56 a 20-year zero with a face value of \$1000. You assumed semi-annual compounding, which gave you an interest rate of 6%. Six months later the interest rate increased to 7% . The present value is now:

$$\frac{1000}{1.035^{39}} = 261.41$$

The value of your investment dropped by $306.56 - 261.41 = 45.15$. You will get \$1000 if you keep the bond for another 19.5 years, but if you sell it now, you lose \$45.15, a return of:

$$\frac{-45.15}{306.56} = -14.73\%$$

for a half-year or -29.46% per year. Therefore, *the value of a bond decreases when interest rates increase.*

Now, let's look at the sensitivity of the Price of a Zero to Interest Rate Drops. Assume semi-annual compounding, and that you just bought the zero for \$306.56. Then, six months later, the interest rate decreased to 5%. The present value now is:

$$\frac{1000}{1.025^{39}} = 381.74$$

The value of your investment went up by $381.74 - 306.56 = 75.18$. You will get \$1000 if you keep the bond for another 19.5 years. if you sell it now, you make a \$75.18 profit, a return of:

$$\frac{75.18}{306.56} = 24.5\%$$

for a half-year or 49% per year (using the bond convention). In conclusion, *the value of a bond increases when interest rates decrease.*

13.2 Present Value of Coupon Bonds

Each cash flow gets discounted according to the number of periods in the future that it will be paid out. We assume a constant interest rate r, per half year. The PV of the first coupon of C dollars is $\frac{C}{1+r}$, that of the second coupon is $\frac{C}{(1+r)^2}$, etc. The general formula is given by:

$$\text{bond price} = \sum_{t=1}^{2T} \frac{C}{(1+r)^t} + \frac{\text{PAR}}{(1+r)^{2T}} \tag{13.2}$$

where PAR = par value and T is the maturity (in years).

Let's see the intuition of that equation on a simple example: a bond with par of \$1,000 paying \$20 every 6 months, with a maturity of 2 years i.e., four 6-month periods. Let's assume the relevant rate is 2% per year, i.e. 1% per six-month period. So the discount factor in each period will be $1 + 1\% = 1.01$, raised to a power equal to the number of periods. The cash flows and each of their present values can be visualized as follows. The first payment is assumed to be paid exactly 6 months from now and the last cash flow will equal par plus the final coupon:

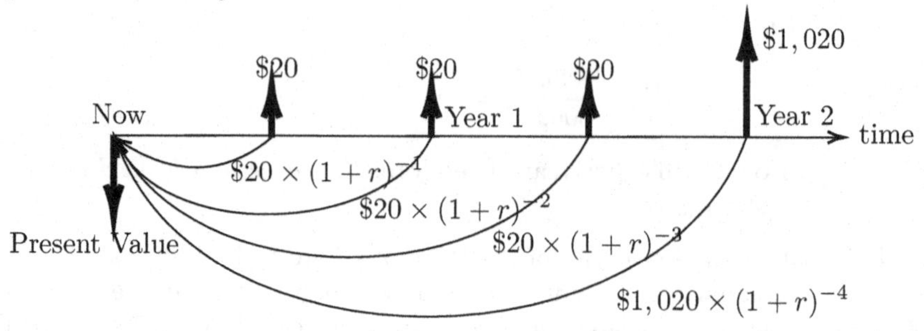

The present valuation can be tabulated as follows. For each period (in numbers of semesters), the cash flows are listed. The discount rate here is constant, and the discount factors equal one plus that rate, raised to the power of the top row. The cash flows are divided by the discount factors, which gives us each contribution to the total present value:

Period	1	2	3	4	4
Cash flow	\$20	\$20	\$20	\$20	\$1,000
Rate	0.01	0.01	0.01	0.01	0.01
Discount Factor	1.01	1.01^2	1.01^3	1.01^4	1.01^4
Present Value	20/1.01	$20/1.01^2$	$20/1.01^3$	$20/1.01^4$	$1000/1.01^4$

The present value of the bond, then, is the sum of the present values of all 5 cash flows:

$$\frac{20}{1.01}+\frac{20}{1.01^2}+\frac{20}{1.01^3}+\frac{20}{1.01^4}+\frac{1000}{1.01^4} \tag{13.3}$$

which equals 1039.02.

Of note: Another expression for the price of a bond is as follows:

$$\text{bond price} \quad = \quad \frac{C}{r}+\left(\text{PAR}-\frac{C}{r}\right)(1+r)^{-2T} \tag{13.4}$$

Equation (13.4) is derived using the fact that, for $a \neq 1$:

$$1+a+a^2+\ldots+a^n=\frac{1-a^{n+1}}{1-a}$$

and then setting $a = 1/(1+r)$.

Equation (13.2) and the table above give us a great opportunity to apply a lot of what we've learned in R[1]. We can apply the same process we used to construct the table: we build a vector of copies of the rate using **rep()**:

```
rep(0.01, 4)
```

```
[1] 0.01 0.01 0.01 0.01
```

and we add 1 to all the elements of the created vector:

```
1 + rep(0.01, 4)
```

```
[1] 1.01 1.01 1.01 1.01
```

Separately, we prepare a sequence of integers going from 1, the first period, to 4, the fourth and last 6-month period, using **seq()**:

```
seq(1, 4, 1)
```

```
[1] 1 2 3 4
```

Then the discount factors are the elements of the first sequence raised to each of the values in the second sequence:

```
(1 + rep(0.01, 4))  ^  seq(1, 4, 1)
```

```
[1] 1.010 1.020 1.030 1.041
```

Each coupon payment is discounted using these discount factors. We could build a vector of coupons using **rep()** like we did for rates, but dividing a number by a vector already results in a vector. We can thus write something like:

```
20 / ( (1 + rep(0.01, 4))  ^  seq(1, 4, 1) )
```

```
[1] 19.80 19.61 19.41 19.22
```

and we can then add all these present values together using **sum()**, plus that of the final par payment (par divided by 1.01^4). To wrap it up, we can create a function that takes par amount, coupon, flat discount rate and the number of periods as inputs and produces the total present value of all cash flows (and thus the bond's fair price) as its output:

```
bondValuation = function(par, coupon, rate, periods) {
  discount.factors = 1 + rep(rate, periods)
  compounding.exponents = seq(1, periods, 1)
  discount.factors = discount.factors ^ compounding.exponents
  cash.flows = coupon / discount.factors
  final.cash.flow = par / discount.factors[periods]
```

[1]When running the examples of this textbook, don't forget to systematically load the **tidyverse** package.

```
  cash.flows = append(cash.flows, final.cash.flow)
  total.present.value = sum(cash.flows)
  total.present.value
}
```

For example, we calculated in Equation (13.1) the price of a 20-year zero assuming a flat discount rate of 6% annually. We can verify that our function produces the same fair value:

```
bondValuation(1000, 0, 0.03, 40)
```

```
[1] 306.6
```

In Equation (13.3), we had also calculated the present value of another bond, this time with a coupon. We can verify we are getting the same result:

```
bondValuation(1000, 20, 0.01, 4)
```

```
[1] 1039
```

13.3 Exercises

13.3.1 Alternative Formula for the Present Value of a Coupon Bond

Write a function calculating the fair value of a coupon bond based on Equation (13.4).

■ SOLUTION

```
bondValuation.alternate = function(par, coupon, rate, periods) {
  present.value = coupon/rate +
    (par - coupon/rate)/(1+rate)^periods
  present.value
}
```

We can verify that prices match those we calculated earlier:

```
bondValuation.alternate(1000, 0, 0.03, 40)
```

```
[1] 306.6
```

```
bondValuation.alternate(1000, 20, 0.01, 4)
```

```
[1] 1039
```

■ END OF SOLUTION

13.3.2 Modified Duration

Modified duration is the percentage change in the value of a bond for a small chage in rates – say, an increase of one tenth of a percentage point. Write a function that takes par, coupon, a constant semi-annual rate and a number of periods and returns the modified duration.

■ SOLUTION

```
modifiedDuration = function (par, coupon, rate, period) {
  initial.value = bondValuation(par, coupon, rate, period)
  final.value   = bondValuation(par, coupon, rate+0.001, period)
  pct.change    = (final.value - initial.value) / initial.value
  pct.change
}
```

We can test it out on the two bonds we took earlier as examples: a 20-year zero with par value of $1000 and a yield to maturity of 6%, and a 2-year bond with par of $1,000 paying $20 every 6 months and a yield-to-maturity of 2%. Their durations are:

```
modifiedDuration(1000, 0, 0.03, 40)
```

```
[1] -0.03807
```

```
modifiedDuration(1000, 20, 0.01, 4)
```

```
[1] -0.003839
```

So the 20-year zero would lose 3.8% while the 2-year bond would lose 0.4%.

■ END OF SOLUTION

13.3.3 Yield to Maturity

Write a function that, given the par, market price, maturity (or number of periods) and coupon of a bond, provides the yield to maturity of that bond.

■ SOLUTION

We use the **optimize()** function to find the discount rate that minimizes the difference between, on the one hand, the bond valued at that discount rate and, on the other hand, the stated market price. Ideally that difference will be zero, so we seek to minimize the square of that difference:

```
ytm = function(par, coupon, periods, price) {
  f = function(x) {
    (bondValuation(par, coupon, x, periods) - price)^2
  }
  optim = optimize(f, c(0, .2))
```

```
  2*optim$minimum
}
```

We use the same two bonds and verify that the 20-year zero with par value of $1000 selling for 306.6 dollars has a yield to maturity of 6% and that the 2-year bond with a par of $1,000 paying $20 every 6 months and a market value of $1,039 has a yield-to-maturity of 2%:

```
ytm(1000, 0, 40, 306.6)
```

```
[1] 0.05996
```

```
ytm(1000, 20, 4, 1039)
```

```
[1] 0.02001
```

■ END OF SOLUTION

14

Principal Component Analysis

In all the analyses so far, we implicitly or explicitly suspected what variables could have an influence on the others. For example, in Chapter 12 on regression we explicitly tested the hypothesis that one variable explained another, and in Chapter 13, we explicitly valued bond as a function of prevailing interest rates.

In many situations however, we do not know what variable or variables explain what we observe, or even worse, we do not know what, in the data, matters most. **Principal Component Analysis** (**PCA**) is a method that helps on both fronts: it first suggests what in the data matters most by identifying where most of the **variance** is, i.e., where changes in the data are most significant. It also offers possible explanatory variables for that variance, with the limitation that this explanatory variable is a combination of the input variables. In fact, we want keep one key property in mind: PCA is only a change of coordinate system, from your initial set of variables to a linear combination of "better" variables. Let see that on an example.

14.1 Directions of Most Variance

Let's consider two variables and let's plot them against each other in Figure 14.1. One variable is the monthly return of Ford's stock and the other is the monthly return of General Motors; each dot on the plot thus represents one month.

The direction that sees the largest changes, or more technically that has the largest variance, is the **first principal component** (PC1). The direction orthogonal to the first PC is the second principal component – but in a 2-dimensional problem, there is only one direction that can be orthogonal to PC1. We insist that PC2 is orthogonal to PC1 even though Figure 14.1 can distort the axes.

The three-dimensional case is illustrated in Figure 14.2, where the returns of Google are added. Oftentimes, plotting your data (here, 3 numerical variables) will look like a football. The long side of the football clearly is the first principal component: it is along that dimension that most of the changes in the data,

DOI: 10.1201/9781003320555-14

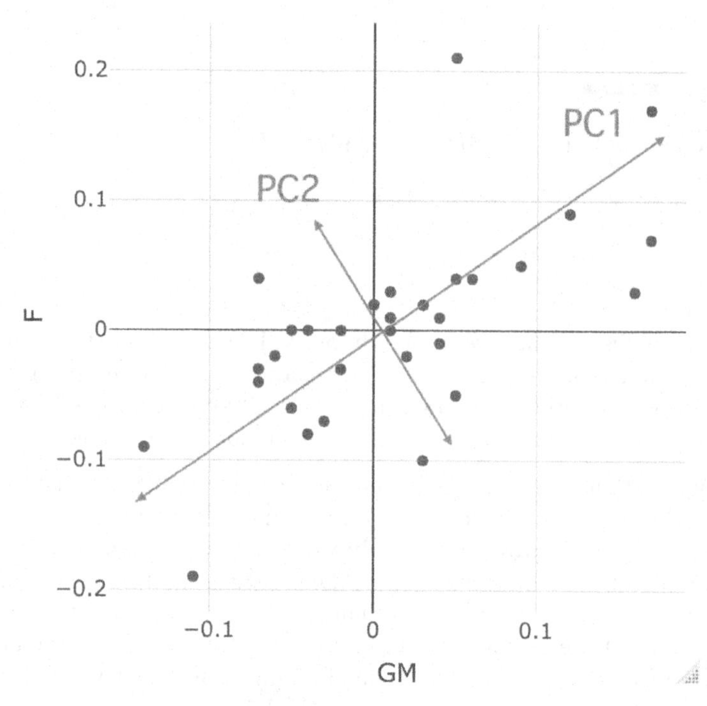

FIGURE 14.1 Direction of most variance in the case of two variables

from one end to the cloud of points to the other end, happens. But then, and in contrast to 2-D, there is an infinite number of directions orthogonal to the first PC. Among them, the one that has the largest variance is the second component (PC2). Then, the third principal component is the (single) dimension that is orthogonal to both PC1 and PC2. And the reasoning continues for dimensions greater than 3.

Because PCA can be applied to any set of numerical values, the data can have very different ranges, even possibly expressed in different units. To be able to compare these data "apple to apple," the first step in performing a principal component analysis is to normalize the data, that is, very specifically, to convert them to **z-scores**. As detailed on page 35, this is done by subtracting their mean (what is called **centering**) and dividing them by their standard deviation (called **scaling**). Because z-scores are unit-less, z-scoring allows us to compare the influence of variables expressed in different units.

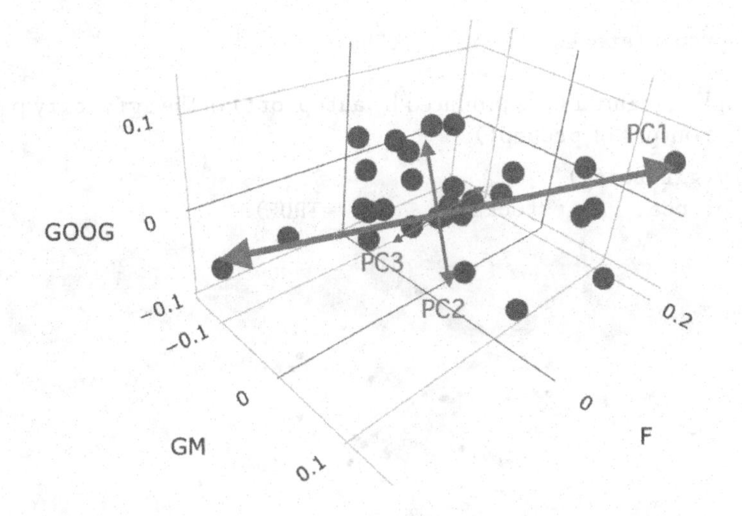

FIGURE 14.2 Direction of most variance in the 3-dimensional case

14.2 Application to a Full Example

Let's see how we find these principal components. The sample file `stocks5.csv` we first read on page 14 contains 8 stocks and 31 monthly returns. We read it again as follows:

```
stocks = read.csv(file = 'stocks5.csv')
```

(As a matter of fact, we already used these data in this chapter without mentioning it: Figures 14.1 and 14.2 plot some of these data using function **`plot_ly()`** of the **plotly** package.)

Thanks to PCA, we are going to see if common drivers influence the returns of these stocks, or if return drivers are specific to some type of stocks – for example, depending on their industry.

Because the data set contains 8 stocks, the starting point is a 8-dimensional space, and each point corresponds to one month – specifically, the coordinates of each point are the returns of the stocks that month.

Here, we cut corners and don't do any z-scoring because all the variables are returns (hence unit-less) and are assumed to be centered on 0 with approximately the same volatility. Fortunately, the **`prcomp()`** function (and a similar function called **`princomp()`**) performs this centering by default and performs scaling if we specify so:

```
pca = prcomp(stocks, scale. = TRUE)
```

The graph in Figure 14.3 is produced by **autoplot()** of the **ggfortify** package using the output of **prcomp()**:

```
library(ggfortify)
autoplot(pca, data=stocks, loadings=TRUE)
```

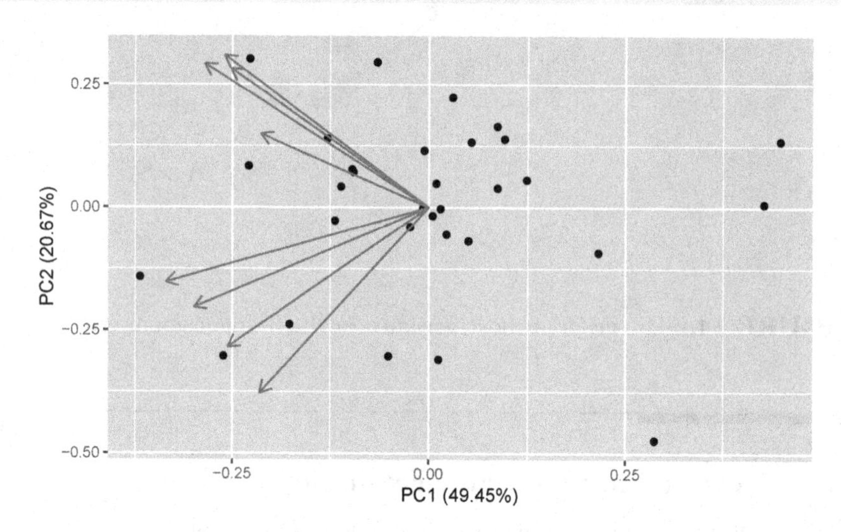

FIGURE 14.3 Plotting the first 2 principal components, and the projections of the 8 stocks in this coordinate system

The axes of this graph are the first two principal components. Each red arrow is a stock; more precisely, the coordinates of the end points of each arrow are the correlations, called **loadings**, to the respective principal components.

Figure 14.3 clearly shows that PCA finds that all stocks are carried in the same direction along PC1. That direction happens to be toward the left, but that's just random. The financial interpretation of that direction followed by all stocks is unambiguous: it is the sensitivity of all stocks to the broad market, what we had called the **beta** to the market on page 173.

Figure 14.3 also shows that PCA separates stock into two groups, one driven positively by the second principal component PC2 and the other heading south. At this point, we don't know what these stocks are and we can't interpret this second component – but we'll have answers to these questions soon. Before we get there, we should assess how good of a model we have, and specifically how much of the dispersion in the data is explained by each principal component.

14.3 How Much Variance is Explained by Each Principal Component?

As the axis legends in Figure 14.3 indicate, the first principal component explains 49.45% of all variance. We can verify that by examining the content of the variable `pca`, in particular its field named **`sdev`** that contains the standard deviation of each principal component. (The **`str(pca)`** command gives you all the fields in `pca`.) And since variance is the square of standard deviation, we verify how much of total variance each principal component explains, in decreasing order:

```
vars = pca$sdev^2
(varspc = vars/sum(vars))
```

```
[1] 0.494521 0.206676 0.102390 0.068015 0.064686
[6] 0.031957 0.024804 0.006952
```

We can verify that the variance of PC1, almost 4%, is 49.45% of total variance. The second principal component explains 20.668%, etc. So the first two principal components explain about 70% of total variance. How much variance is explained by each principal component is often visualized using a **scree plot** - essentially, a plot of variances sorted in decreasing order:

```
plot(varspc)
```

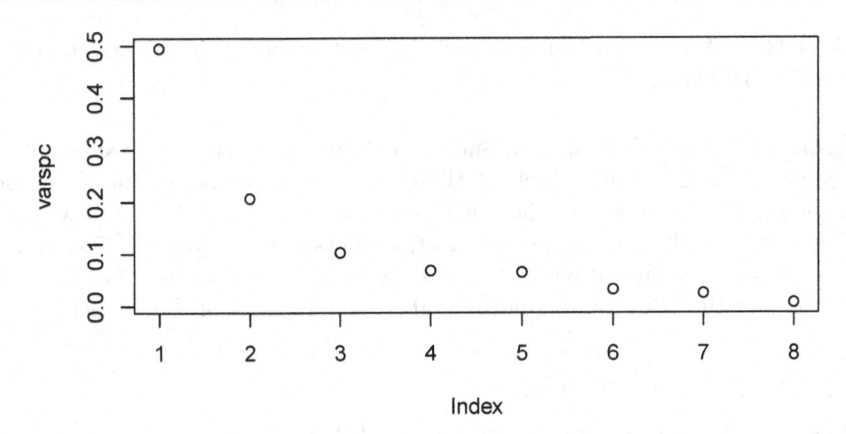

FIGURE 14.4 Scree plot of the PCA

Moreover, we see in Figure 14.3 that the stocks are clearly separated into two groups – even though PCA knows nothing about the identity of these stocks. So many questions still come to mind: Can we decompose these groups further?

Also, can we explain more than 70% of variance? We are going to answer both questions next.

Plotting "Less Principal" Components: PC2 and PC3

Displaying the second and third principal components can be done using the **x** and **y** parameters of **autoplot()**.

```
autoplot(pca, data=stocks, loadings=TRUE, x=2, y=3)
```

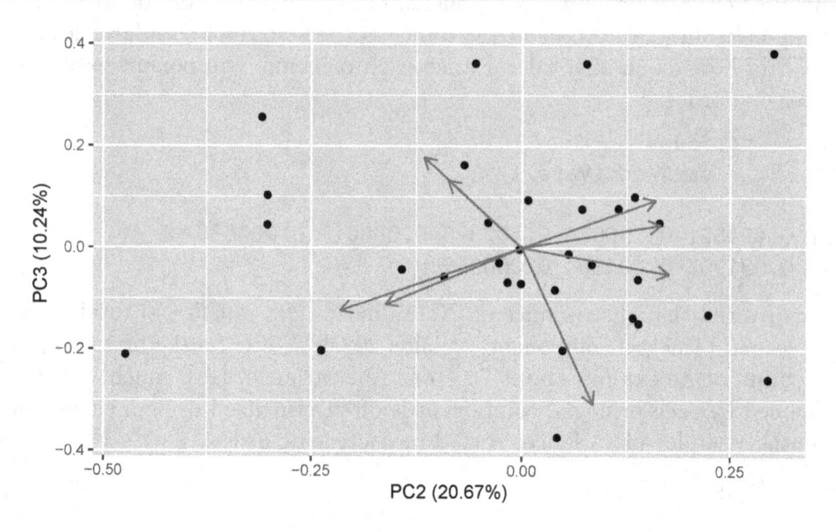

FIGURE 14.5 Second and third principal components, and projection of the 8 stocks on that coordinate system

Figure 14.5 shows that the 2nd and 3rd component separate the stocks into 4 groups. We can't tell which of these stocks correspond to those on the previous slide – except that those with positive loadings on PC2 are the same of course. Now, PC3 clearly separates the stocks with negative PC2 loadings into 2 groups, which will be of interest once we reveal the names of each stock. (Remember that PCA knows nothing about the identity of these stocks.)

Now With Stock Names

By setting **loadings.label** to **TRUE**, we ask R to display the name of each stock:

```
autoplot(pca, data=stocks, loadings=TRUE, loadings.label=TRUE)
```

As illustrated in Figure 14.6, principal component analysis finds a statistical difference between, on the one hand, technology stocks (Google, with ticker GOOG; Apple, ticker AAPL; Amazon, ticker AMZN; and Salesforce, ticker

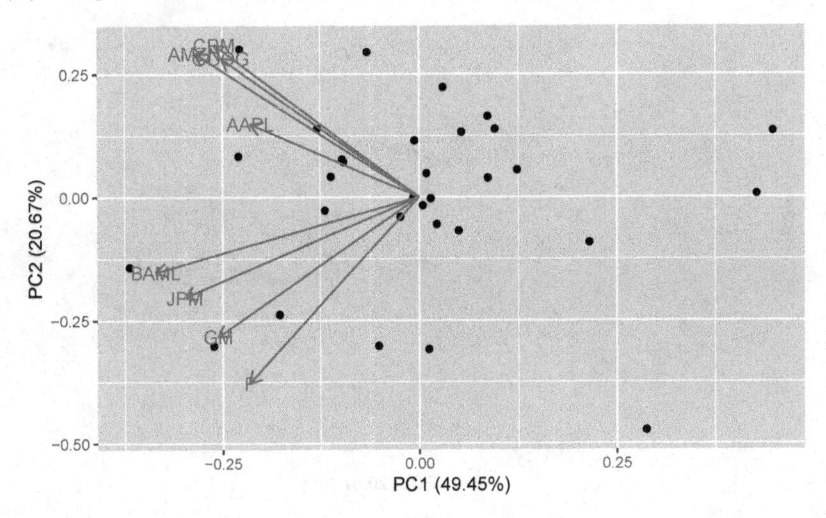

FIGURE 14.6 PCA finds patterns in returns that correspond to different industries

CRM) and, on the other hand, all the other stocks: Ford (F), General Motors (GM), JP Morgan (JPM) and Bank of America Merrill Lynch (BAML). That is, without knowing the company's names, let alone the industries they operate in, PCA finds patterns in their returns that separate them according to the industry they are in. This holds also for the second and third principal components, plotted in Figure 14.7 thanks to the code below:

```
autoplot(pca, data=stocks, loadings=TRUE, loadings.label=TRUE,
         x=2, y=3)
```

So PCA finds a statistical difference among technology stocks, with Apple standing out. It also separates the other stocks into two groups that we can easily identify as auto stocks (GM and Ford) and banks (JPM and Bank of America Merrill Lynch).

14.4 Chapter-End Summary

Principal Component Analysis is a key tool in exploratory data analysis to identify what drives most of the data in numerical datasets. If you think of such datasets as cloud of points in a multi-dimensional space, with one point per observation, then what probably matters most to you is the direction along which most of the variation takes place. (Of course, different variables can have very different ranges of values, and even be expressed in different

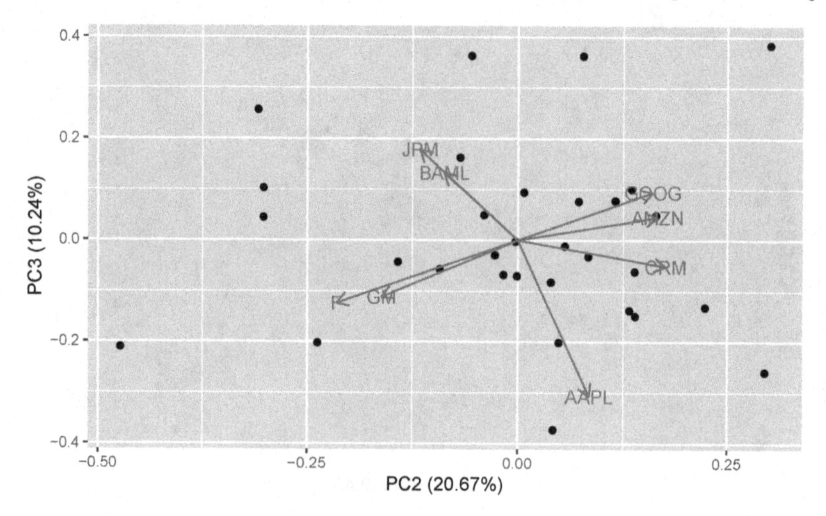

FIGURE 14.7 Principal Components 2 and 3, with Stock Names

units, so variables are first normalized using z-scoring.) We saw how to quickly identify and plot that direction of most variation, then the second direction driving most of the remaining variations, etc. We also saw that the directions, although not always immediately intuitive in and of themselves, can help cluster observations in meaningful way – such as, in the case of stocks, depending on the industry these firms operate in.

14.5 Exercises

14.5.1 PCA on Rates

Perform PCA on the rates provided in the `fredgraph.xlsx` file. How do you interpret the principal components?

14.5.2 PCA on ACWI

Take the 10 stocks with the largest capitalizations in the ACWI benchmark and perform a principal component analysis. Plot the first 3 principal components using for instance the **plotly** package.

15

Options

An **option** gives the right to buy or sell a financial instrument (oftentimes, a stock or a basket of stocks) at a specific price, called the **strike**. If that right applies only on *one specific day*, then the option is called a **European option**; if the option owner can exercise that right on any day until a specific date, the **expiry** or **maturity**, then the option is a so-called **American option**. Moreover, if the option gives the right to *buy* at a given price, it is named a **call**; if it gives the right to sell at the agreed-upon price, it is named a **put**.

15.1 European Options

The fact that European options can only be exercised at their expiration date makes them amenable to pricing using a closed-form formula called the **Black-Scholes** equation. In contrast, American options (which we will investigate on page 301) are typically handled through simulations.

One formulation of the **Black-Scholes** equation to calculate the price C at time t of a call option maturing at T on a stock whose current value is S is:

$$C(S,t) = e^{-r(T-t)}[FN(d_1)] - KN(d_2)]$$

F equals $Se^{(r-q)(T-t)}$ and $N()$ is the standard normal cumulative distribution function, implemented in R in the `pnorm()` function. Moreover:

$$d_1 = \frac{1}{\sigma\sqrt{T-t}}\left[\ln\left(\frac{S}{K}\right) + \left(r - q + \frac{1}{2}\sigma^2\right)(T-t)\right] \tag{15.1}$$

and:

$$d_2 = d_1 - \sigma\sqrt{T-t}$$

where K is the strike price, σ is the annualized volatility (defined on page 167), r the annualized risk-free rate and q the continuously compounded **dividend yield**, as defined on page 272.

Because $F = Se^{(r-q)(T-t)}$, we have:

$$C(S,t) = Se^{-r(T-t)}e^{(r-q)(T-t)}N(d_1) - Ke^{-r(T-t)}N(d_2)$$

DOI: 10.1201/9781003320555-15

and, after simplification:

$$C(S,t) = Se^{-q(T-t)}N(d_1) - Ke^{-r(T-t)}N(d_2)$$

We can simplify further by setting $\tau = T - t$ and write a function that calculate the price of a call option using Black-Scholes:

```
BlackScholes.Call = function(S, K, sigma, r, q, tau){
  d1 = (log(S/K) + (r - q + sigma^2/2)*tau) / (sigma*sqrt(tau))
  d2 = d1 - sigma * sqrt(tau)
  price = S * exp(-q*tau) * pnorm(d1) - K*exp(-r*tau)*pnorm(d2)
  return(price)
}
```

We can now price a call option with strike 100 when the current price is 105, annualized volatility 30%, the risk-free rate 3%, the dividend yield 1%, with 10 days to go before expiry:

```
BlackScholes.Call(S = 105, K = 100, sigma = 0.3, r = 0.03,
                  q = 0.01, tau = 10/365)
```

```
[1] 5.482
```

Similarly, the value P of a put is given as:

$$P(S,t) = e^{-r(T-t)}[KN(d_2) - FN(d_1)]$$

hence:

$$P(S,t) = Ke^{-r(T-t)}N(d_2) - Se^{(r-q)(T-t)}e^{-r(T-t)}N(d_1)$$

and finally:

$$P(S,t) = Ke^{-r\tau}N(d_2) - Se^{-q\tau}N(d_1)$$

Note that d_1 and d_2 are as before, so the code for **d1** and **d2** stays the same and the following piece of code calculates the price of a put:

```
BlackScholes.Put = function(S, K, sigma, r, q, tau){
  d1 = (log(S/K) + (r - q + sigma^2/2)*tau) / (sigma*sqrt(tau))
  d2 = d1 - sigma * sqrt(tau)
  price = K*exp(-r*tau)*pnorm(-d2) - S*exp(-q*tau)*pnorm(-d1)
  return(price)
}
```

A put with the same parameters as the earlier call would then be valued at:

```
BlackScholes.Put(105, 100, 0.3, 0.03, 0.01, 10/365)
```

```
[1] 0.4288
```

The **RQuantLib** package was introduced on page 5. It will prove invaluable again to calculate the fair value of and to derive properties on options.

```
library(RQuantLib)
```

But more simply, the **RQuantLib** library offers a function, aptly named **EuropeanOption()**, to calculate the value of such an option using Black-Scholes. For example, the price of the right to buy, for \$100 and in 10 days, a stock whose value is currently \$105 can be done as follows. Notice that time, such as time to expiry, is usually expressed in models and equations as fractions of a year, since everything is annualized.

```
(eurOption = EuropeanOption(type="call",
                            underlying = 105,
                            strike=100,
                            dividendYield = 0.01,
                            riskFreeRate = 0.03,
                            maturity = 10/365,
                            volatility = 0.3))
```

```
Concise summary of valuation for EuropeanOption
  value    delta    gamma     vega    theta      rho
 5.4914   0.8440   0.0455   4.1829 -24.1951   2.3091
```

From the output of this command, which is the content of object **eurOption**, we understand that each entry can be read separately using the usual dollar-sign notation:

```
eurOption$value
```

```
[1] 5.491
```

We can verify that this value is very close to the value calculated above using our own **BlackScholes.Call()** function. But does it make intuitive sense? If we think about it, the stock is already worth \$105, and we are asking to buy it for \$100, so it stands to reason that buying that right should cost us about \$5. If we hand out that money now, the seller of the option can invest the 5 dollars at the prevailing interest rate (hence our passing **riskFreeRate** to **EuropeanOption()**). On the other hand, the value of the stock can appreciate a bit more, hence the importance of the **volatility** parameter.

These results of course hold only for one current price of the stock, and a natural question is how the value of the call option changes as a function of the current price. One way to do this is to call **EuropeanOption** repeatedly using a **for** loop, for different values of the stock price:

```
stock.prices = seq(90, 110, by=1)
option.prices = rep(0,length(stock.prices))
for (i in 1:length(stock.prices)) {
```

```
  eo = EuropeanOption(
    type = "call",
    underlying = stock.prices[i],
    strike = 100,
    dividendYield = 0.01,
    riskFreeRate = 0.03,
    maturity = 10 / 365,
    volatility = 0.3
  )
  option.prices[i] = eo$value
}
```

Note that the different values are stored into an array that was defined before entering the loop – the idea being to first allocate storage in memory, then fill the array out using the successive calculations.

We can then plot these prices as a function of the current stock price thanks to the code below, resulting in Figure 15.1.

```
plot(stock.prices,
     option.prices,
     type = "o",
     main="")
```

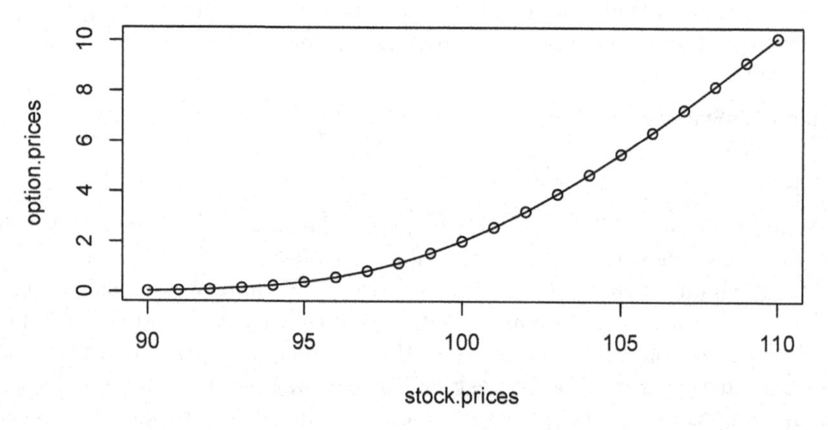

FIGURE 15.1 Option Value a.f.o. Different Stock Prices (Mat = 10 days)

We observe that the value of a call option is very small when the current price is much less than the option's strike (which is fixed at $100 in this example). (How much less depends on several factors, which we are important and are discussed later.) A value close to zero makes sense, since the odds are long that the stock price will reach the strike and thus that the call will be exercised in the remaining time to expiry. Because the current price is much less than the strike, the call option is said to be **out of the money**.

At the other end of the spectrum of stock prices, we observe that the value of the option increases fast when the stock's price is significantly above the $100 strike. For a stock price of $110, the call's value is essentially 10 dollars – and that makes sense: A cheaper value would mean arbitrageurs would buy the call, sell the stock for $110, and pocket the difference (assuming prices don't change much until expiry).

Actually, applying the same reasoning, the option value should be close to zero when the stock price is $89 or $90 or $91: the option's value should not change much when it is so deep out of the money. That sensitivity to changes in the stock's price is called the option's **delta**, and delta is close to zero for a deep out-of-the-money call (i.e., when the stock price is much lower than the strike). We can experimentally verify that delta is small by calculating the approximate slope on the left side of the graph, dividing the rise by the run for our first two data points:

```
(OTM.delta = (option.prices[2] - option.prices[1]) /
   (stock.prices[2] - stock.prices[1]))
```

```
[1] 0.0252
```

Likewise, when the stock price is significantly above the call's strike, the call is **in the money** and its value increases by $1 for each $1 increase in the stock's price – which makes sense when we think of the profit an arbitrageur would do if it were otherwise. This 1-for-1 match mean that the **delta** is 1 when the option is deep in-the-money. We calculate it on the far right side of the graph:

```
n = length(option.prices)
(ITM.delta = (option.prices[n]-option.prices[n-1]) /
    (stock.prices[n]-stock.prices[n-1]))
```

```
[1] 0.9675
```

Finally, the sensitivity of the option's value to the stock price is half way through, or 0.5, when the stock's price is close to the strike. (In that case, the call is said to be **at the money**.) We can again check that delta is 0.5 for an at-the-money call by calculating the slope of the curve going through two points close to the strike:

```
nmid = round(n/2)
(ATM.delta = (option.prices[nmid+1] - option.prices[nmid]) /
    (stock.prices[nmid+1] - stock.prices[nmid]))
```

```
[1] 0.4743
```

Note that delta calculated above is not 0.5 precisely because the two current stock prices we considered in our calculation are not symmetrically spaced around the strike price of $100.

Another consideration is that this "current" price changes, every second, as we get closer to the option expiry. In other words, it makes sense to also look at the evolution of option values as a function of remaining time. This sensitivity of option prices to the passage of time is called **theta**, as we will discuss shortly.

This means that at least two parameters have to be considered in our discussions now: the current price, and the time remaining until expiry. And there are more: The alert reader will have noticed that we assumed (arbitrary?) values for volatility, dividend yield and risk-free rate. Each of these parameters impacts option pricing and should be, ideally, considered as well.

Fortunately, **RQuantLib** offers a function that accepts exactly 2 vectors of parameter values and returns a matrix of option valuations[1]. We first set a few maturities; for the sake of illustration, we will consider 1, 3, 5 and 10 days, or rather their respective fractions of a year (because the **RQuantLib** analytics use years as the unit of time):

```
(maturities = round(c(1/365, 4/365, 5/365, 10/365, 20/365), 4))
```

```
[1] 0.0027 0.0110 0.0137 0.0274 0.0548
```

Then, the **EuropeanOptionArrays** function returns a matrix of option values for the different stock prices and maturities we specify:

```
european.options.mats =
  EuropeanOptionArrays("call",
                       underlying = stock.prices,
                       strike = 100,
                       dividendYield = 0,
                       riskFreeRate = 0.01,
                       maturity = maturities,
                       volatility = 0.3)
```

To know where **EuropeanOptionArrays()** stores the different calculations, we use the **names()** function to get the full list of fields:

```
names(european.options.mats)
```

```
[1] "value"
[2] "delta"
[3] "gamma"
[4] "vega"
[5] "theta"
[6] "rho"
```

We notice that, in addition to values, **EuropeanOptionArrays()** returns the corresponding **greeks**, which are different sensitivities of the price of an option to small changes in the value of one of the parameters. These sensitivities

[1]Unfortunately, the package does not let us vary more than 2 parameters in one shot.

are classically denoted with letters of the Greek alphabet, hence their names. We will discuss these sensitivities shortly, but the fact that they are already calculated by **EuropeanOptionArrays()** will come handy later.

We have to keep in mind that the value of the options will be different for each strike price and for each volatility level. Hence, a matrix of valuations are returned. In this example, the first 3 rows of that matrix can be seen as follows:

```
head(european.options.mats$value, 3)
```

```
      0.0027     0.011   0.0137  0.0274 0.0548
90 2.921e-12 0.0003427 0.001394 0.03053 0.2046
91 2.977e-10 0.0012418 0.004025 0.05541 0.2904
92 1.824e-08 0.0039916 0.010571 0.09611 0.4031
```

This table shows the different option prices for different remaining maturities across the columns and for different stock prices down the rows.

We can plot these different values as what option traders call a **surface** of option values. To do that, we perform a **3D plot** using functions **plot_ly()** and **add_surface()** of the **plotly** package:

```
library(plotly)
plot_ly() %>%
  add_surface(x=~stock.prices,
              y=~maturities,
              z=~european.options.mats$value)
```

After possibly some rotation of the plot, you should be getting something that looks like the plot in Figure 15.2.

However, a 3D plot does not make clear what's happening when maturities change. To do that, we are going to plot on the same 2D graph all the curves corresponding to the different maturities, so as to highlight their differences.

From an R standpoint, the first step is to create a chart, even a blank one, that reserves space for the curves and has the axes at the right scales. We can also give the graph its desired title. To do that, we plot one of the curves, knowing we are going to overwrite that plot.

```
plot(stock.prices,
     european.options.mats$value[,1],
     type = "l",
     main="Option Value vs. Stock Price")
```

Technical Detail Actually, since we are going to overwrite that curve, we don't even have to plot it – so we could also have the line type as "nil" by setting **type=** to **"n"** instead of **"l"**, as shown below.

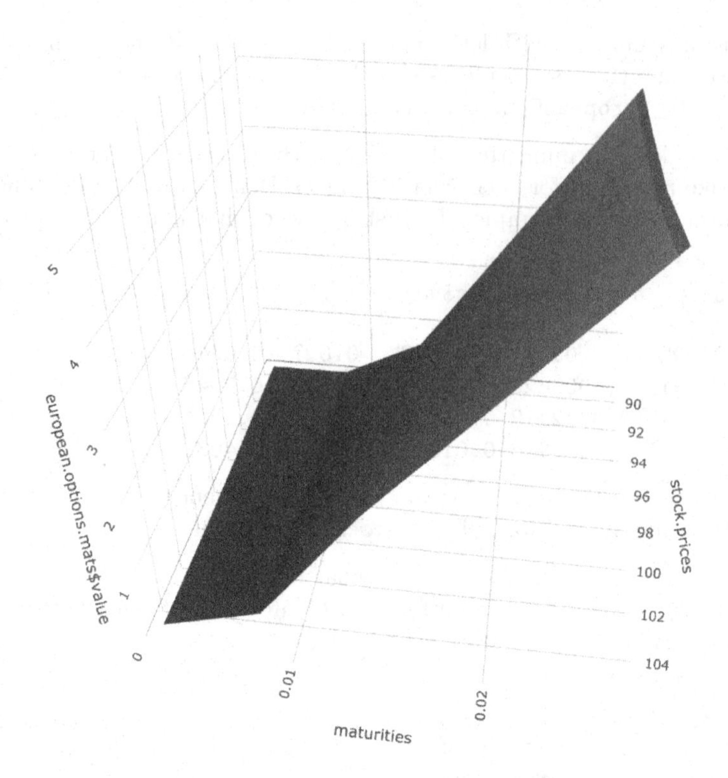

FIGURE 15.2 Surface of option prices as a function of maturities and strike price

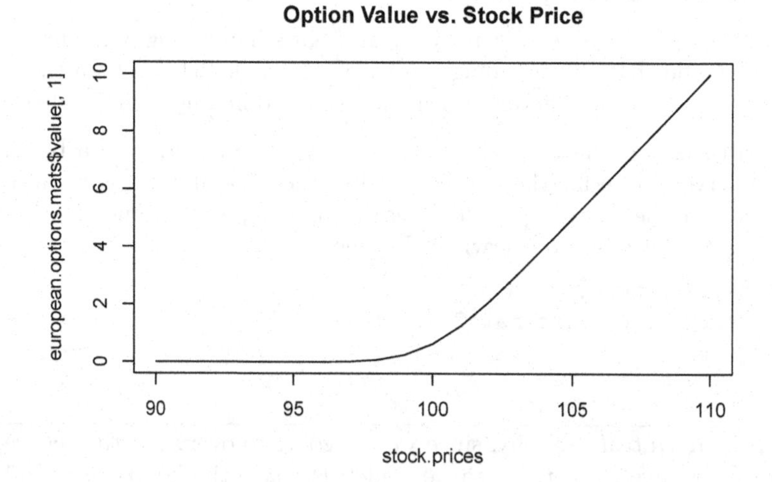

FIGURE 15.3 Plotting the output of EuropeanOptionArrays()

Now comes the second step of our R construction. We can now plot Figure 15.4, using the **lines()** function, with one curve for each of the maturities. Remember: per the comment above, the first **plot()** is here only to set the scales right. For explanations on the **lty=** named argument, please see page 19.

```
plot(stock.prices,
     european.options.mats$value[,1],
     type = "n")
colors = topo.colors(length(maturities))
for (i in 1:length(maturities)) {
  lines(stock.prices,
        european.options.mats$value[,i],
        col=colors[i])
}
legend(x = "topleft",
       legend = c("1 day","4 days","5 days","10 days","20 days"),
       lty=1,
       col = colors)
```

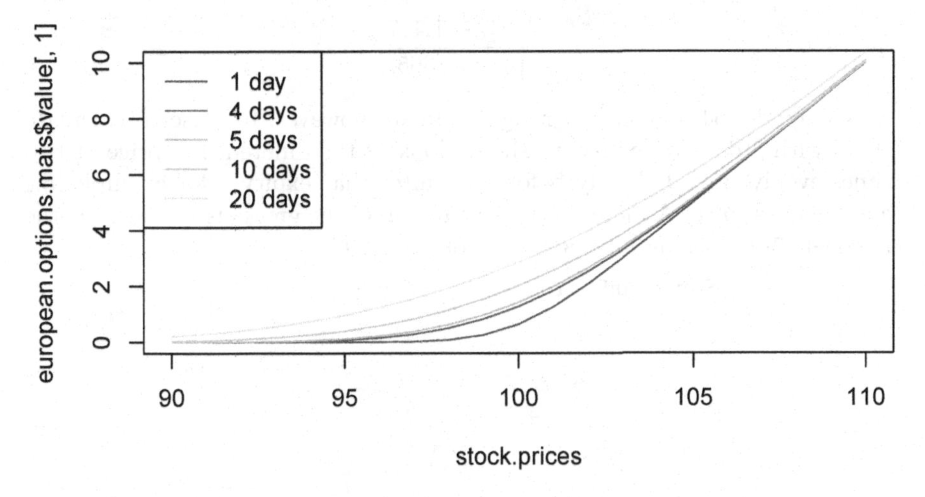

FIGURE 15.4 Option Value for Different Remaining Maturities

Note that **european.options.mats** now provides us with the full matrix of option prices for different stock prices and maturities, so Figure 15.1 is just one of the curves in Figure 15.4.

Figure 15.4 shows that the price of an option follows a round curve when there is still some time until option expiry, such as 20 days to go, but has a "hockey stick" shape when we are close to expiry, such as one day left. This stands to reason: with some time to go, a current stock price below the option strike still leaves room for a rally that would propel the stock price above the strike. With one day to go however, such odds get longer, and the option becomes worthless. On the other hand, for a current stock price above the strike, i.e. when the option is in-the-money, the stock can rally further and give the option holder an even larger profit. With one day to go though, the profit becomes more certainly equal to exactly the difference between the current price and the strike – the relationship becomes a line of slope 1 that goes through the strike point on the x-axis.

We read on the graph that, for a strike of $100, the option value drops from about 1.4 dollars 5 days before expiry to about 1.2 dollars, 4 days from maturity. We can verify these values by reading the relevant entries in the data:

```
european.options.mats$value[11,2]
```

```
[1] 1.267
```

```
european.options.mats$value[11,3]
```

```
[1] 1.417
```

The sensitivity of that option to the passage of time, or **theta**, is thus close to

$$\frac{1.267 - 1.417}{(5-4)/365},$$

which equals -54.8. That's a rough estimate however, because to calculate theta with 5 days to expiry we should look at the difference in price at two times evenly spread shortly before and after that expiry – for example, 5.1 and 4.9 days to go. A more precise calculation of the greeks is provided by the **EuropeanOption** function, as we saw on page 292:

```
(eurOption = EuropeanOption(type="call",
                   underlying = 100,
                   strike=100,
                   dividendYield = 0.01,
                   riskFreeRate = 0.03,
                   maturity = 5/365,
                   volatility = 0.3))
```

```
Concise summary of valuation for EuropeanOption
   value    delta    gamma     vega    theta      rho
  1.4239   0.5101   0.1128   4.6994 -51.7310   0.6887
  divRho
 -0.7085
```

As can be seen, our estimate was slightly off and a better estimate of **theta** is:

```
eurOption$theta
```

```
[1] -51.73
```

The matrix of option values also allows us to make a much more precise visualization of how **delta** changes as a function of maturity. Figure 15.5 matches the approximate delta calculations we made earlier. In particular, we can confirm that, close to expiry, delta is close to zero for stock prices far below the strike, close to 1 when the stock is well above the strike, and 0.5 when the stock price is around the option strike.

```
plot(stock.prices,
     european.options.mats$delta[,1],
     type = "n")
colors = topo.colors(length(maturities))
for (i in 1:length(maturities)) {
  lines(stock.prices,
        european.options.mats$delta[,i],
        col=colors[i])
}
legend(x = "topleft",
       legend = c("1 day","4 days","5 days","10 days","20 days"),
       lty=1,
       col = colors)
```

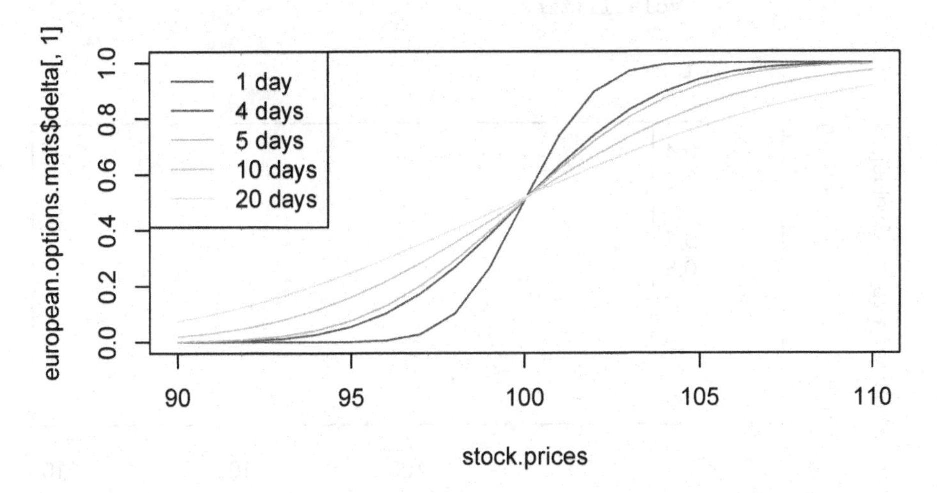

FIGURE 15.5 Delta a.f.o. Current Price, for Different Maturities

We can refine our analysis one step further. Indeed, a more subtle situation is that faced daily by option traders: When considering to buy (or sell) an option on a given stock, they have to determine the proper valuation of that option for different strike prices and different assumptions on volatility. So now, volatility is going to be one of the parameters passed to **EuropeanOptionArrays()**:

```
volatilities = seq(0.4, 0.8, by=0.1)
european.options = EuropeanOptionArrays(
```

```
  "call",
  underlying = stock.prices,
  strike = 100,
  dividendYield = 0,
  riskFreeRate = 0.01,
  maturity = 7 / 365,
  volatility = volatilities
)
```

We can then visualize in Figure 15.6 the impact of different volatility assumptions using the code below:

```
plot(stock.prices,
     european.options$value[,1],
     type = "n")
colors = topo.colors(length(volatilities))
for (i in 1:length(volatilities)) {
  lines(stock.prices, european.options$value[,i], col=colors[i])
}
legend(x = "topleft",
       legend = volatilities,
              lty=1,
       col = colors)
```

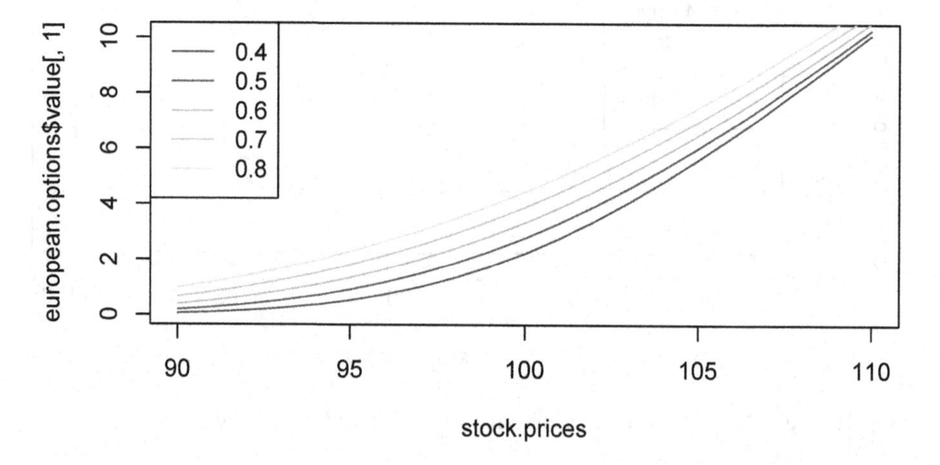

FIGURE 15.6 Option Value a.f.o. Current Price, for Different Volatilities

As we can see in Figure 15.6, the value of the option increases with volatility, irrespective of the current stock price: the curve that corresponds to an annualized volatility of 80%, (in yellow depending on your book version) is higher than the curve for a vol of 70% (in green), itself higher than the blue curve for a vol of 60%. That makes sense: for deep out-of-the-money call options where the stock price is much lower than the strike, a higher volatility means a higher probability that the stock still has time to reach the strike, giving the option more value. For an at-the-money option, the exact prices for the specified volatilities can be read as follows:

```
european.options$value[nmid, 1:length(volatilities)]
```

```
  0.4   0.5   0.6   0.7   0.8
1.758 2.304 2.853 3.402 3.952
```

So for an increase of annualized volatility from 40% to 80%, or an increase of 0.4, the option value increases by 3.952 less 1.758 or 2.195 dollars. The **vega** of the option, then, is the change in option value for a small change in volatility, so in this case 2.195 divided by 0.4, or 5.487. Note that our change in volatility, of 40 percentage points, was not "small" by any stretch of the imagination. A much more precise estimate of vega is again provided by the `EuropeanOption` function:

```
eurOption$vega
```

```
[1] 4.699
```

Implied Volatility

The **implied volatility** is the volatility that makes the Black-Scholes valuation of an option equal to the price quoted on the market. Calculating it can be done using `which.min()`. For example, let's say the underlying stock is worth $106 today, and the price quoted on the market for a call at a strike of 100 is $7.20. The volatility that makes the market price equal to the theoretical price is can be found by enumerating possible volatilities:

```
volatilities = seq(0.4, 0.8, by=0.01)
stock.prices = seq(106, 107, by=1)
european.options = EuropeanOptionArrays(
  "call",
  underlying = stock.prices,
  strike = 100,
  dividendYield = 0,
  riskFreeRate = 0.01,
  maturity = 7 / 365,
  volatility = volatilities
)
```

Then we find which of these vols makes the value closest, in absolute terms, to the quoted 7.20 dollars:

```
which.min(abs(european.options$value[1,] - 7.20))
```

```
0.59
  20
```

The first value is the volatility we are looking for, 59%, and the second is the index of that value in the **volatilities** vector.

Of course, a cleaner way to find the implied vol is to use the **optimize()** function we saw on page 74. We need, however, to repackage the **EuropeanOption** function to extract only its **value** field and then calculate the absolute value of the difference with the current stock price:

```
europ.wrapper = function(vol, u, s, d, r, m, p) {
  tmp=EuropeanOption(
    type = "call",
    underlying = u,
    strike = s,
    dividendYield = d,
    riskFreeRate = r,
    maturity = m,
    volatility = vol
  )
  return(abs(tmp$value - p))
}
```

We can then call **optimize()** on a sensible interval of possible vols. The other arguments are passed by **optimize()** to the function it optimizes, **europ.wrapper()**:

```
optimize(europ.wrapper,
         interval = c(0.4, 0.8),
         u = 106,
         s = 100,
         d = 0,
         r = 0.01,
         m = 7 / 365,
         p = 7.20)
```

```
$minimum
[1] 0.5883


$objective
[1] 6.115e-05
```

The first value, 58.8%, is the desired implied vol, indeed close to the 59% we found earlier. The second is how far the calculated option value is from the quoted price, assuming that implied vol.

15.2 American Options

As mentioned earlier, the fact that European options can only be exercised at their expiration date makes them relatively easier to model. In contrast, American options can be exercised at any time until their expiry. That additional flexibility has to have value, so an American option always has to have a higher price than its European counterpart, all else being equal. But that additional flexibility comes with a conceptual cost, which is an increased difficulty in modeling. Indeed, there are no formula or equation to value an American option, only numerical methods, and none of them is the de-facto standard in the industry.

One of these numerical method is called **Crank-Nicolson**. Using that method, we get the following value for an example option and some of its greeks. Notice that here again maturity is expressed in years:

```
amOption = AmericanOption(
  type="call",
  underlying = 105,
  strike = 100,
  dividendYield = 0.01,
  riskFreeRate = 0.03,
  maturity = 10 / 365,
  volatility = 0.3,
  engine = "CrankNicolson"
)
summary(amOption)
```

```
Detailed summary of valuation for AmericanOption
 value  delta  gamma   vega  theta    rho divRho
5.4928 0.8442 0.0455     NA     NA     NA     NA
with parameters
NULL
```

Two observations are in order here. First, we notice that most greeks are not calculated. This is not an error in our code or in the function, but a conscious design decision by the creators of **QuantLib**: these greeks are tedious to compute using numerical methods, implying that their calculation considerably slows down the time to execute the **`AmericanOption()`** function. Given that these greeks are not needed often, the authors of QuantLib preferred to leave their calculations to the user.

The second observation is to compare the valuation we obtained to that calculated on page 289 for a European option with otherwise identical parameters:

```
Concise summary of valuation for EuropeanOption
```

```
  value    delta    gamma      vega    theta      rho
 5.4914   0.8440   0.0455   4.1829 -24.1951   2.3091
 divRho
-2.4616
```

As we can verify, the value of the American option is (here, slightly) higher than that of its European equivalent.

15.3 Embedded Optionality in Callable Bonds

A **callable bond** is a bond in which the borrowing company, or **issuer**, keeps the right to pay back to the investor the amount it owes and thus to stop paying interests on the loan. Since the issuer has the right but not the obligation to do so, it falls under the definition of an **option**. Moreover, a bond is typically callable only on specific dates, meaning that the option is European, not American.

However, that option does not trade independently but is instead part of the callable bond: this is called an **embedded option**. Valuing an embedded option thus presents an additional challenge: we have to value the entire "package," and describing the callable bond can be tedious. Let's see that on an example using the **RQuantLib** package:

```
library(RQuantLib)
```

Assume we need to price a callable bond issued on September 16, 2021 and maturing on Sept 16, 2022. This bond has quarterly coupon payments and has a face value of $100. It is callable quarterly as well.

We will first build the **callability schedule**: it is a dataframe whose columns are "Price" (the price at which the bond can be called), "Type" and "Date". The example was made simple enough that we have only 4 quarters from bond issuance to maturity:

```
number.of.quarters = 4
Price    = rep(as.double(100),number.of.quarters)
Type     = rep(as.character("C"), number.of.quarters)
Date     = seq(as.Date("2021-09-15"),
           by = '3 months',
               length = number.of.quarters)
callability.schedule = data.frame(Price, Type, Date)
```

We then specify the different properties of the bond: the face amount, the issue and maturity dates, and the callability schedule we just constructed. Surprisingly, the way the library is made, the **coupon rate** (i.e., the ratio of the coupon value in dollars divided by the face amount) is specified separately. In this example, the coupon rate is chosen to be 4%, or $4 for a face value of $100 paid in four quarterly coupons of approximately $1 (the dollar amounts of the different coupon payments are different, as we will see later, because the number of days when the bond market is opened differs in each quarter).

```
bond.params   = list(faceAmount    = 100,
                     issueDate     = as.Date("2021-09-16"),
                     maturityDate = as.Date("2022-09-16"),
                     callSch       = callability.schedule)
coupon = c(0.04)
```

We then specify the accounting conventions used in the pricing of the bond. Fixed Income in general is extremely particular on the way days are counted. Do weekends count in calculating interests? Do holidays or market closes count? How many days does it take for a transaction to settle and thus for the investor to get their money back? All of these details change the return on a bond investment change by a few **basis points**, or hundredths of a percent, or even fractions of a basis point. But don't forget that these small percentages apply to gigantic amounts of money invested in corporate and government debt, so they do make a difference in actual dollars.

In any case, here is an example of a complete specification in case you need to go that granular. If these parameters do not mean anything to you, then you probably don't need to worry about them.

```
date.params   = list(settlementDays = 3,
                     calendar = "UnitedStates/GovernmentBond",
                     dayCounter = "ActualActual",
                     period = "Quarterly",
                     businessDayConvention = "Unadjusted",
                     terminationDateConvention= "Unadjusted")
```

Finally, we need to indicate which method we want to use to price the embedded option. We have to realize that the issuer will call the bond if the interest rate at which it could issue new debt is less than what it had promised to the investors of the current bond: if that's the case, then the company issues new bonds for which it will pay a lower interest, and use the proceeds of that sale to call the old bonds. In other words, the probability of a callable bond being called depends primarily on the path, going forward, of the bond's yield relative to prevailing interest rates. Depending on how the bond yield changes, up and down and up and down every day, one may or may not reach a situation where calling the bond is beneficial to the issuer. Going into the modeling of the path of rates is beyond the scope of this book, but one can imagine that a tree of possible scenarios opens up and that the probability of events following each of the branches in the tree depends on assumptions on the volatility and direction of rates. There are several such models, and in this example we are using the **Hull-White** tree methodology that takes two parameters (alpha for the propensity of the rate to revert to previous levels and its volatility sigma) to describe the movement of a specific bond's yield. We will be assuming values of 3% and 1%, respectively, for these two parameters, and the reader is referred to the technical literature for details on how to choose the values of these two parameters. The **term=** parameter is more intuitive, however, and is simply the assumed *constant* prevailing interest rate (i.e., the term structure is flat). In the example below, prevailing interest rates are assumed to be flat at 5.5% over the period under consideration. We also specify how many intervals of time we want over the period – the higher the number meaning the more branches in the tree of scenarios, yielding potentially more accurate numbers but causing calculation time to increase fast.

```
HullWhite.params = list(term = 0.055,
                        alpha = 0.03,
                        sigma = 0.01,
                        gridIntervals = 40)
```

Now that we have described the bond and specified which model we prefer to use to forecast possible paths in interest rates, we can value the bond as of a specific date using the **CallableBond()** function of the **RQuantLib** package. Note that the maturity date (specified earlier as September 16, 2022) has to be in the future relative to the valuation date.

```
setEvaluationDate(as.Date("2021-11-23"))
```

```
[1] TRUE
```

```
CallableBond(bond.params, HullWhite.params, coupon, date.params)
```

```
Concise summary of valuation for CallableBond
 Net present value :  98.44
       clean price :  97.629
       dirty price :  98.44
    accrued coupon :  0.81096
             yield :  0.070829
        cash flows :
       Date     Amount
 2021-12-16   0.99726
 2022-03-16   0.98630
 2022-06-16   1.00822
 2022-09-16   1.00822
 2022-09-16 100.00000
```

The **clean price** is the theoretical price of a coupon bond not including any coupon payments that are due but yet unpaid – what is called **accrued interest**. That is, the clean price doesn't include the partial coupon owed to the investor since the last coupon payment, yet the price quoted on financial news sites typically is a clean price. Dirty price is the theoretical price of the bond including the accrued interest. In our example, we can verify that the clean price of 97.6292 plus the accrued coupon of 0.811 equals 98.4401, which is exactly the dirty price.

Finally, note that the **net present value** (**NPV**) of the bond should be very close, but is not exactly equal, to the dirty price. That's because the NPV calculates the present value of the cash flows as of the valuation date, while the dirty price takes into account the small delay between the transaction date and the settlement date at which the investor receives the money. That delay was set at 3 days when we defined **dateparams**. We can also note that the amount of each coupon is slightly different each quarter, as we explained earlier.

15.4 Exercises

15.4.1 Black-Scholes

The d_1 term of Black-Scholes is shown in Equation (15.1). Assuming $S = 90$, $K = 100$, $r = 0$, $q = 0.03$ and a time to maturity $T - t$ of 0.05 year, plot d_1 and $N(d_1)$ as a function of σ. Assume volatility σ cannot exceed 60%.

■ SOLUTION

```
d1.afo.s =
  function(s) {
     (log(90/100) + (-0.03 + 0.5*s^2)*0.05)/sqrt(0.05)/s
  }
```

```
x = seq(0, .6, length=100)
d = d1.afo.s(x)
plot(x, d, type="l")
```

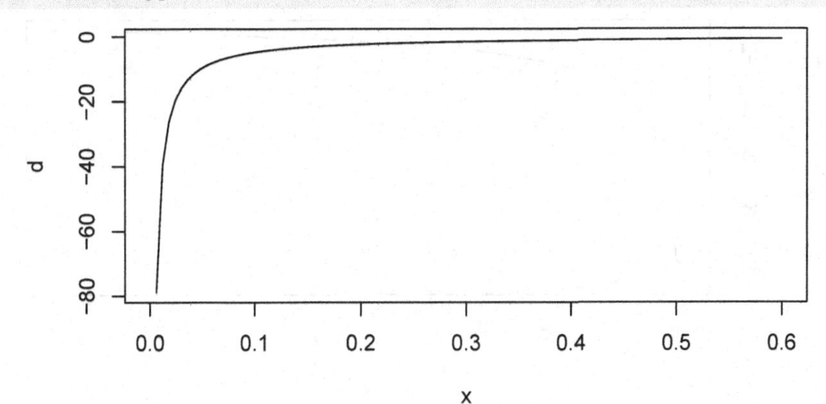

For $N(d_1)$:

```
x = seq(0, .6, length=100)
nd = pnorm(d)
plot(x, nd, type="l")
```

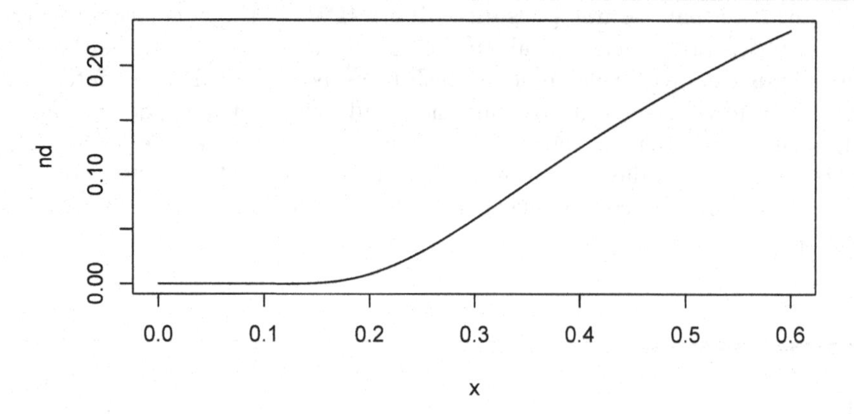

■ End of solution

15.4.2 Plot d1 as a Function of Time

Likewise, plot d_1 as a function of time. Assume that the time to expiry $T - t$ does not exceed one fourth of a year, that volatility is 20%, and the same values as in the previous exercise for the other parameters. ■ Solution

```
d1.afo.dt =
  function(dt) {
    (log(90/100) + (-0.03 + 0.5*0.2^2)*dt)/sqrt(dt)/0.2
  }
```

```
x = seq(0, 0.25, length=100)
y = d1.afo.dt(x)
plot(x, y, type="l")
```

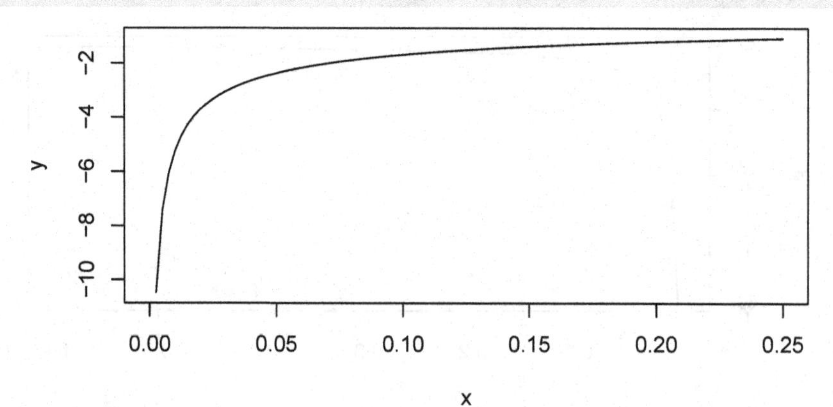

■ End of solution

16

Value at Risk

The powerhouse of risk management is **Value at Risk**, or **VaR**. It is defined in combination with a probability: VaR tells you how much you stand to lose with that probability, called **significance level**. There might be a small chance, say 1 out of 20, that your investment loses a small amount, but a smaller probability, say 1%, that it loses a lot. Value at risk quantifies the losses you can expect at these different significance levels. Note you could also say that there is a 95% **confidence level** that your investment will not lose more than that small amount, but if you want to be super sure, you can say that you're 99% (100 - 1) *confident* that any loss would not be larger than the bigger loss. When we'll need to use math, we will denote the significance level with α and the confidence level will be $1 - \alpha$.

VaR is also determined for a specific period: it stands to reason that the probability of a loss in the next hour is much less than over the next year. As an example, if the VaR on an asset is \$100 million at a one-week, 95% confidence level, there is a only a 5% chance that the value of the asset will drop more than \$100 million over any given week. As said, VaR is itself defined in terms of probabilities: The VaR at a 95% confidence level is the loss that has a $100 - 95 = 5$ percent chance of happening. This is of course a very high level definition of VaR; for a more mathematical presentation, the reader is referred to financial statistics textbooks[1].

To get the intuition of that definition, assume you have two worst cases:

- A large loss, say \$100,000, with probability 1%;
- A smaller loss, say \$10,000, with probability 5%;
- Then $VaR(5\%) = \min(10,000, 100,000) = \$10,000$ since both probabilities (of losses greater than either amounts) are less than or equal to 5%;

[1] As a quick summary: Suppose $\mathcal{L}$ is a loss over the holding period T. Then $VaR(\alpha)$ at significance level α is the αth upper quantile of $\mathcal{L}$:

$$P(\mathcal{L} \geq VaR(\alpha)) = \alpha$$

and

$$VaR(\alpha) = \min\{x : P(\mathcal{L} \geq x) \leq \alpha\}.$$

DOI: 10.1201/9781003320555-16

- But $VaR(1\%) = \min(100,000) = \$100,000$ since there is only one event, that of a loss of 100,000, that has a probability less than or equal to 1%

A few simple matters to clarify: first, Value at Risk and other metrics can be expressed in dollars or in percentage (return) terms. It doesn't really matter, but unless there is no ambiguity on the reference value of your investment, a return is best. Second, we care about losses, which are "negative profits." When it comes to losses, returns are negative, too. But we will casually talk of losses being "greater than" to mean they are more deeply negative.

More importantly, the example above hints to VaR's shortcomings: First, VaR tells how much you stand to lose – *or more*! And it doesn't tell you *how much more*. The risk of a large loss that we mentioned earlier is still lurking, despite having a smaller probability. VaR can thus be deceptive in that you may feel good about the loss you risk 5% of the time but hides from you the painful loss that can still happen 1% of the time.

And maybe more importantly, VaR is not **sub-additive**. What does sub-additive mean?

Let's review the example given by Wikipedia: Suppose you consider investing in either of two bonds, with independent probability of default. Let's say the default probability is 4% for both, so there is a 96% probability of no loss. Let's also assume that, upon default, these two bonds A and B have a recovery rate of 70% – meaning you get only 70% back. If you buy either bond, then the VaR at 5% significance level, or $VaR(5\%)$, is *zero* since the probability of a loss is less than the requested probability threshold of 5%.

Now, assume instead you invest in both bonds, 50% in each. Your goal is to diversify – as you should be encouraged to do! The probability that only A defaults is $96\% \times 4\% = 3.84\%$, with a total portfolio loss of $50\% \times 30\% = 15\%$. The probability that only B defaults is $96\% \times 4\% = 3.84\%$, with a total loss of $50\% \times 30\%$, or 15%. Then, there is also a probability of (4% times 4% =) 0.16% that both bonds default, for a total loss of 30%. So the probability of a 15%-or-worse loss is $3.84\% + 3.84\% + 0.16\% = 7.84\%$[2]. (By the way, we can verify that the probability that either or both bonds default is $1 - (96\%)(96\%) = 7.84\%$.) This probability is higher than the significance level you wished. A risk manager would report that $VaR(7.8\%) = 15\%$, so $VaR(5\%)$ is at least 15%. So the risk taken by investing in 2 bonds is greater than investing in a single bond. This unsavory property is called **subadditivity**. Despite this shortcoming, VaR is unfortunately still used often in the context of portfolios (the issue does not matter when assessing a single investment) but **conditional VaR**, also known as **expected loss**, fixes that.

[2]So in this two-bond example, we would write that $P(\mathcal{L} \geq 15\%) = 7.84\%$.

16.1 Parametric VaR

As said, Value at Risk measures the potential loss in value of an asset or portfolio over a defined period for a given significance level. Losses are not binary, however, and correspond to negative returns among a distribution of returns. That distribution is either approximated by a "classic" distribution, such as the normal distribution, or it's not – in which case the VaR calculation is based directly on the past returns. Because "classic" distributions are often described by very few parameters (for example, mean and standard deviation for the normal distribution), the first type of VaR calculation is called **parametric VaR**. Otherwise, since we do not need to make assumptions neither on which parameters matter nor on their values, the second family is called **nonparametric VaR**.

Let's assume that returns on an investment are normally distributed. Then the worst return in α percent of cases is the αth percentile of the normal distribution – so for a 95%-confidence VaR, we look for the 5th percentile of the distribution. A classic result of basic statistics is that 5% of the area under the curve of the standard normal distribution (corresponding to an aggregate probability of 5%) consists of all the points to the left of $z = -1.645$. That z-score is related to a return r by

$$z = \frac{r - \mu}{\sigma},$$

assuming returns on the investment follow a normal distribution of mean μ and standard deviation σ. So if the mean of that investment's returns is 6% and if the standard deviation of these returns is 10%, then the return r_{VaR} that corresponds to $z = -1.645$ is such that

$$-1.645 = \frac{r - 0.06}{0.10}.$$

Thus, $r_{VaR} = -0.1645 + 0.06 = -0.1045$ or -10.45%. That value r_{VaR} is VaR. There are no other calculations to perform. All the work was in determining the mean and the standard deviation.

Let's work this out on an earlier example of a portfolio of three stocks:

```
getSymbols("AAPL", from="2018-01-01", to="2021-12-06")
```

```
[1] "AAPL"
```

```
getSymbols("JNJ", from="2018-01-01", to="2021-12-06")
```

```
[1] "JNJ"
```

```
getSymbols("JPM", from="2018-01-01", to="2021-12-06")
```

```
[1] "JPM"
```

```
aapl.daily = periodReturn(AAPL, period="daily")
jnj.daily  = periodReturn(JNJ,  period="daily")
jpm.daily  = periodReturn(JPM,  period="daily")
holdings.returns = cbind(aapl.daily, jnj.daily, jpm.daily)
```

If we assume the portfolio is rebalanced daily, then we can assume that the weights are constant; here, we even assume they are equal:

```
equal.weights = c(1/3, 1/3, 1/3)
```

The daily portfolio returns are the weighted average of the holdings' returns:

```
port.returns =  holdings.returns %*% equal.weights
```

Then we calculate the z-score (here, at a 5% significance level), the standard deviation and the average return. VaR, in percentage terms, for a horizon of one day is then the return whose distance from the mean return is the z-score times the daily volatility:

```
(zscore    = qnorm(0.05, 0, 1))
```

```
[1] -1.645
```

```
(port.vol   = sd(port.returns))
```

```
[1] 0.0147
```

```
(port.mean  = mean(port.returns))
```

```
[1] 0.0007996
```

```
(VaR.param  = port.mean + zscore * port.vol)
```

```
[1] -0.02337
```

So the 1-day horizon VaR at a 95% confidence level calculated by the parametric method is 2.34%.

Parametric methods can assume distributions other than normal: for example, t-distributions can be used. Either way, the parameters of the chosen family of distributions have to be calibrated on the historical returns – a process called distribution fitting that we discussed on page 191. Once the general shape of distributions is decided, the method can extrapolate what are the worst cases could be – i.e., we calculate VaR based on the left tail of the fitted distribution even if no event corresponding to that left tail has happened yet. This is implicitly what we assume a moment ago when we took a z-score, because z-scores assume a normal distribution. But we can make the distribution fitting explicit:

```
library(MASS)
(f = fitdistr(port.returns, "normal"))
```

```
      mean          sd
  0.0007996   0.0146895
 (0.0004671) (0.0003303)
```

We are of course finding the same mean and volatility, and the rest of the VaR calculations is as above.

16.2 Nonparametric VaR

For nonparametric estimations, the loss distribution is not assumed to follow a parametric family such as the normal or t-distributions. Instead, this method assumes that only returns observed in the past can happen in the future, and the 5% worst return that can be observed in the future is the 5% worst return that already happened. For that reason, this method is also called **historical VaR**.

The calculation boils down to:

- Sorting all observed returns in decreasing order;
- Find the worst α-th percentile, where α is the significance level;
- The return at the α-th percentile is the VaR.

Fortunately, all this is done in R with a single function: `quantile()`. We illustrate that on the S&P 500 data from the **Ecdat** package. As seen earlier, we load the return time series using the `data()` function:

```
data(SP500, package="Ecdat")
```

As always, we look at the structure:

```
str(SP500)
```

```
'data.frame':    2783 obs. of  1 variable:
 $ r500: num  -0.01173 0.00245 0.01105 0.01905 -0.00557 ...
```

We observe that the data frame contains one vector named `r500` of 2783 observations of daily returns of the S&P 500. We place these returns in a new variable and calculate the lowest 5th percentile:

```
returns = SP500$r500
```

```
alpha = 0.05
quantile(returns, alpha)
```

```
      5%
-0.01514
```

We see that **quantile()** produces a **named vector** of one element. Its numerical value can be extracted using **as.numeric()**:

```
(VaR = as.numeric(quantile(returns, alpha)))
```

```
[1] -0.01514
```

The result is that this worst 5th percentile, which is the VaR at the 95% confidence level, is 0.0151, or 1.51% (expressed as a positive number by convention).

Applying this non-parametric method to our running example of three stocks is left as Exercise 16.8.5.

16.3 Calculating VaR Using the Covariance Matrix

The parametric and non-parametric methods assume that the weights given to each holding will be the same going forward, as of the day we are calculating VaR. But some times weights are about to change, or a new holding is about to enter the portfolio, or a large position that used to influence the portfolio's historical returns has now been exited. The **covariance method** to calculate VaR simplifies these issues.

The variance σ_P^2 of a portfolio equals:

$$\sigma_P^2 = w^T \Sigma w$$

where w are the holding weights and Σ is the covariance matrix of their returns. To compare the result of this method to that of the two earlier calculations, we keep the same equal weights:

```
(cm = cov(holdings.returns))
```

```
                 daily.returns daily.returns.1
daily.returns        0.0004302       0.0001244
daily.returns.1      0.0001244       0.0001884
daily.returns.2      0.0001972       0.0001270
                 daily.returns.2
daily.returns          0.0001972
daily.returns.1        0.0001270
daily.returns.2        0.0004283
```

```
port.variance = t(equal.weights) %*% cm %*% equal.weights
port.stddev   = sqrt(port.variance)
```

The portfolio's average return equals the weighted average of the holding's average returns, or the **sumproduct** of the weights and the average returns:

```
holdings.avg    = colMeans(holdings.returns)
port.avg.return =  sum(equal.weights * holdings.avg)
```

Finally, the VaR calculated by the covariance method is the average portfolio return less z-score times the standard deviation calculated from the covariance matrix:

```
(VaR.cov  = port.avg.return + zscore * port.stddev)

         [,1]
[1,] -0.02337
```

We find the same 1-day 95% VaR as we did using the parametric method.

16.4 Conditional Value at Risk

Expected Shortfall, also known as **Expected Tail Loss** or **Conditional Value At Risk**, is an improvement over VaR. Expected shortfall tells you the probability that you lose $VaR(\alpha)$ *or more*, and it tells you *how much more* that loss would be, on average. So it fixes one shortcoming of VaR: it does tell how much more than VaR we stand to lose – weighted by the probabilities of each event. Mathematically, expected shortfall is defined as the expected value of losses $\mathcal{L}$ worse than the value at risk:

$$ES(\alpha) \quad = \quad \mathbf{E}[\mathcal{L}|\mathcal{L} \geq \text{VaR}(\alpha)]$$

where **E** is the expected value function seen on page 169. Intuitively, ES(α) is the average of VaR(u) over all the u's that are less than or equal to α. Thanks to that definition, expected shortfall takes care of sub-additivity. We won't prove that statement but will give the intuition of why expected shortfall correctly captures the risk of the single-bond example discussed earlier: In that example, we saw that $VaR(5\%)$ equals 0 because the bonds have a default probability of 4%, which is less probable than the 5% worst-case significance level. However, VaR jumps to 30% for $0\% < \alpha \leq 4\%$. In contrast, the expected shortfall at a 5% significance level would be the average of values-at-risk over the interval:

$$ES(5\%) = \frac{(5\% - 4\%) \times 0 + (4\% - 0\%) \times 30\%}{5\%} = 24\%$$

This doesn't prove that ES(5%) for the two-bond portfolio is less than ES(5%) for one bond, which is what sub-additivity means, but at least this hopefully explains why the risk of a one-bond portfolio is more accurately reflected using expected shortfall.

16.5 Calculating VaR Using PerformanceAnalytics

One intuitive way to visualize VaR is to use the **chart.Histogram()** function (package **PerformanceAnalytics**). We encountered that function on page 88, but we are now using it with the **add.risk** method. Please see its output in Figure 16.1.

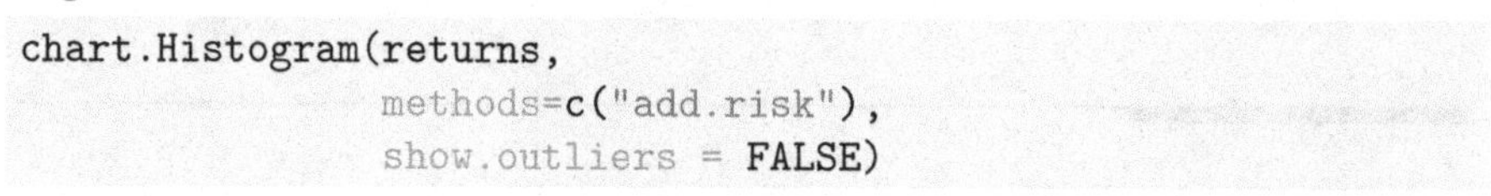

```
chart.Histogram(returns,
                methods=c("add.risk"),
                show.outliers = FALSE)
```

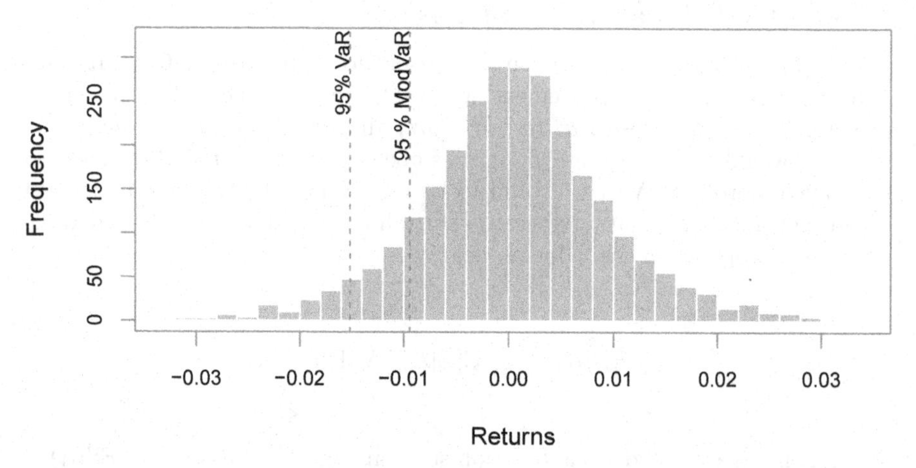

FIGURE 16.1 Using chart.Histogram() to visualize VaR

Notice that the function displays VaR but also **modified VaR**. Many implementations of VaR assume a symmetrical distribution, which is flawed for assets with significantly non-normal (e.g., skewed) distributions. Extensions were thus suggested that directly incorporate the higher moments (skew and kurtosis) of the return distribution into the VaR calculation using a Cornish-Fisher expansion (a special case of Taylor expansion). The resulting VaR is referred to as "Cornish-Fisher VaR" or "Modified VaR," and it produces the same results as standard VaR when the return distribution is normal – and thus may be used as a direct replacement.

The **PerformanceAnalytics** package includes functions to directly calculate VaR using the `VaR()` function. For instance, reading again the returns of the S&P 500 index as we did earlier, we calculate the VaR at a 95% confidence level as follows:

```
returns = SP500$r500
VaR(returns, method="gaussian", p = 0.95)
```

```
      [,1]
VaR -0.01745
```

Note that we requested that a normal (Gaussian) distribution be fitted to the history of returns. This VaR of 1.75% is not too far off from the 0.0151, or 1.51%, calculated earlier.

The same package also offers the `ETL()` function to calculate the **expected tail loss**:

```
ETL(returns, method="gaussian")
```

```
      [,1]
ES -0.02199
```

As expected, the expected loss of 2.2% is larger than VaR since it includes the probability-weighted losses in excess of VaR.

16.6 Calculating VaR Using Tidyquant

On page 166, we introduced `tq_performance()` from the **tidyquant** package as a general function to calculate multiple returns-based statistics. This function also calculates VaR, but it is a bit more particular about its input. (Similarly, `tq_performance()` can also calculate expected shortfall.)

Being meant as a general performance tabulation wrapper, the function takes the periodic returns of the investment of interest, which it calls `Ra` internally, and the returns (over the same period and with the same periodicity) of a "benchmark," which it calls `Rb`. A benchmark is not needed here, so we set `Rb` to `NULL`. The function to be calculated here is specified as being `VaR()`.

```
aapl.tq %>%
    tq_transmute(select     = adjusted,
                 mutate_fun = periodReturn,
```

```
                   period      = "monthly") %>%
    tq_performance(Ra = monthly.returns,
                   Rb = NULL,
                   performance_fun = VaR)
```

```
# A tibble: 1 x 1
       VaR
     <dbl>
1 -0.0879
```

So the VaR of Apple's stock, based on the historical data we have, is 8.793%.

By default, VaR and CVaR are calculated for a 95% **confidence level**, but that can be changed using the **p** parameter. Below, we set the confidence level to 99%:

```
aapl.tq %>%
    tq_transmute(select      = adjusted,
                 mutate_fun = periodReturn,
                 period      = "monthly") %>%
    tq_performance(Ra = monthly.returns,
                   Rb = NULL,
                   performance_fun = VaR,
                   p  = 0.99)
```

```
# A tibble: 1 x 1
      VaR
    <dbl>
1 -0.113
```

The value is less, as it should: the 1% (1-99%) worst case has to be a loss greater than the 5% (1-95%) worst case.

While we are at it, other metrics can also be computed by changing **performance_fun**. For example:

```
aapl.tq %>%
    tq_transmute(select      = adjusted,
                 mutate_fun = periodReturn,
                 period      = "monthly") %>%
    tq_performance(Ra = monthly.returns,
                   Rb = NULL,
                   performance_fun = SharpeRatio)
```

```
# A tibble: 1 x 3
  `ESSharpe(Rf=0%,~ `StdDevSharpe(Rf=~ `VaRSharpe(Rf=0~
              <dbl>              <dbl>            <dbl>
1             0.186              0.284            0.204
```

You'll notice that the **SharpeRatio()** function calculates three risk-adjusted measures of return. **VaRSharpe** equals the return in excess of the risk-free rate divided by VaR, while **ESSharpe** equals that excess return divided by the expected shortfall. The middle value, labeled **StdDevSharpe**, is the standard Sharpe ratio defined in Equation (9.2). The risk-free rate used in this calculation can also be specified, and it has to be expressed in the same time period as the returns: for example, the risk-free rate should be monthly here since the returns are monthly. Below, we assume an annualized risk-free rate of 3% so we divide it by 12 to make it monthly (that's the approximate but standard convention to get a monthly risk-free rate):

```
aapl.tq %>%
    tq_transmute(select     = adjusted,
                 mutate_fun = periodReturn,
                 period     = "monthly") %>%
    tq_performance(Ra = monthly.returns,
                   Rb = NULL,
                   performance_fun = SharpeRatio,
                   Rf = 0.03 / 12)
```

```
# A tibble: 1 x 3
  `ESSharpe(Rf=0.2~ `StdDevSharpe(Rf=~ `VaRSharpe(Rf=0~
              <dbl>              <dbl>            <dbl>
1             0.160              0.245            0.175
```

We can verify that all three ratios are lower. For example, the standard Sharpe ratio went from 0.284 to 0.245: When interest rates are higher, the bar for investments is higher. That's because when the risk-free rate is 3%, a rational investor would not take the risk of investing in stocks to get the same 3%. The investment has to provide a higher return to deserve back the 0.284 it had with interest rates at 0.

How sensitive is VaR to α?

Value at Risk depends on the significance level α. It is therefore important to check how VaR would change depending on α – maybe picking too low a significance level hides extreme tails, for example. To assess this sensitivity to α visually, we can have α vary and plot the respective VaRs, as follows – resulting in the plot in Figure 16.2.

```
alpha = seq(0.005, 0.07, by=0.005)
VaR = -as.numeric(quantile(returns, alpha))
plot(alpha, VaR, type="l", ylab="VaR(alpha)")
```

As Figure 16.2 shows, if the significance level is too stringent, then only the largest losses are captured, which explains why the curve is concave up.

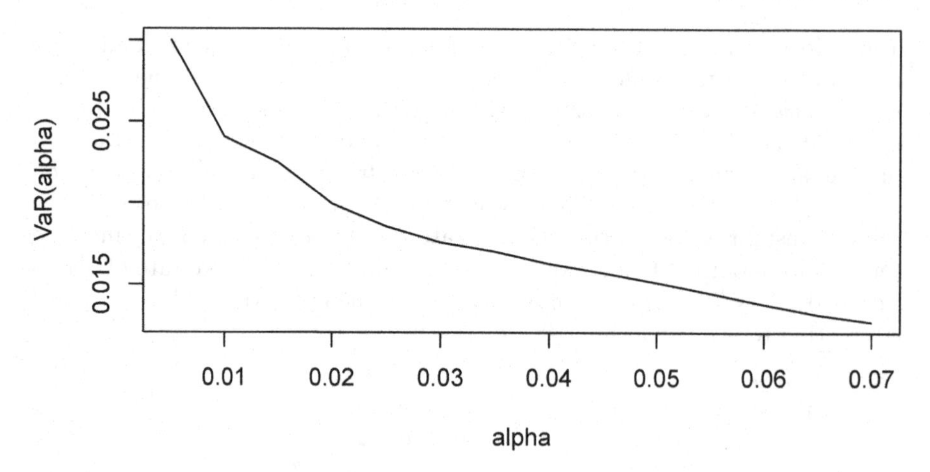

FIGURE 16.2 Sensitivity of VaR as a function of the significance level

16.7 Chapter-End Summary

VaR tells how much you stand to lose, *or more*; and it doesn't tell you *how much more*. Moreover, because VaR is not sub-additive, it can create an incentive *not* to diversify your portfolio. Expected shortfall, also known as Conditional Value at Risk (CVaR), solves both issues.

We reviewed three ways to calculate VaR: the first one is based on past returns; implicitly, it assumes that the worst returns observed in the past are also the worst that can happen in the future. Parametric methods, in contrast, fit a standard distribution like a normal distribution to the historical returns and extrapolate from the fitted distribution what the worst cases are. In other words, parametric methods calculate VaR based on the left tail of the fitted distribution, even if no event in that left tail has happened yet. Finally, the third method relies on the covariance matrix of the assets currently in the portfolio – which matters most when the portfolio's holdings have changed over time.

Both VaR and CVaR can easily be calculated using standard financial packages, but we should remember that VaR is quite sensitive to the value chosen for its significance level (a choice that is often arbitrary) and on the time horizon.

16.8 Exercises

16.8.1 How Sensitive is VaR to α, Revisited

We saw how to study the sensitivity of the VaR on the S&P500 index to the significance level. Find another method based on the **qnorm()** function.

■ SOLUTION

```
mu = mean(returns)
sig = sd(returns)
alpha = seq(.0025,.25,by=.002)
VaR = -mu + sig*qnorm(1-alpha)
plot(alpha, VaR, type="l", ylab="VaR(alpha)")
```

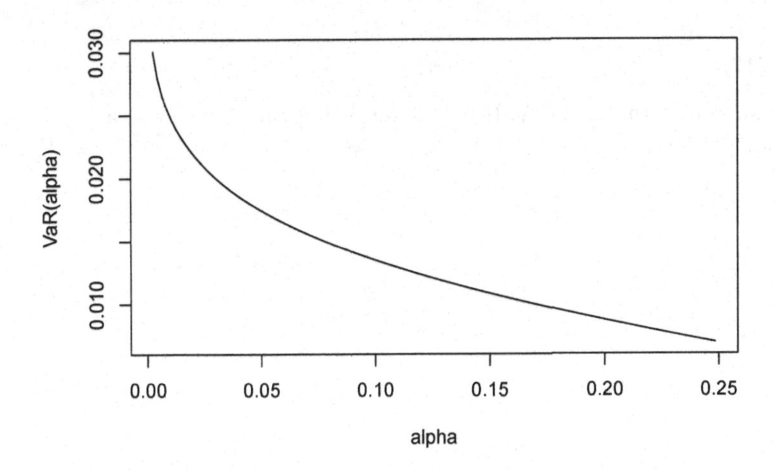

FIGURE 16.3 Sensitivity plot produced in Exercise 15.4.1

■ END OF SOLUTION

16.8.2 Comparing VaR Methods

The **VaR()** function can use four **methods**: "modified", "gaussian", "kernel" and "historical". Compare the results they provide on the S&P 500 at some fixed confidence level.

16.8.3 Comparing CVaR Methods

The **ETL()** function can use three **methods**: "modified", "gaussian" and "historical". Compare the results they provide on the S&P 500 at some fixed confidence level.

16.8.4 Rolling VaR

Value at Risk calculations depend on the returns provided as input, and those are typically the daily or monthly returns observed over the last few years. If that **rolling window** is fixed at 36 monthly returns, calculate and graph the Gaussian 5%-significance level VaR on the S&P 500 over the past five years. How much has VaR changed over that period?

16.8.5 Non-parametric VaR

Apply the non-parametric method to calculate the VaR(5%) at a one-day horizon to the three-stock portfolio seen in this chapter.

■ SOLUTION

```
quantile(port.returns, 0.05)
```

```
      5%
-0.02125
```

This is close to the 2.34% VaR calculated using the parametric and covariance methods.

■ END OF SOLUTION

17

Time Series Analysis

A **time series** is just a fancy name for historical data at regular time intervals. (Or relatively regular: work days or days when the stock market is open are not regularly spaced.)

A time series is also a data structure in R, and it is essentially a vector of values paired with a vector of dates and times. Time series objects are created using the `ts()` function, and we saw an example of its use on page 173.

What is meant by time series *analysis* is the discovery of a pattern in the values and the creation of a mathematical model that allows us to make predictions on future values.

Most of these predictions rely on a few assumptions. A key assumption is that the values revert to some average level – i.e., that a plot of these values will show they trend along a flat, horizontal line. Such a **stationary** series can show random oscillation, but those oscillation should be around some fixed level in a phenomenon called **mean-reversion**. In contrast, if the series wanders without returning repeatedly to some fixed level, then the series should *not* be modeled as a stationary process.

Another assumption is that errors (made by the model in predicting values around the trend line) do not get worse over time. If these assumptions are not satisfied, then some transformation should be attempted to correct the issue.

The mathematics behind this modeling can be intimidating, so it is important to keep the intuitions in mind. It is also helpful to remember that all these models are doing is trying to find a *linear* expression, somehow, for future values. We are going to review each of the following steps in details, but here are the intuitive steps, painted with a broad brush, in building a time series model.

First, a time series can have an upward or downward trend. If that trend is a curve, then by definition it's not a line – and fitting a linear model on it will be misleading. It will fail our first assumption, too: values are not reverting to a stable level. If the trend is linear however, then we are closer to the assumption of a horizontal mean-level line, but we need to flatten the values. There are two options: Option 1, we calculate ourselves the **first difference** of the values, i.e. the difference between one value and the next one in the time series. Option

DOI: 10.1201/9781003320555-17

2: We'll use a so-called **integrated** model – which simply means it will do the differencing for us to remove the up- or downward trend.

Moreover, if values tend to decrease or increase, then erratic movements in the data can appear to grow larger over time. In that case, a **variance stabilizing response transformation** might help.

The models we are going to study and build are called **ARIMA**:

- **AR** stands for **autoregressive**, which simply means that the values at one time depends on previous values. How many past values matter has to be determined, and is usually denoted with p. In other words, a value is predicted based on a regression of the previous p values in the time series.
- **I** stands for **integrated**; the corresponding parameter d equals 0 if we request no differencing and 1 if we request first differencing (and higher integers for higher differencing, not discussed in this book).
- **MA** stands for **moving average**, but the term is kind of a misnomer. It actually means that the predicted value depends on the previous *errors* on the previous points. How many past errors we are trying to capture (and catch up on) is denoted with Parameter q.

So an ARIMA model will be fully specified as ARIMA(p, d, q). Note that if d and q are zero, then we simply have an AR(p) model; and if p and d are zero, then we say we have a MA(q) model.

But enough with terminology. We'll see that finding the proper ARIMA model is relatively easy anyway – and how it can even be automated. The process can be summed up as follows:

1. Visualize the time series
2. "Stationarize" the time series
3. Plot the so-called ACF and PACF (to be defined shortly) to help find the ARIMA parameters
4. Build the ARIMA model on in-sample data
5. Make a forecast on out-of-sample data

Let's look concretely what this all means: We load the Mishkin data set of inflation rates (found in the `Ecdat` package and stored in the **RDA** format), and plot them in Figure 17.1.

```
load("Mishkin.rda")
inflRates = Mishkin[,1]
plot(inflRates,ylab="Inflation Rate",type="l",xlab="Year")
```

The main observation is that there is some upward drift. That's what should prompt us to calculate first differences. This is done easily in R using the `diff()` function. Plotting the differences, in Figure 17.2 clearly shows a much more stable time series:

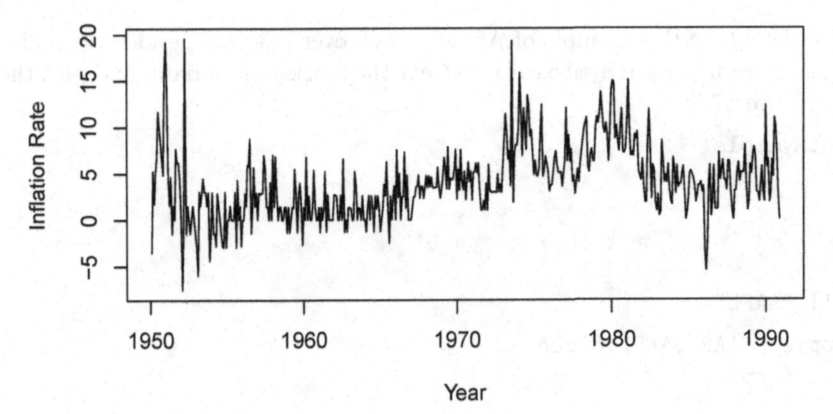

FIGURE 17.1 Inflation Rates

```
plot(diff(inflRates),
     ylab="Change in Inflation Rate",
     xlab="Year",
     type="l")
```

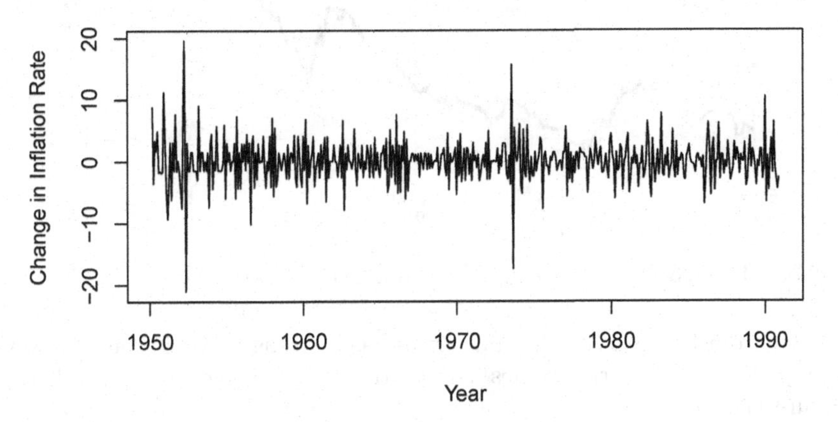

FIGURE 17.2 First-Differenced Inflation Rates

A visual inspection of Figure 17.2, i.e. of the plot of *changes* in inflation rates, suggests stationarity. The data also seems to have constant variance: with the exception of very brief spikes, the ups and downs of the differenced time series seem to have constant magnitude. If that wasn't the case, a **variance stabilizing transformation** such as log or **Box-Cox** could be attempted, in particular if the data showed some curved trend. Otherwise, a different class of models that explicitly captures change in variance over time would be needed. An example of such models are **Auto-Regressive Conditional Heteroskedasticity** or **ARCH** models.

Now let's take the example of Apple's stock over a 3-year period. We pull the time series using **getSymbols()**, extract the market-close data, and plot them in Figure 17.3:

```
getSymbols("AAPL",
           from = "2018-04-01",
           to = "2021-04-09",
           get = "stock.prices")
```

```
[1] "AAPL"
```

```
Apple = AAPL$AAPL.Close
plot(Apple)
```

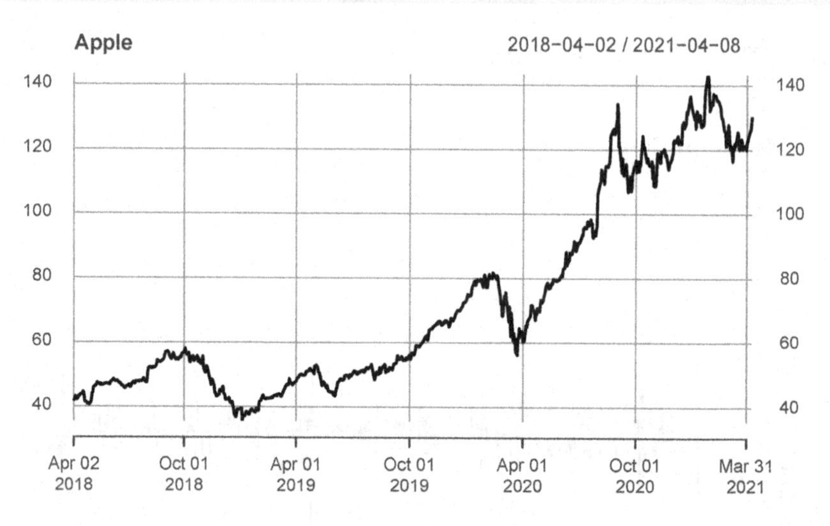

FIGURE 17.3 Apple's Stock Price from 2018 to early 2021

Clearly from Figure 17.3, the time series has a strong upward trend. However, let's plot the time series values against themselves lagged by 1, resulting in Figure 17.4:

```
plot(as.vector(Apple[1:length(Apple)-1]),
     as.vector(Apple[2:length(Apple)]))
```

Figure 17.4 shows a strong positive linear relationship. Thus, what we called an **autoregressive** model is promising.

However, we still haven't solved the issue of the strong upward trend of Apple's stock, which contradicts our first assumption. Here again, a trick to remove the trend is to use first-differences. Plotting these differences in absolute values, from one day to the next, results in Figure 17.5. The index on the x-axis is simply the number of days since the beginning of the time series.

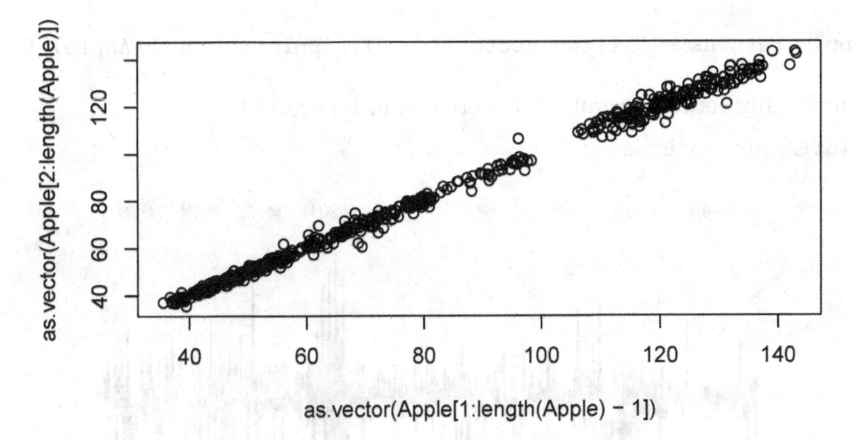

FIGURE 17.4 Plotting Apple's stock price against itself with a lag of 1 shows a strong relationship

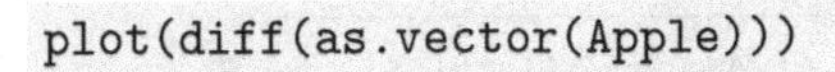

```
plot(diff(as.vector(Apple)))
```

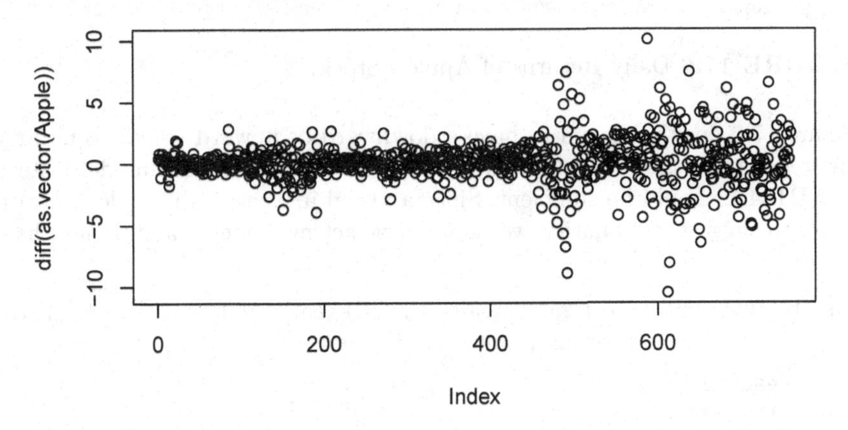

FIGURE 17.5 First-differencing of Apple's stock

The time series now has a nice horizontal trend.

However, we notice a large increase over time in these daily differences. This should not surprise us: since the stock has dramatically increased in *absolute* dollar price, *absolute* changes from one day to the next also increased. Our second assumption is breached. We need to normalize these absolute price differences, and this is precisely what returns are here for. Let's calculate all the returns over the entire period by dividing the daily differences by the values at the beginning of each day – except the last one:

```
Apple.returns = diff(as.vector(Apple))/Apple[1:length(Apple)-1]
```

These daily returns are plotted over time in Figure 17.6:

```
plot(Apple.returns)
```

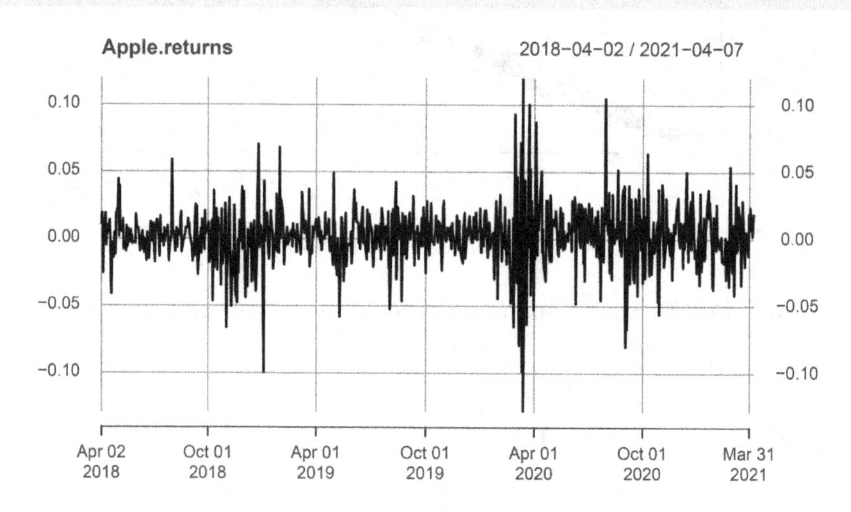

FIGURE 17.6 Daily Returns of Apple's Stock

Figure 17.6 does not seem to have a downward or upward trend, so no first-difference term should be needed; in other words, an **ARMA** model instead of **ARIMA** should be sufficient. Since a visual inspection of Apple's returns seems to suggest stationarity, we should have achieved Step 2 and we now have something we can model.

Finally, we will investigate again the 10-year rates contained in the `fredgraph.xlsx` dataset and plot them in Figure 17.7:

```
xl = read.xlsx("fredgraph.xlsx", startRow=14)
plot(xl$DGS10, type="l")
```

We see some reversion to the mean but also a lot of variability around it that hides a trend (and also indicates that variance is not constant). For that reason, we calculate the first differences and plot them in Figure 17.8:

```
rate.changes.10yr = diff(xl$DGS10)
plot(rate.changes.10yr, type="l")
```

Here again, we end up with a sequence of numbers that is visually much more stationary That will insure higher quality in our modeling.

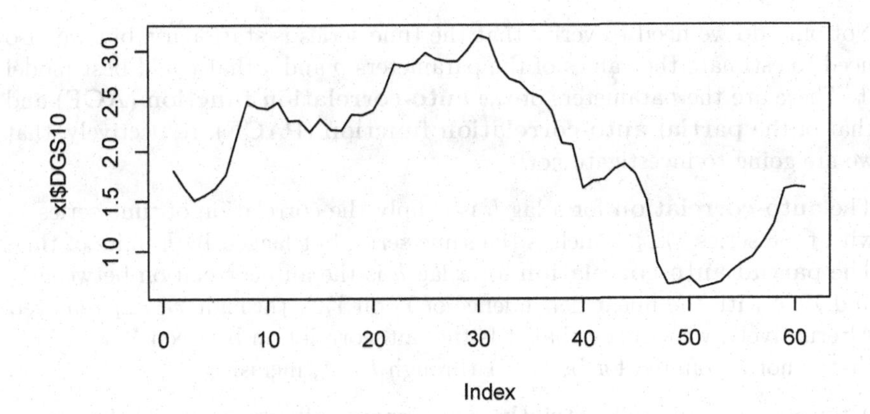

FIGURE 17.7 10-year Interest Rates

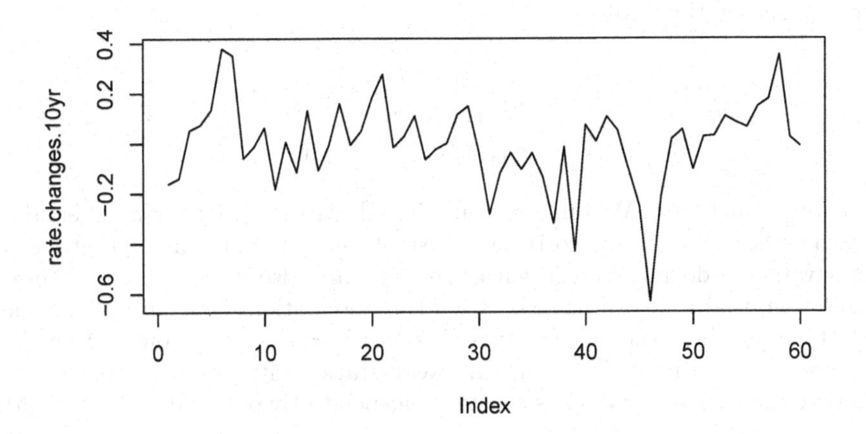

FIGURE 17.8 First-Differenced 10-year Interest Rates

17.1 ACFs and PACFs

Not only do we need to verify that the time series is stationary, but we also need to estimate the values of the parameters p and q that could best model it. These are the parameters of the **auto-correlation function** (**ACF**) and that of the **partial auto-correlation function** (**PACF**), respectively, that we are going to investigate now.

The **auto-correlation** for a lag h is simply the correlation of time series Y_t with time series Y_{t-h}, which is the same series but lagged by h units of time. The **partial auto-correlation** for a lag h is the autocorrelation between Y_t and Y_{t-h} with the linear dependence of Y_t on Y_{t-1} through Y_{t-h+1} removed. Alternatively, we can say that it is the autocorrelation between Y_t and Y_{t-h} that is not accounted for by lags 1 through $h-1$, inclusive.

A time series is said to be **strictly stationary** if all aspects of its behavior are unchanged by shifts in time. In particular, this implies that the distribution of each of the $X_1, X_2, ..., X_n$ is the same. In other words, the probability distribution of a sequence of observations does not depend on when they started.

Strict stationarity is a very strong assumption, because it requires that "all aspects" of stochastic behavior be constant in time. Often, it will suffice to assume less, namely, weak stationarity.

A time series is **weakly stationary** if the mean, variance and covariance are unchanged by time shifts:

$$E(Y_t) = \mu$$
$$Var(Y_t) = \sigma^2$$
$$Cov(Y_t, Y_s) = \gamma(|t-s|)$$

for some function γ. We are not assuming other distributional characteristics (such as skewness and kurtosis) are constant, but we are assuming that mean and variance do not change with time. We are also saying, per the third condition, that the covariance between two observations depends only on the lag between them (i.e., on the time distance $h = abs(t-s)$) but *not* on the values t or s individually. Assuming weak stationarity, the auto-correlation between Y_t and Y_{t+h}, which is constant independently of t, is denoted by $\rho(h)$.

White noise is the simplest example of a stationary process. The weakly stationary sequence $Y_1, Y_2, ...$ is **weak white noise** if:

- For $h = 0$: $\gamma(0) = \sigma^2$ and $\rho(0) = 1$
- For $h \neq 0$: $\gamma(h) = 0$ and $\rho(h) = 0$

But how can we test that $\rho(0) = 1$ and $\rho(1) = \rho(2) = ...\rho(k) = 0$? The solution is to plot these different correlations as a function of h and see if they are different from 0. This is where ACFs come into play.

The **autocorrelation function** can be plotted using the `acf()` function. For example, for inflation rates:

```
inflation.autocorrelations = acf(inflRates)
```

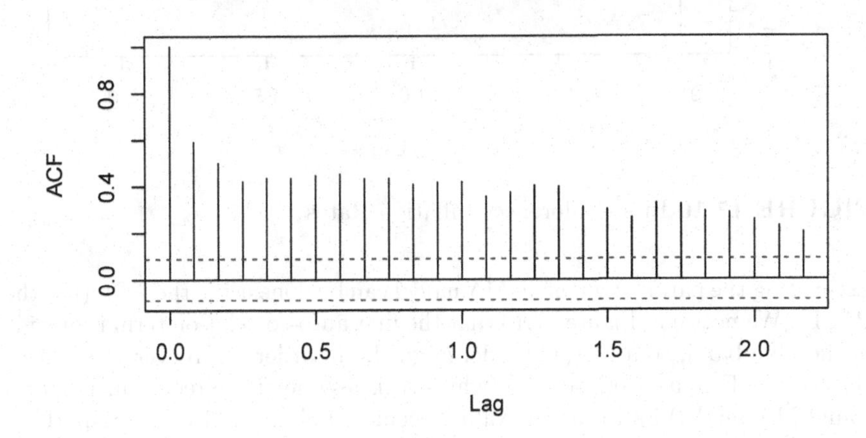

FIGURE 17.9 ACF of the Inflation Rates time series.

In the case of Figure 17.9 for inflation rates, we observe that the ACF decays to zero slowly. This makes us suspect non-stationarity, which agrees with our initial visual inspection of Figure 17.1.

Each time we see a very slow decay to zero in an ACF, we can forget about stationarity. Actually, if the ACF (or the PACF) does not tail off but instead stays close to 1 over many lags, then first-differencing will be needed before we look again at the ACF and PACF of the differenced data.

And indeed, we had already established that the first difference of inflation rates, plotted in Figure 17.2, was a much better candidate. We can confirm that by analyzing its ACF, plotted in Figure 17.10:

```
acf(diff(inflRates))
```

Plots of ACFs typically include test bounds, and you may have noticed dotted lines (shown in blue, depending on your version of the book). These lines give the values beyond which the autocorrelations are significantly different from zero. The null hypothesis that one autocorrelation is zero is rejected if its plot is outside the bounds.

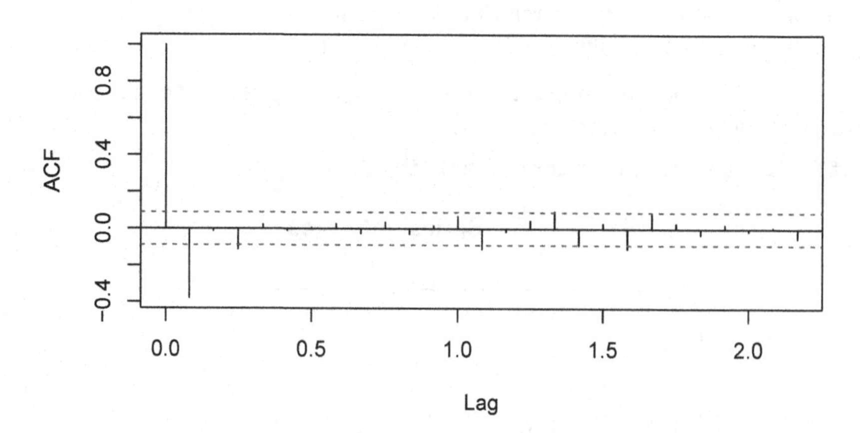

FIGURE 17.10 First-Differences Inflation Rates

Estimating the parameter q of an MA model can be done using the ACF (not the PACF). We notice in Figure 17.10 that the first auto-correlation term is outside of the blue band, so a MA(1) model might be in order. Moreover we observe that the ACF trends toward zero (which satisfies one of the requirements for a sound MA model) and that the values seem to follow an alternating pattern, indicating that the single coefficient of the MA(1) model should probably be negative.

But ACF plots should be inspected simultaneously with the Partial ACFs, or PACFs. Partial autocorrelations that are significantly different from 0 indicate lagged terms of the series that are useful predictors of the value at time t. In other words, time series that can be described by an AR model have theoretical PACFs with non-zero values at the AR terms (the single spike in the PACF plot), and zero elsewhere. (As said earlier, the ACF should trend toward zero in some fashion.) Let's start with inflation rates changes. Its PACF, plotted in Figure 17.11, is obtained using the **`pacf()`** command:

```
pacf(diff(inflRates))
```

(Note that both the ACF and the PACF can be produced in one short using command **`acf2()`** of the **`astsa`** package.)

The multiple spikes (on the negative side) in the PACF Figure 17.11 indicate an AR model of degree at least 1 but up to 6. The 4th point is clearly outside of the dotted blue lines, but points 5 and 6 start to be less significant, so we can consider an AR(4) model. Given that the ACF plot had earlier suggested an MA(1) model and that differencing is already done, an ARIMA(4,0,1) is our best guess so far for the first differences of inflation rates.

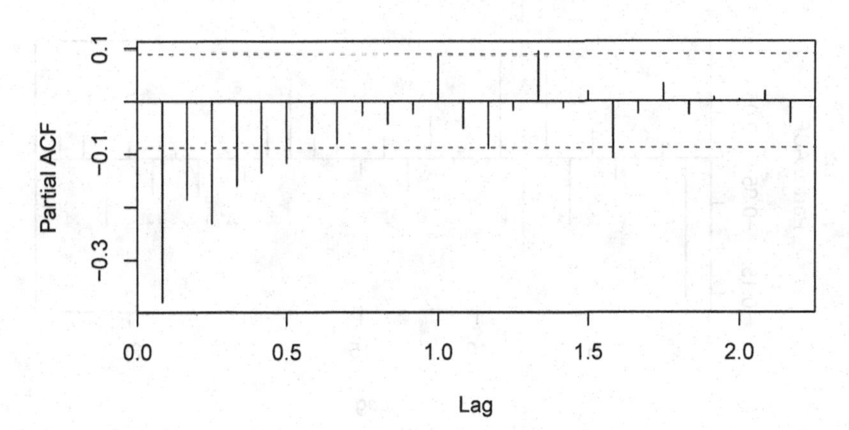

FIGURE 17.11 PACF of First-Differenced Inflation Rates

In the case of Apple, the ACF of the daily returns plotted in Figure 17.6 is calculated as follows, resulting in Figure 17.12:

```
acf(Apple.returns)
```

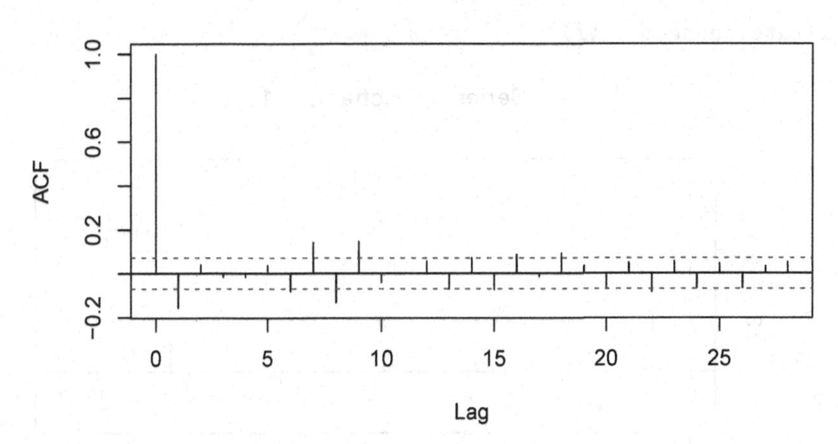

FIGURE 17.12 ACF for Apple's Returns

The first autocorrelation is barely outside of the dotted line so is barely significant. We can ignore it and conclude that a MA model is not relevant here. However, the PACF for Apple's returns is as follows:

```
pacf(Apple.returns)
```

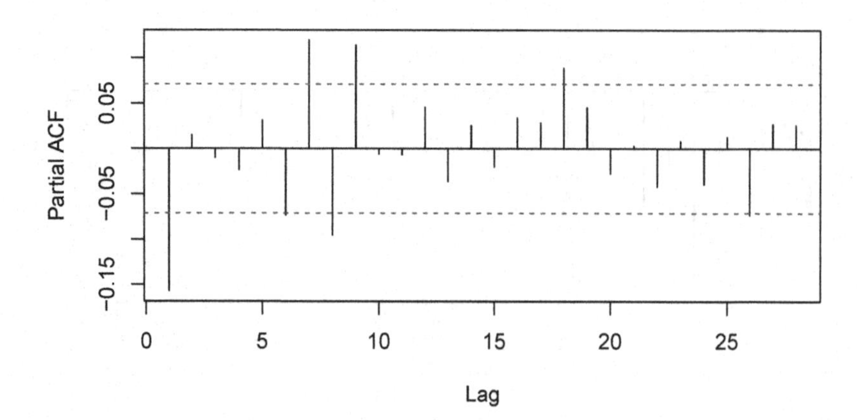

FIGURE 17.13 PACF for Apple's Returns

The first auto-correlation is clearly significant; this suggest an AR(1) model. Equivalently, since a moving average model would be of degree 0, the model can be denoted ARMA(1,0) or ARIMA(1,0,0).

Finally, let's look in Figure 17.14 at the ACF for the changes in 10-year interest rates we had plotted in Figure 17.8:

```
acf(rate.changes.10yr)
```

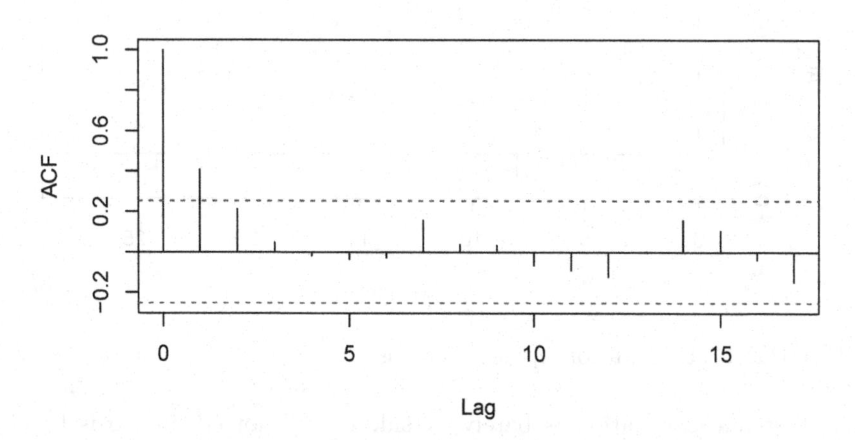

FIGURE 17.14 ACF for changes in 10-year rates

The ACF suggests an MA(1) model. From the ACF plot above, a positive value of that coefficient is in order because the lagged correlation is positive;

had the lagged correlations alternated between positive and negative values while tapering off to zero as lags increase, then a negative coefficient would be in order.

We now look at the PACF, shown in Figure 17.15:

```
pacf(rate.changes.10yr)
```

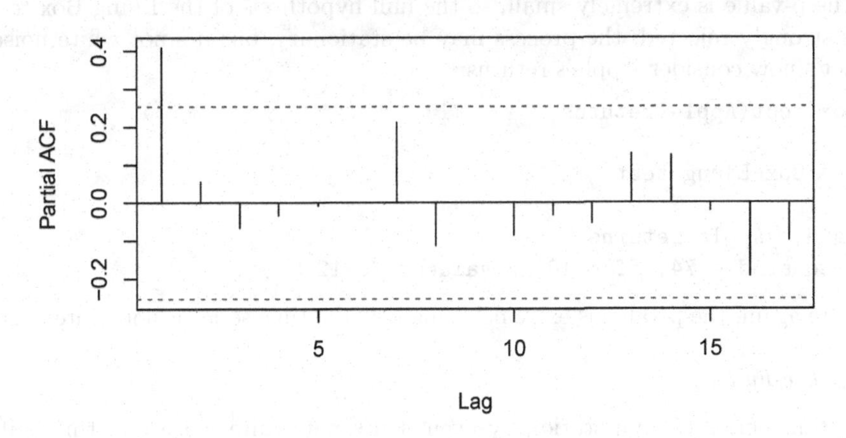

FIGURE 17.15 PACF for Changes in 10-year Rates

The partial autocorrelation of lag 1 seen in Figure 17.15 is significantly outside of the dotted lines but none of the others, so an AR model of order 1 is in order. Our best guess for a model of the daily changes in rates is thus ARMA(1,1), i.e. ARIMA(1,0,1). Moreover, the value of the AR(1) coefficient can be read from Figure 17.15: it is close to 0.4.

Ljung-Box

We have a small problem though... At the usual significance level 5%, one can expect to see about 1 out of 20 autocorrelations outside of the test bounds *simply by chance.* So if the autocorrelations are tested one at time, then there is a high chance of concluding that one or more is nonzero. An alternative to using the bounds to test autocorrelations one at a time is to use a simultaneous test. A **simultaneous test** is one that tests whether a group of null hypotheses are all true, versus the alternative that at least one of them is false. A relevant test here is the **Ljung-Box Test** of white noise.

The null hypothesis of the Ljung–Box test is[1]:

$$H_0 : \rho(1) = \rho(2) = ...\rho(k) = 0$$

[1]You'll notice that R calls it the Box-Ljung test. I don't know the history of that test but I believe the most usual name is Ljung-Box.

for some integer k. We apply it on the first differences on inflation rates:

```
Box.test(diff(inflRates), lag = 10, type = "Ljung-Box")
```

```
	Box-Ljung test

data:  diff(inflRates)
X-squared = 80, df = 10, p-value = 5e-13
```

The p-value is extremely small, so the null hypothesis of the Ljung-Box test is strongly rejected: the process may be stationary, but it's not white noise. Let's now consider Apple's returns:

```
Box.test(Apple.returns, lag = 10, type = "Ljung-Box")
```

```
	Box-Ljung test

data:  Apple.returns
X-squared = 74, df = 10, p-value = 7e-12
```

Here again, the p-value is extremely small, so the time series is not white noise.

Dickey-Fuller

A time series can be stationary even if it's not white noise and thus fails the Ljung-Box test. The **Dickey-Fuller test** is the most common test for stationarity. Its null hypothesis is that the time series is *not* stationary, that is, that some of the roots are greater than 1. This test can be performed using the `adf.test()` function of the `tseries` package:

```
library(tseries)
adf.test(diff(inflRates))
```

```
	Augmented Dickey-Fuller Test

data:  diff(inflRates)
Dickey-Fuller = -13, Lag order = 7, p-value =
0.01
alternative hypothesis: stationary
```

The p-value is relatively low, and actually we are being warned that the actual value is even smaller. According to this test, we can conclude that the time series is stationary. Now let's apply Dickey-Fuller on Apple's returns:

```
adf.test(Apple.returns)
```

```
Warning in adf.test(Apple.returns): p-value smaller
than printed p-value

	Augmented Dickey-Fuller Test
```

```
data:  Apple.returns
Dickey-Fuller = -8.1, Lag order = 9, p-value =
0.01
alternative hypothesis: stationary
```

A low p-value here again allows us to reach the same conclusion.

17.2 But What Are These Autoregressive (AR) and Moving Average (MA) Models?

The simplest autoregressive model is the model of degree 1, denoted AR(1). A process Y_t with constant parameters μ and ϕ is AR(1) if:

$$Y_t = (1 - \phi)\mu + \phi Y_{t-1} + \epsilon_t \tag{17.1}$$

Equation (17.1) is a regression of the Y's against themselves, hence the term "auto-regressive." And specifically, each Y_t at time t is regressed against ("is predicted by", or "is a function of", if you prefer) Y_{t-1} at time $t-1$, hence the order of 1. The intercept of the regression is $(1 - \phi)\mu$ and the slope is ϕ. What is less obvious, is that μ is in fact the mean of the time series Y. Even less intuitively, parameter ϕ determines the amount of feedback from one time step to the next, with a larger absolute value of ϕ resulting in more feedback. Critically:

- The process is stationary if $|\phi| < 1$.
- If $|\phi| > 1$, the process is "explosive" and plotting it will leave no doubt that it is not stationary.
- The case $|\phi| = 1$ is the tricky case, we'll come back to it shortly.

More generally, an *AR*(p) model is such that:

$$Y_t = \beta + \phi_1 Y_{t-1} + .. + \phi_p Y_{t-p} + \epsilon_t$$

How does that relate to the PACFs we discussed earlier? In Figure 17.13, we notice an alternating pattern that trends down to zero, so the AR(1) model has a negative coefficient – more specifically, approximately -0.16 from the graph. Given that the lag-1 autocorrelation is also the coefficient of the lagged variable in an AR(1) model, a model for Apple's return Y_t^{Apple} at time t would look like

$$Y_t^{Apple} = \beta + -0.16 Y_{t-1}^{Apple} + \epsilon_t$$

As an aside, it can be shown that Y_t is stationary if and only if all the roots of the polynomial $1 - \phi_1 x - ... - \phi_p x^p$ have absolute values greater than one.

Let's now turn to MA models. You'll notice that Equation (17.1) has an error term at the end, ϵ_t. That error term is the part not fully explained by simple linear regression, but it still contains information. An **MA model** of order q, denoted MA(q), is a regression of Y_t against those lagged error terms:

$$Y_t = a + \epsilon_t + b_1\epsilon_{t-1} + \ldots + b_q\epsilon_{t-q} \tag{17.2}$$

The ACF autocorrelations relate to the coefficients $b_1..b_q$ in a complicated way. The simplest relationship is for $q = 1$, and in this case the auto-correlation for a lag of 1, ρ_1 is:

$$\rho_1 = \frac{b_1}{1 + b_1^2} \tag{17.3}$$

So, reading ρ_1 from an ACF plot can give us b_1 if we solve for $\rho_1 b_1^2 - b_1 + \rho_1 = 0$. However in practice, we simply use (17.3) to know if b_1 is nonzero, since clearly a null value for ρ_1 (or a value not statistically different from zero) means a null value for b_1. The sign of ρ_1 gives the sign of b_1, too. The relationships for other lags are more complicated but explain why, earlier, we looked at the longest lag for which the ACF was outside the dashed blue lines of statistical significance: That determined the longest lag for which ρ_q and thus b_q would be nonzero, and thus what the appropriate order q of the model would be.

17.3 Fitting a Model

So far, we've seen how to determine the lags p and q of AR and MA models, respectively. But that did not give us the *values* of the $\phi_1, .., \phi_p$ and $b_1, .., b_q$ coefficients. Deriving the numerical values of model parameters is called **model fitting**, and one way to fit an ARIMA model in R is to use functions **arima()** or **auto.arima()**, as we are going to see.

For example, we saw that a good model for Apple's returns would be ARIMA(1,0,0). The **arima()** function gets us the parameter of the AR(1) part of the model:

```
arima(Apple.returns, order=c(1,0,0))
```

```
Call:
arima(x = Apple.returns, order = c(1, 0, 0))

Coefficients:
```

```
         ar1  intercept
      -0.157      0.002
s.e.   0.036      0.001

sigma^2 estimated as 0.000487:  log likelihood = 1820,  aic = -3634
```

The value of the autoregression coefficient is given as -0.1572, which is very close to the value of -0.16 we eyeballed earlier.

Another useful function is **auto.arima()** from the **forecast** package:

```
library(forecast)
```

Calling **auto.arima()**, with the optional **trace=** parameter set to true, produces a long output but doing so on one example will help us understand what's happening. Moreover, the trace is relatively short in the case of Apple's returns.

```
auto.arima(Apple.returns, trace=TRUE)
```

```
 Fitting models using approximations to speed things up...

 ARIMA(2,0,2) with non-zero mean : -3627
 ARIMA(0,0,0) with non-zero mean : -3617
 ARIMA(1,0,0) with non-zero mean : -3633
 ARIMA(0,0,1) with non-zero mean : -3633
 ARIMA(0,0,0) with zero mean     : -3614
 ARIMA(2,0,0) with non-zero mean : -3631
 ARIMA(1,0,1) with non-zero mean : -3631
 ARIMA(2,0,1) with non-zero mean : -3629
 ARIMA(1,0,0) with zero mean     : -3629

 Now re-fitting the best model(s) without approximations...

 ARIMA(1,0,0) with non-zero mean : -3634

 Best model: ARIMA(1,0,0) with non-zero mean

Series: Apple.returns
ARIMA(1,0,0) with non-zero mean

Coefficients:
         ar1   mean
      -0.157  0.002
s.e.   0.036  0.001

sigma^2 estimated as 0.000488:  log likelihood=1820
AIC=-3634   AICc=-3634   BIC=-3620
```

We see that numerous models were tried out, from ARIMA(2,0,2) to ARIMA(1,0,0), but all were considered integrated already (the middle parameter, d, is always 0 in these attempts). Eventually, an ARIMA(1,0,0) model is established as the best option, and its first coefficient is confirmed to be -0.16.

Inflation Rates

Let's now find the parameters of an ARIMA model on the first-differenced inflation rates:

```
arima(diff(inflRates), order=c(4,0,1))
```

```
Call:
arima(x = diff(inflRates), order = c(4, 0, 1))

Coefficients:
        ar1    ar2     ar3     ar4     ma1  intercept
      0.189  0.013  -0.121  -0.019  -0.834      0.000
s.e.  0.066  0.054   0.052   0.054   0.047      0.024

sigma^2 estimated as 8.46:  log likelihood = -1219,  aic = 2452
```

But to verify our intuition on an appropriate model, let's perform an automatic calibration of an ARIMA model on changes in inflation rates:

```
auto.arima(diff(inflRates))
```

```
Series: diff(inflRates)
ARIMA(4,0,1)(2,0,0)[12] with zero mean

Coefficients:
        ar1    ar2     ar3     ar4     ma1   sar1
      0.192  0.012  -0.129  -0.020  -0.839  0.089
s.e.  0.068  0.054   0.053   0.056   0.050  0.049
        sar2
     -0.030
s.e.  0.051

sigma^2 estimated as 8.51:  log likelihood=-1217
AIC=2450   AICc=2450   BIC=2484
```

We see something unexpected: the optimizer finds some seasonality in the data. In this case, we know that this does not make sense so we force the seasonality off:

```
auto.arima(diff(inflRates), seasonal=FALSE)
```

```
Series: diff(inflRates)
```

```
ARIMA(1,0,1) with zero mean

Coefficients:
         ar1      ma1
       0.238  -0.877
s.e.   0.055   0.027

sigma^2 estimated as 8.59:  log likelihood=-1222
AIC=2449   AICc=2449   BIC=2462
```

Changes in 10-year Rates

Earlier, we had estimated that ARIMA(1,0,1) would be the best model for **rate.changes.10yr**. Let's calibrate the parameter values for such a model:

```
arima(rate.changes.10yr, order=c(1,0,1))
```

```
Call:
arima(x = rate.changes.10yr, order = c(1, 0, 1))

Coefficients:
         ar1      ma1  intercept
       0.493  -0.101     -0.005
s.e.   0.235   0.260      0.035

sigma^2 estimated as 0.0247:  log likelihood = 25.82,  aic = -43.65
```

However, the automated procedure finds a different model, ARIMA(1,0,0), as the best fit:

```
auto.arima(rate.changes.10yr)
```

```
Series: rate.changes.10yr
ARIMA(1,0,0) with zero mean

Coefficients:
         ar1
       0.408
s.e.   0.117

sigma^2 estimated as 0.0252:  log likelihood=25.74
AIC=-47.48   AICc=-47.27   BIC=-43.29
```

Note that the value of the autoregression coefficient is indeed very close to 0.4 as we had read on page 333 from Figure 17.15.

17.4 Forecasting

At this point, we think we have a good model for our time series. But the best test for a model is its predictive power – i.e., whether it allows us to perform forward-looking forecasts.

To do that, we are going to redo the calibration of a model on Apple's stock stopping a few days early. This part of the time series (most of it actually) will be the **train set** on which **in-sample** calibration of the model is performed. The rest, the **test set**, will be used to compare the model's predictions to what actually happened in the last few days.

So we start with creating the in-sample data set:

```
number.of.days = length(Apple.returns)
days.out.of.sample = 20
in.sample = Apple.returns[1:(number.of.days-days.out.of.sample)]
```

We then build a model (automatically using **auto.arima()** again) and request a **forecast()** for the 20 days kept aside as the test set. We can then plot these forecasts:

```
model = auto.arima(in.sample)
future.returns = forecast(model,
                          h=days.out.of.sample,
                          level=c(99)) # 99% confidence level
plot(forecast(future.returns))
```

The forecast values shown in Figure 17.16 are hard to read. We can verify that the first few days show some volatility, but then the forecasted values stabilize around the forecasted mean. (But at least, the plot confirms we still have an ARIMA(1,0,0) model!)

To make predictions easier to appreciate, let's plot the predicted **price** of Apple's stock on the out-of-sample days. First, we extract the point estimate of the **returns** at each time step in the column called **mean** (the name of that column can be found using the command **str(future.returns)**). These returns are compounded following Equation (9.1) seen on page 159.

```
compounded.forecasts = 1+future.returns$mean
(compounded.forecasts = cumprod(compounded.forecasts))
```

```
 [1] 1.003 1.005 1.007 1.008 1.010 1.012 1.013 1.015
 [9] 1.017 1.018 1.020 1.022 1.024 1.025 1.027 1.029
[17] 1.030 1.032 1.034 1.036
```

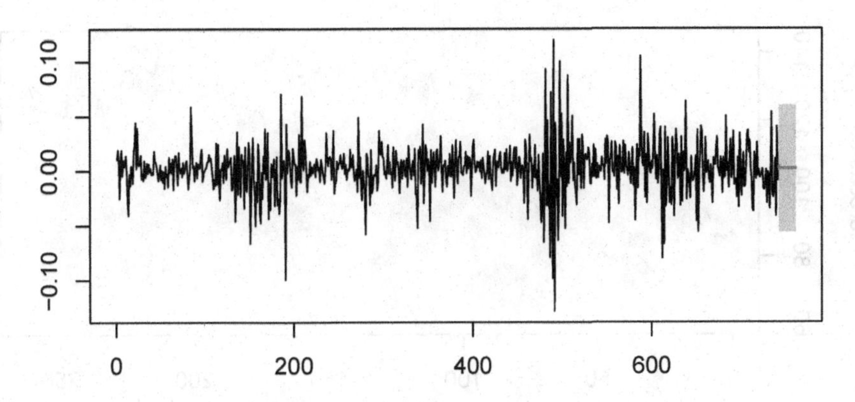

FIGURE 17.16 Out-of-sample Forecasts of Apple's Returns

From these compounded returns, the forecasted stock prices simply equal the value of the stock on the last day of the in-sample data set multiplied by each of the successive compounded returns:

```
apple.vec = as.vector(Apple)
stock.prices.in.sample =
  apple.vec[1:(length(apple.vec) - days.out.of.sample)]
stock.prices.forecasted =
  stock.prices.in.sample[length(stock.prices.in.sample)] *
  compounded.forecasts
```

We are now going to plot two time series: the actual prices over the entire period and, for the final few out-of-sample days, the forecasted values. The output is shown in Figure 17.17.

```
forecasted =
  c(apple.vec[500:(length(apple.vec)-days.out.of.sample)],
    stock.prices.forecasted)
plot(forecasted,
     type="l",
     main="Actual Stock Price (red) vs Prediced (black)")
lines(apple.vec[500:length(apple.vec)], col="red")
```

The black line segment in Figure 17.17 shows that the out-of-sample ups and downs of Apple's stock prices are not correctly predicted, and that their sheer *volatility* isn't either. But the general trend is correct – even though the most recent period, approximately from Day 210 to Day 240, had exhibited a down trend.

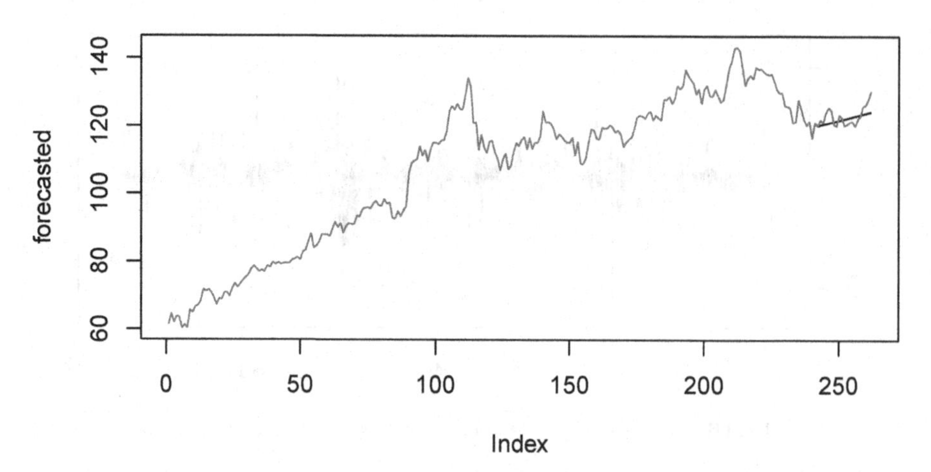

FIGURE 17.17 Apple stock price forecasted by the model

17.5 First Differencing, or Integrated Model?

We have seen the entire "workflow" of modeling time series: We started with a visual inspection, then considered stabilizing transformation, then plotted and interpreted ACFs and PCFs to decide the best ARIMA(p, d, q) model, then calibrated the model's parameters, and finally made and tested predictions. Now that we have done all that, we should come back to a few points. First is the difference between calculating first-differences "ourselves" vs. requesting a model with $d > 0$.

We saw that all 3 time series (inflation rates, Apple's stock price and 10-year interest rates) exhibit trends either up or down (instead of oscillating around a mean) or do not clearly exhibit stationarity. As discussed, one way to remove a linear trend is to use so-called first difference, which is simply to subtract each value from the next one in the time sequence.

This first-differencing is precisely what the **integration** part in ARIMA performs. We can verify that the results of an ARIMA model of degrees (1,1,0) applied on inflation rates, i.e. AR(1), 1st-differencing, and MA(0), are the same as an ARMA(1,0) model, i.e. without integration, applied to the "already-integrated" first-difference series. The former is:

```
arima(inflRates, order=c(1,1,0))
```

```
Call:
arima(x = inflRates, order = c(1, 1, 0))
```

```
Coefficients:
          ar1
       -0.385
s.e.    0.042

sigma^2 estimated as 10.4:  log likelihood = -1268,  aic = 2541
```

And the latter is:

```
arima(diff(inflRates), order=c(1,0,0))
```

```
Call:
arima(x = diff(inflRates), order = c(1, 0, 0))

Coefficients:
          ar1  intercept
       -0.385      0.004
s.e.    0.042      0.105

sigma^2 estimated as 10.4:  log likelihood = -1268,  aic = 2543
```

As you can verify, the results are identical. In particular, the AR1 coefficient is the exact same.

17.6 A Digression: The Intuition of the ACF Values

There's another point we did not elaborate much on: What exactly are ACFs? Where do these values come from?

The answer is: There are almost what their name says, i.e. a series of correlations of a time series with itself, at different lags. Let's consider again Figure 17.9 that plots the ACF for **inflRates**. The first 3 autocorrelations (if we exclude the correlation of the series with itself at lag 0, which of course is 1) are:

```
inflation.autocorrelations$acf[2:4]
```

```
[1] 0.5913 0.5001 0.4194
```

Here's how we can find the autocorrelation at a lag of 1. We first extract the first $n-1$ values and the last $n-1$ values, and regress the latter against the former:

```
n = length(inflRates)
xs = inflRates[1:n-1]
```

```
ys = inflRates[2:n]
cor(xs, ys)
```

```
[1] 0.5942
```

So we are *close* to the ACF value for a lag of 1, but not quite there yet. The correct estimate of autocorrelation for a discrete process with known mean and variance is calculated as follows[2]:

```
mu = mean(inflRates)
sd = sd(inflRates)
sum((xs - mu)*(ys - mu))/((n - 1)*sd^2)
```

```
[1] 0.5913
```

So the intuition is there: each kth value of the ACF is the correlation of the time series with itself, but lagged by k and with an adjustment to make the estimate unbiased.

[2]See, e.g., https://en.wikipedia.org/wiki/Autocorrelation#Estimation.

18

Machine Learning

Machine Learning is the branch of algorithmics devoted to automatically discovering patterns in data and making predictions from these patterns. As I am sure you already know, this field has grown tremendously in popularity in the last decade or so. Given the (already wide) scope of this book, we will have to limit ourselves to the machine learning techniques most frequently used in finance – even if that's a very subjective criterion. In particular, we will limit ourselves to methods where the input data are quantitative. However, keep in mind that machine learning on unstructured data such as text has been applied to finance, for example to the automatic analysis of the frequency of certain words in the quarterly filings of corporations. On the other hand, the output (the prediction) of the algorithms we will investigate can be categorical or continuous. In the group producing categorical outputs, **classification algorithms** decide to which possible discrete category (which **class**) each input data point corresponds to. We are going to see five such classification methods: KNN, logistic regression, decision trees, regression trees and K-means clustering.

18.1 Supervised Algorithms

When an algorithm learns from **train data** (also called **in-sample** data) before being vetted on **test data**, it is called a **supervised** algorithm. KNN, logistic regressions and decision trees are examples of supervised algorithms. (In contrast, **unsupervised** algorithms detect patterns or similarities –for example, based on "closeness"– and continuously improve their pattern detection as they are fed more data.)

Let's say that a supervised algorithm has to classify data into "Positive" (the data point satisfies a certain condition) or "Negative." Once the algorithm has learned from (we often say it was **calibrated** on) the training data set, it is tested **out-of-sample** on the test data set. Since we know the correct answers for the test data, we can measure the performance of the algorithm: we can assess how well it has learned on the train data to make correct predictions on the test data. Specifically, we count how many predictions were **false**

DOI: 10.1201/9781003320555-18

negatives or **false positives** – what percentage of test data is incorrectly labeled as "Negative" when they should be "Positive," or inversely.

These statistics on the merits of a classification algorithm are typically summarized in a **confusion matrix**. A confusion matrix is a 2×2 table showing, in each column, how many data points were predicted to be (classified as) Positive or Negative and, in each row, how many of these data points are actually Positive or Negative. So a confusion matrix will look like this:

	Predicted	
Actual	Positive	Negative
Positive	12	2
Negative	3	18

In this example, 35 data points were in the test data set. 15 were classified as Positive by the algorithm: 12 correctly so, and those are called **true positives**; their count is denoted TP. 3 were incorrectly predicted as Positives, so they are **false positivess**, which we denote with $FP = 3$; 20 were classified as Negative: 18 correctly (**true negative**, and $TN = 18$) and 2 incorrectly (**false negative**, and $FN = 2$).

In R, a confusion matrix can be built quickly using the `table()` function seen on page 43. Imagine your classification algorithm predicted values as A or B, and your output had five data points:

```
predicted.values = c("A","B","A","A","B")
actual.values    = c("B","B","A","A","A")
```

We see that 2 actual A's were correctly predicted as A's so, if we decide (it's only a convention after all) that A's are the positives, $TP = 2$. But one A got classified as a B, so $FN = 1$, and a B got incorrectly classified as an A, so $FP = 1$. Finally, one B was correctly labeled, so $TN = 1$. But the confusion matrix can be constructed as simply as in the following command:

```
table(actual.values, predicted.values)
```

```
             predicted.values
actual.values A B
            A 2 1
            B 1 1
```

Or, as yet another example:

```
predicted.values = c("Positive","Negative","Positive","Positive")
actual.values    = c("Negative","Negative","Positive","Positive")
table(actual.values, predicted.values)
```

```
             predicted.values
actual.values Negative Positive
     Negative        1        1
```

```
Positive          0        2
```

Note that the row and column order is the alphabetical order, so "Negative" is the left column and "Positive" is the right one, like "A" came before "B" in the earlier example.

Now, to assess the quality of the classification, we define three metrics – **sensitivity, specificity** and **accuracy**:

$$\text{sensitivity} = \frac{TP}{TP + FN} \tag{18.1}$$

$$\text{specificity} = \frac{TN}{TN + FP} \tag{18.2}$$

$$\text{accuracy} = \frac{TP + TN}{TP + TN + FP + FN} \tag{18.3}$$

Based on the alphabetical order of Negative then Positive, False Negatives are therefore reported on the second row and in the first column; so, given a confusion matrix **cm**, FN equals **cm[2,1]**. The three other numbers can be found easily, and sensitivity, specificity and accuracy can be implemented as the following functions:

```
sensitivity = function(cm) {
  return(cm[2,2]/(cm[2,2]+cm[2,1]))
}
specificity = function(cm) {
  return(cm[1,1]/(cm[1,2]+cm[1,1]))
}
accuracy = function(cm) {
  return((cm[1,1]+cm[2,2])/(cm[1,1]+cm[1,2]+cm[2,1]+cm[2,2]))
}
```

We are now ready to discuss specific algorithms. To compare their performance, we will apply them on the same data, specifically the `MSCI AC World.csv` data set.

18.2 KNN

The **KNN** algorithm is a supervised classifier algorithm that decides whether a data point belongs to one of two categories based on the majority of the k already-known data points that are closest to the new data point. To make sure we always can get a majority, k is always odd.

The first obvious question to investigate is what "closest" means. The input data points are vectors of numerical values, for example different accounting metrics on a company. So we can define "distance" between two data points (two stocks, for example) as the distance between, say, their price-to-book and their market capitalizations.

How do we define distance? Quite simply: it's the **Euclidean distance**. In general, if your data set has two columns A and B and the values in these columns for two records x and y are denoted x_A, x_B, y_A and y_B then the Euclidean distance between the two records is:

$$d(x, y) = \sqrt{(x_A - y_A)^2 + (x_B - y_B)^2}. \tag{18.4}$$

However, we still have a problem: if A is price-to-book and B is market cap, then x_A and y_A are unit-less numbers in the few ten's maybe and x_B and y_B are in dollars with magnitudes going to the trillions. How can we add them together? For (18.4) to make sense, the data have to be normalized (z-scored) before calculating their distance.

One might ask also what the best k is. There is no general answer, but consider the two extremes: If $k = N$, then there is one big group, so the model is under-fitted. And inversely, if $k = 1$, then the response is influenced by a single input, so the model is over-fitted. I sometimes hear suggestions that k should be close to the square root of the total number of observations in the data set, but that doesn't make sense for large data sets. You are better off trying a few reasonable values and see which one gives you the best in-sample specificity, sensitivity and accuracy.

The KNN library we will use is the **class** package:

```
library(class)
```

We first read and inspect the data:

```
stocks = read.csv("ACWI.csv")
```

```
str(stocks)
```

```
'data.frame':   2974 obs. of  13 variables:
 $ Ticker           : chr  "AAPL" "MSFT" "AMZN" "FB" ...
 $ Name             : chr  "Apple" "Microsoft" "Amazon" "Facebo
 $ GICS.Sector      : chr  "Information Technology" "Informatio
 $ Country          : chr  "United States" "United States" "Uni
 $ Ending.Weight    : num  3.47 2.81 2.29 1.21 1.11 1.1 0.83 0.
 $ Market.Cap.      : chr  "2,193,582" "1,899,924" "1,744,112"
 $ Dividend.Yield   : chr  "0.62" "0.85" "--" "--"  ...
 $ Price.to.Cash.Flow: chr  "30.73" "31.17" "33.63" "22.64" ...
 $ Price.to.Book    : chr  "33.40" "14.13" "21.03" "7.22" ...
 $ Price.to.Sales   : chr  "8.41" "13.55" "4.58" "10.92" ...
```

```
$ ROE                : chr  "73.69" "40.14" "27.44" "25.42" ...
$ ReturnPast52Weeks : chr  "78.98" "40.72" "40.15" "58.80" ...
$ Buy                : logi  TRUE FALSE FALSE TRUE FALSE FALSE
```

The basis of the learning is the **Buy** column: each stock is labeled a "buy" if it did better than the mean of all stocks in that period. (The average return of all ACWI stocks in that period was 52.22%.) That column represents the decision we want the "machine" to "learn."

We then perform the usual cleaning and conversions to numerical values using **as.numeric()**, as we did for example on page 9. The cleaning includes removing all rows with **NA**s thanks to the **na.omit()** function.

The data must be normalized, which means that the z-score of each variable must be computed. As you remember, the **z-score** of a value is the distance of that value from its mean in units of standard deviation, so it is equal to the difference between its value and the mean, divided by the standard deviation. We can do this in one step using the **scale()** function of the **dpylr** library:

```
stocksDividend.Yield       = scale(stocks$Dividend.Yield)
stocksPrice.to.Cash.Flow = scale(stocks$Price.to.Cash.Flow)
stocksPrice.to.Book        = scale(stocks$Price.to.Book)
stocksPrice.to.Sales       = scale(stocks$Price.to.Sales)
stocksROE                  = scale(stocks$ROE)
stocksReturnPast52Weeks   = scale(stocks$ReturnPast52Weeks)
```

We then take a random sample of the dataset to be used as the training data. The **sample()** function makes our life easier: it creates a vector of random values. Here, we request to randomly pick either the number 1 or the number 2 as many times as there are records in **stocks**. The fact we have two choices is specified by the first argument of **sample()**, and the number of random picks is specified by the second parameter.

```
ind = sample(2, nrow(stocks), replace=TRUE, prob=c(0.75, 0.25))
```

The output of **sample()** will decide whether each element of **stocks** should be included in the training set (when **sample()** returns 1) or in the test data set (when **sample()** returns 2). We have to specify with what probabilities each element falls in what bucket – above, we specified that there is a 75% chance that a 1 turns up and a 25% probability to get a 2.

Based on the index **ind**, we can then create two data subsets, the training and test sets, with the labels (the correct classes) being kept separately:

```
stocks.training    = stocks[ind==1, 7:11]
stocks.test        = stocks[ind==2, 7:11]
stocks.trainLabels = stocks[ind==1, 13]
stocks.testLabels  = stocks[ind==2, 13]
```

We can then use the **knn()** function from the **class** package:

```
prediction = knn(train = stocks.training,
                 test  = stocks.test,
                 cl    = stocks.trainLabels,
                 k     = 3)
```

The results are in the variable we called **prediction**. We then build the corresponding **confusion matrix**:

```
(confusionMatrix = table(Actual_Value = stocks.testLabels,
                         Predicted_Value = prediction))
```

```
            Predicted_Value
Actual_Value FALSE TRUE
       FALSE   300  103
       TRUE    123   76
```

And we calculate the summary metrics:

```
sensitivity(confusionMatrix)
```

```
[1] 0.3819
```

```
specificity(confusionMatrix)
```

```
[1] 0.7444
```

```
accuracy(confusionMatrix)
```

```
[1] 0.6246
```

Specificity and accuracy are pretty good. Sensitivity less so.

However, we don't know if a portfolio invested in all the stocks as buys would have beaten investing blindly in all the stocks of the test dataset, on average. We first extract the stocks that were in the test set:

```
pred = as.logical(prediction)
stocks.testset = stocks[ind==2,]
```

Then we calculate the average return for those that were predicted as buys:

```
returns.prediction = stocks.testset$ReturnPast52Weeks[pred]
mean(returns.prediction)
```

```
[1] 60.12
```

In contrast, the average (hence, equally weighted) return of the stocks in the whole test set is significantly lower, confirming the efficacy of the algorithm:

```
all.returns = stocks.testset$ReturnPast52Weeks
mean(all.returns)
```

```
[1] 45.23
```

Note that the return of the test set, 45.23%, does not have to equal that of entire ACWI benchmark.

18.3 Logistic Regression

Logistic regressions are also classifiers, so their output is also a discrete variable, i.e., a category. But its salient feature is that the output is decided based on a continuous probability function calibrated to the input data set.

For example, let's assume we are interesting in trying to understand how a portfolio manager picks their stocks. We suspect that dividend yield is the main criterion but we'd like to test that assumption by predicting if, based on our knowledge of which other stocks are or are not in the portfolio, a new stock would be included. The intuition of logistic regression is that the binary in-sample information we have (whether each stock is in the portfolio or not, and the dividend yield of each stock) is smoothed into a probability curve that depends on dividend yield. Any new stock can thus be plotted on the probability curve given its dividend yield. Finally, if the corresponding probability is greater than a certain threshold, we hypothesize that the stock would be in the portfolio.

Let's see that concretely: We have information on 18 stocks whose dividend yields range from 2% to 8%. Six stocks are in the portfolio, which is indicated by a 1 in the **portf** vector below. Vector **div.yld** gives the dividend yields of each stock. From that information, can we tell how the dividend yield of a stock affects the probability that this stock will be included in the portfolio?

```
div.yld = c(2.03, 2.55, 2.79, 3.12, 3.31, 3.50, 3.75, 3.75, 4.25,
            4.32, 4.75, 4.96, 5.72, 5.89, 6.67, 6.85, 7.67, 8.12)
portf   = c(0,    0,    0,    0,    0,    0,    0,    0,    1,
            0,    1,    0,    0,    0,    1,    1,    1,    1)
```

We use the **glm()** function of the **stats** package with the **binomial** method to build the logistic model of the value of **portf** as a function of **div.yld**. Note the tilde operator, and that the two variables are combined in a dataframe we call **stock.picks**:

```
stock.picks = data.frame(div.yld, portf)
model = glm(portf ~ div.yld,
            data   = stock.picks,
            family = binomial)
```

The **model** that got created is a function mapping dividend yields to a probability of inclusion in the portfolio. We will see shortly how to interpret the coefficients:

```
model
```

```
Call:  glm(formula = portf ~ div.yld, family = binomial,
           data = stock.picks)

Coefficients:
(Intercept)      div.yld
      -6.99         1.26
```

To see what that function looks like, we are going to apply it, using **predict()**, to 500 possible dividend yields ranging from 2 to 9 (percent) :

```
newdata         = data.frame(div.yld = seq(2, 9, len=500))
newdata$portf = predict(model, newdata, type="response")
```

We can now plot these predicted values to trace the logistic regression curve. The code appears below and the plot in Figure 18.1.

```
plot(portf ~ div.yld, data=stock.picks)
lines(portf ~ div.yld, newdata)
```

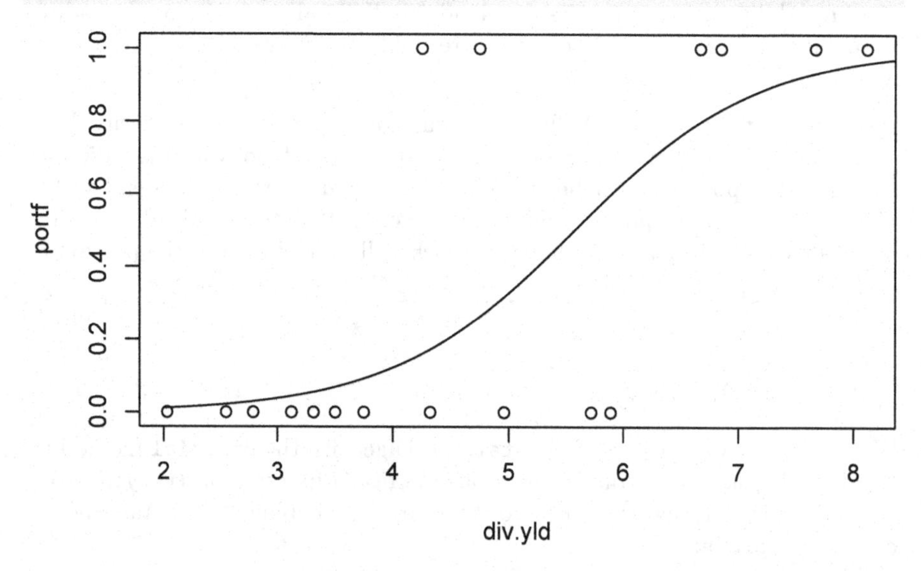

FIGURE 18.1 Plot of the logistic curve

Mathematically, the smooth curve in Figure 18.1 is a sigmoid, and more precisely it is the plot of a logistic function. The standard logistic function is defined as

$$\sigma(t) = \frac{1}{1 + e^{-t}}.$$

However, t can be a linear function of a single (or more!) explanatory variable x, so $t = a + bx$ and the generalized logistic function is:

$$p(x) = \sigma(a + bx) = \frac{1}{1 + e^{-a-bx}}.$$

Note that both σ and p have values in $(0, 1)$.

The output provided the coefficients for Intercept = -6.99 and div.yld = 1.26. If d is the dividend yield, then the probability p of being selected by the PM and be in her portfolio:

$$p = 1/(1 + e^{(-(-6.99+1.26\times d))})$$

These coefficients are entered in the logistic regression equation to estimate probability p. For example, if a stock has a dividend yield of 3%, entering the value $d = 3$ in the equation gives the estimated probability of being in the portfolio:

```
d = 3
(p = 1/(1+exp(-(-6.99 + 1.26 * d))))
```

```
[1] 0.03879
```

Let's work on a more comprehensive and more realistic application based on our running ACWI example. We prepare the training data set with an explicit column for the correct decisions (labels):

```
stocks.training$Buy.or.Sell = stocks.trainLabels
```

Just to clarify where we are standing at this point, let's look at the first 3 records of this training data set:

```
head(stocks.training, 3)
```

```
  Dividend.Yield Price.to.Cash.Flow Price.to.Book
1           0.62              30.73         33.40
2           0.85              31.17         14.13
8           1.67              18.04          8.41
  Price.to.Sales   ROE Buy.or.Sell
1           8.41 73.69        TRUE
2          13.55 40.14       FALSE
8          11.62 29.84        TRUE
```

A logistic model is then established using the **glm()** function:

```
model = glm(Buy.or.Sell ~ Dividend.Yield +
                          Price.to.Cash.Flow +
                          Price.to.Book +
                          Price.to.Sales +
```

```
                    ROE,
                    data=stocks.training,
                    family='binomial')
```

We now predict on the test set using the **predict()** function of the **stats** package:

```
prediction = predict(model, stocks.test, type='response')
stocks.test$predicted = ifelse(prediction>0.4, TRUE, FALSE)
```

And we calculate the confusion matrix:

```
(confusionMatrix = table(Actual_Value = stocks.testLabels,
                         Predicted_Value = prediction>0.4))
```

```
            Predicted_Value
Actual_Value FALSE TRUE
       FALSE   314   89
       TRUE    103   96
```

We can calculate the usual metrics on this confusion matrix:

```
sensitivity(confusionMatrix)
```

```
[1] 0.4824
```

```
specificity(confusionMatrix)
```

```
[1] 0.7792
```

```
accuracy(confusionMatrix)
```

```
[1] 0.6811
```

We may want to try a different threshold though. Let us make it 0.3 now:

```
stocks.test$predicted03 = ifelse(prediction>0.3, TRUE, FALSE)
```

The confusion matrix is then:

```
(confusionMatrix = table(Actual_Value = stocks.testLabels,
                         Predicted_Value = prediction>0.3))
```

```
            Predicted_Value
Actual_Value FALSE TRUE
       FALSE   170  233
       TRUE     37  162
```

And the summary statistics are:

```
sensitivity(confusionMatrix)
```

```
[1] 0.8141
```

```
specificity(confusionMatrix)
```

```
[1] 0.4218
```

```
accuracy(confusionMatrix)
```

```
[1] 0.5515
```

We see that lowering the threshold increased sensitivity but decreased specificity.

However, we don't know if the return of a portfolio invested in all the predicted stocks would have beaten investing blindly and in equal amounts in all the stocks of the test dataset.

```
returns.prediction =
  stocks.testset$ReturnPast52Weeks[stocks.test$predicted]
mean(returns.prediction)
```

```
[1] 70.03
```

In contrast, the average (hence, equally weighted) return of the stocks in the test set is:

```
all.returns = stocks.testset$ReturnPast52Weeks
mean(all.returns)
```

```
[1] 45.23
```

Moreover, we can assess the sensitivity of the average return to a lower threshold:

```
returns.prediction03 =
  stocks.testset$ReturnPast52Weeks[stocks.test$predicted03]
mean(returns.prediction03)
```

```
[1] 55.22
```

In this example, lowering the threshold would have produced a lower return, but still a higher return than the average stocks in the test set.

18.4 Decision Tree

Intuitively, a **decision tree** present exclusive alternatives reached after a succession of conditions. More technically, it's a recursive partitioning of the space of possible values of the input variables. Each node is a condition on the value of one of the columns, resulting of a split of the universe of possible values. The partitioning has to be exhaustive, meaning that each possible combination of input values has to lead to a conclusion at one of the tree's leaves. So a decision tree is quite an intuitive concept, but how to build that tree is the tough part.

At each node, the split is made (based on both the choice of an independent variable and the choice of a threshold value for that variable) that results in the subsets being most homogeneous. Perfect homogeneity here would mean that the nodes after the split are either all buys or all sells. Homogeneity can be defined in many ways, one of them being the opposite of what is called the **Gini Impurity**.

To build a decision tree, we will use the **rpart** package:

```
library(rpart)
```

This package provides, in particular, the **rpart()** function:

```
model = rpart(Buy.or.Sell ~
                Dividend.Yield +
                Price.to.Cash.Flow +
                Price.to.Book +
                Price.to.Sales +
                ROE,
              data=stocks.training,
              control=rpart.control(maxdepth=3),
              method='class')
```

As a detail, we directed **rpart()** to treat **Buy.or.Sell** as a categorical variable by asking for **method='class'**. Since **Buy.or.Sell** is a factor this would have been the default choice anyway.

> **Technical Detail**
> You may have noticed the **rpart.control()** part. As it name says, it provides additional control parameters that tune the details of the algorithm. In this example, the maximum depth of the tree can be set by the **maxdepth=** parameter – here, a maximum depth of 3. You can also specify the minimum number of data points at a node to attempt a further split using the **minsplit=** parameter. You can also specify that a split must decrease the overall lack of fit by a certain factor before being attempted; that "lack of fit," which is beyond the scope of this textbook, is called the **cost complexity factor** and can be set using the **cp=** parameter.

Let's look at the tree built by **rpart()** on the training dataset:

```
model
```

```
n= 1708
```

```
node), split, n, loss, yval, (yprob)
      * denotes terminal node

1) root 1708 603 FALSE (0.6470 0.3530)
  2) Dividend.Yield>=2.705 579 110 FALSE (0.8100 0.1900) *
  3) Dividend.Yield< 2.705 1129 493 FALSE (0.5633 0.4367)
    6) Price.to.Book< 4.195 667 234 FALSE (0.6492 0.3508) *
    7) Price.to.Book>=4.195 462 203 TRUE (0.4394 0.5606) *
```

Starting from the top of the output, we learn that there are 1,708 cases categorized by the tree. We are then reminded of the syntax of the tree description: each line starts with a node number **node** (a node being a branching point in the tree) followed by the criterion that had to be satisfied to reach that node (**split**), the number of cases that reached that node (**n**), the number of cases that were mis-classified (**loss**), and the class that would be assigned at this node (**yval**). The classification is not final though until we reach a terminal node (a leaf of the decision tree), and terminal nodes are highlighted using the * asterisk symbol[1].

We then have the actual description of the tree. Node 1 is the starting point of the tree, called the **root**. All 1,708 cases go through that node of course. If the decision tree had to make a decision at this point, it would label them all as FALSE. The number following 1708 is the number of error that would have been made if the tree stopped there. We can verify that number from the total of stocks labeled as "buys":

```
length(stocks.training$Buy.or.Sell[
  stocks.training$Buy.or.Sell==TRUE])
```

```
[1] 603
```

Node 1 is followed by Nodes 2 and 3, and as you can verify in the rest of the output, the children nodes of Node n are numbered $2n$ and $2n + 1$ to make things easier. For instance, Node 3 is indeed followed by two rows, slightly indented to make reading easier, for Nodes 6 and 7.

This is clear enough, certainly, but a more entertaining visualization of the decision tree is provided by **rpart.plot()** of the **rpart.plot** package, shown in Figure 18.2.

```
library(rpart.plot)
rpart.plot(model, box.palette="RdBu", shadow.col="gray", nn=TRUE)
```

Figure 18.2 is of course consistent with our description of the content of **model**.

[1]We will not make use in this book of the two numbers that follow, called "yprob".

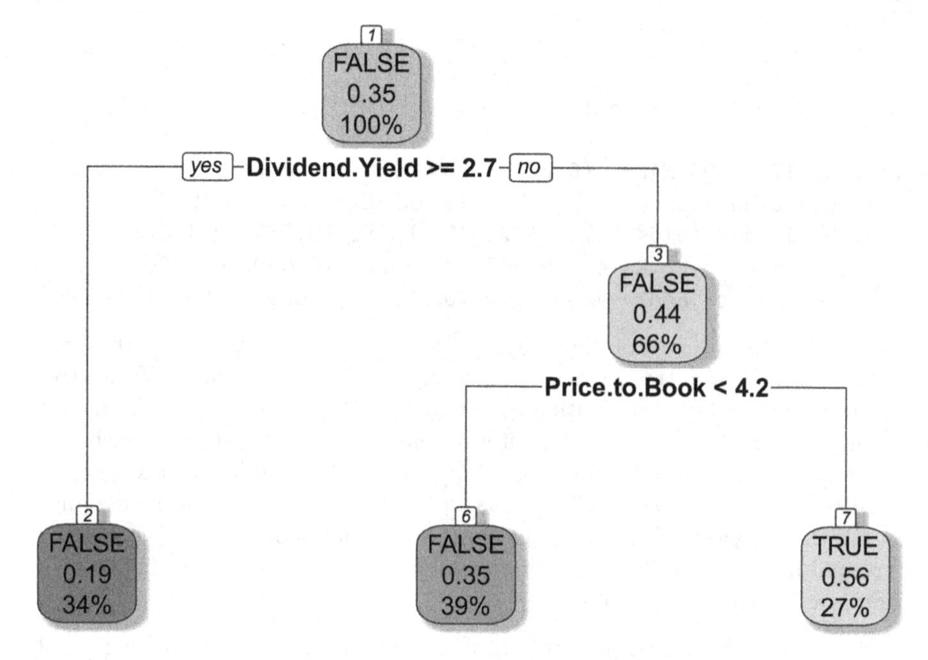

FIGURE 18.2 Visualization of a decision tree using the rpart.plot package

Now that the decision tree has been constructed based on **in-sample data**, we apply it to the *new* **out-of-sample** data, i.e., to the test data set:

```
prediction = predict(model, stocks.test, type='class')
```

We then construct the usual confusion matrix:

```
(confusionMatrix = table(Actual_Value = stocks.testLabels,
                         Predicted_Value = prediction))
```

```
            Predicted_Value
Actual_Value FALSE TRUE
       FALSE   338   65
       TRUE    118   81
```

The usual statistics are as follows:

```
sensitivity(confusionMatrix)
```

```
[1] 0.407
```

```
specificity(confusionMatrix)
```

```
[1] 0.8387
```

```
accuracy(confusionMatrix)
```

```
[1] 0.696
```

These numbers are in-line with those for the other machine learning examples we saw earlier.

The above, however, was based on a simple binary decision whether to invest into a stock or not based on the actual portfolio. But did that portfolio invest in good stocks in the first place? Let's compare the returns that our model would have produced with that of the portfolio. The simple average of the returns of the stocks picked by the algorithm can be calculated as follows:

```
pred = as.logical(prediction)
returns.prediction = stocks.testset$ReturnPast52Weeks[pred]
mean(returns.prediction)
```

```
[1] 75.47
```

The average returns of stocks in the test set is the same as before, since we used the same test set. As a reminder, it is materially lower:

```
all.returns = stocks.testset$ReturnPast52Weeks
mean(all.returns)
```

```
[1] 45.23
```

18.5 Regression Trees (Supervised)

Regression trees are classification algorithms but they are a bit different from the others in that their classes are defined by the algorithm and in that classes represent discrete approximation of a numerical, continuous value.

This is relevant to finance, for example if we want to predict returns. Using a regression tree, only a finite number of returns will be possibly predicted, and each of these discrete return options define a class.

Let's see how it works on our running example. Based on the same index `ind`, we extract the same information plus a categorical variable, which is the GICS sector (GICS sectors were introduced page 12.) Note that we are now including the 6th column of the `stocks` dataset:

```
stocks.training    = stocks[ind==1, c(3,6:11)]
stocks.test        = stocks[ind==2, c(3,6:11)]
stocks.trainLabels = stocks[ind==1, 12]
stocks.testLabels  = stocks[ind==2, 12]
```

```
stocks.training$Return = stocks.trainLabels
```

```
library(rpart)
model = rpart(Return ~
                GICS.Sector +
                Dividend.Yield +
                Price.to.Cash.Flow +
                Price.to.Book +
                Price.to.Sales +
                ROE,
              data=stocks.training,
              control=rpart.control(maxdepth=5,minsplit=30),
              method='anova')
```

The tree can be graphed using **rpart.plot()**, as before, resulting in Figure 18.3.

```
library(rpart.plot)
rpart.plot(model, box.palette="RdBu", shadow.col="gray", nn=TRUE)
```

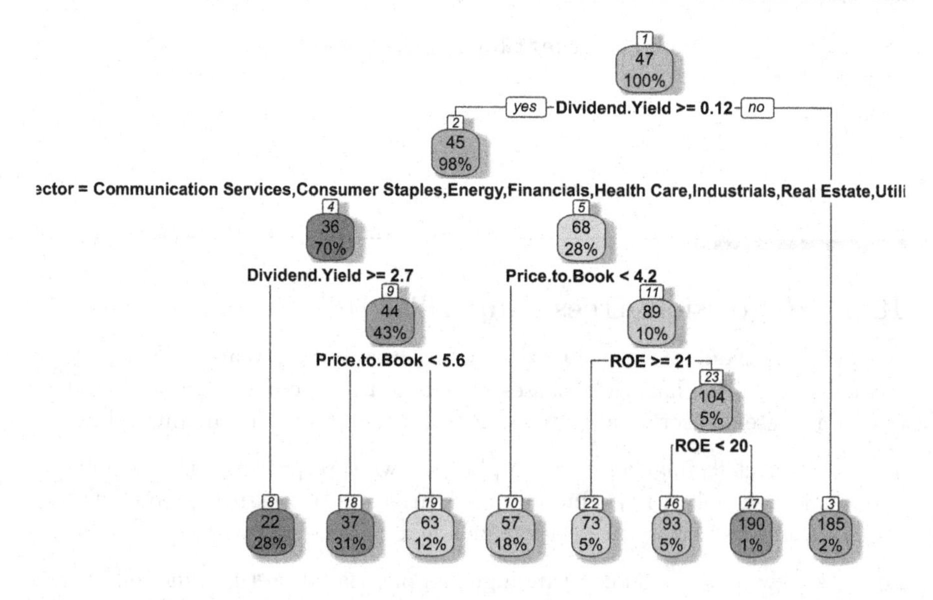

FIGURE 18.3 Regression tree built using the ANOVA method of rpart()

A question of taste: some people find that the decision tree produced by the **prp()** function, as shown in Figure 18.4, is easier to read.

```
prp(model)
```

We now apply the decision tree to *new* data, those of the test data set:

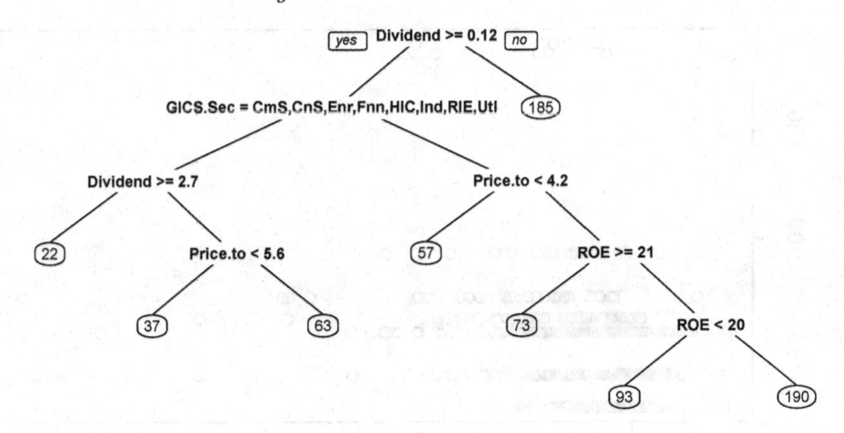

FIGURE 18.4 Decision tree produced by prp()

```
prediction = predict(model, stocks.test)
```

How good is the model?

```
cor(prediction, stocks.testLabels)
```

```
[1] 0.3368
```

Well, the correlation is not great. We can have a better understanding of what is happening if we plot the predicted values against the actual values in the test dataset. The code below results in Figure 18.5.

```
plot(stocks.testLabels, prediction)
```

Figure 18.5 makes it clear that only a finite number of values are terminal values in the nodes of the decision tree, and the largest was 190% while, in contrast, a handful of stocks had returns of 300% or more and thus could not be predicted correctly by the decision tree, reducing the correlation.

18.6 K-Means Clustering

K-means clustering is an **unsupervised** method. K here is the number of clusters that the data will be grouped in. This grouping packs points that are close to each other as possible and such that the distance[2] between (the centers of) groups is as large as possible. In other words, K-means clustering minimizes

[2]The Euclidean distance, as for KNN.

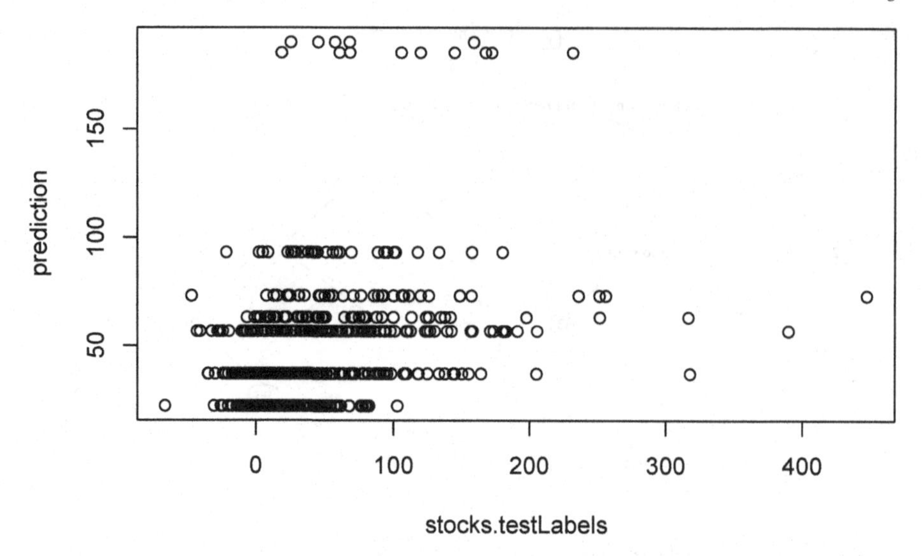

FIGURE 18.5 Plotting the values predicted by the decision tree vs actual values

the within-group dispersion and maximize the between-group dispersion. The algorithm is as follows:

1. Pick K data points at random as the centers of the K clusters.
2. For each other point, calculate their distances to each of the centers; assign that point to the cluster with closest center.
3. Calculate the center (means) of each cluster, and redo the previous step until there are no more changes.
4. Calculate the within-cluster sum of squared distances:

- For each cluster C_j with center m_j, the within-cluster dispersion is

$$W(C_j) = \Sigma_{x \in C_j}(x - m_j)^2$$

- For the entire clustering, the **Total within-cluster sum of squares** defined as:

$$Total.Within.SS = \Sigma_{1 \leq j \leq K} W(C_j)$$

  and often abbreviated as **WSS**. It is a measure of the total variance in the dataset that is explained by the clustering, and you want this number to be as low as possible – remember that K-Means Clustering minimizes the within-group dispersion and maximize the between-group dispersion.

5. Redo all the above N times. By default, $N = 10$ in the **`kmeans()`** function we are going to use.

Let's apply K-means clustering on the constituents of the S&P 500 index (as of Q2'21):

```
sp500 = read.csv("Characteristics Overview.csv")
```

```
sp500$Dividend.Yield     = as.numeric(sp500$Dividend.Yield)
sp500$Total.Return.YTD = as.numeric(sp500$Total.Return.YTD)
sp500$Price..Earnings  = as.numeric(sp500$Price..Earnings)
sp500$Price..Sales     = as.numeric(sp500$Price..Sales)
sp500$Price...Book     = as.numeric(sp500$Price...Book)
sp500$Price..Cash.Flow = as.numeric(sp500$Price..Cash.Flow)
sp500$ROE              = as.numeric(sp500$ROE)

sp500 = na.omit(sp500)
```

As we are now used to, we z-score each of these metrics to make them "unitless" and of the same scale:

```
sp500$Dividend.Yield     = scale(sp500$Dividend.Yield)
sp500$Total.Return.YTD = scale(sp500$Total.Return.YTD)
sp500$Price..Earnings  = scale(sp500$Price..Earnings)
sp500$Price..Sales     = scale(sp500$Price..Sales)
sp500$Price...Book     = scale(sp500$Price...Book)
sp500$Price..Cash.Flow = scale(sp500$Price..Cash.Flow)
sp500$ROE              = scale(sp500$ROE)
```

To make the graphs easier to read, we will focus on just three sectors as defined by GICS, introduced on page 12: Utilities, Energy and Technology. We easily narrow down our dataset using the **filter()** function:

```
three.sectors = sp500 %>%
  filter(GICS.Sector == "Utilities" |
         GICS.Sector == "Energy" |
         GICS.Sector == "Information Technology")
```

Actually, because our charts are going to be a big crowded, let's shorten the sector names using **case_when()**:

```
three.sectors$GICS.Sector = case_when(
  three.sectors$GICS.Sector == "Utilities" ~ "Ut",
  three.sectors$GICS.Sector == "Energy" ~ "En",
  three.sectors$GICS.Sector == "Information Technology" ~ "IT"
)
```

We also focus on two parameters for now: **dividend yield**, defined on page 10, and **return on equity** or **ROE**, also defined on page 11. Plotting this scatter graph using **ggplot** results in Figure 18.6.

```
ggplot(three.sectors,
       aes(x = Dividend.Yield, y = ROE, color = GICS.Sector)) +
  geom_point()
```

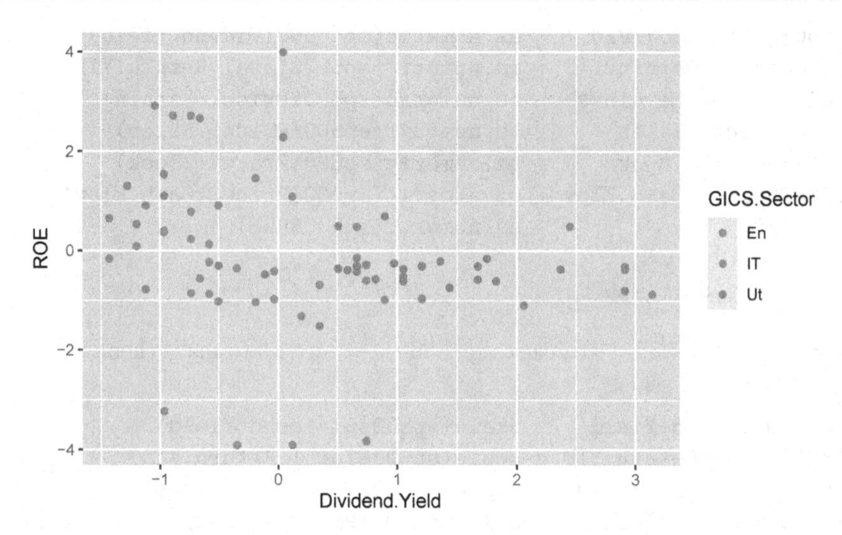

FIGURE 18.6 Scatter graph of dividend yield versus return on equity in 3 sectors of the SP 500 index

Inspecting Figure 18.6, it is visually obvious that IT stocks tend to be grouped in the top left quadrant of the graph, Utilities in the center right, and Energy at the bottom. But K-Means clustering doesn't know that of course, and we will provide the names of the sectors corresponding to each data point only to make scatter plots easier to read. We prepare the dataframe with two columns containing the dividend yields and the ROE values, and label each row with the sector name.

```
data = three.sectors[c("Dividend.Yield","ROE")]
data = as.data.frame(data, row.names = three.sectors$GICS.Sector)
```

A key question now is to decide how many clusters we want the algorithm to categorize the data into. A tempting answer is of course 3, since we know in this application that there are three sectors. But in other applications, we won't know. And even in this case, the GICS classification is somewhat arbitrary and machine learning can reveal categories that are better supported by accounting or financial data.

To find the optimal number of clusters, we can use WSS. WSS depends on the number of clusters, however, and grouping all the points in as many clusters will have a very small TSS: the clustering then explains most of the variance, but how useful is the information that each of your data point belongs to its own cluster? WSS decreases with the number of clusters you request, so there is a trade-off between how granular and precise your grouping is (the number K of clusters) and how much of the data K-Means Clustering explains (WSS).

To visualize the trade-off, one typically plots WSS as a function of K. To do that, we use the **factoextra** package and in particular its **fviz_nbclust()** function to produce Figure 18.7. We explicitly pass **wss** as the function to optimize for each possible value of K.

```
library(factoextra)
fviz_nbclust(data, kmeans, method = "wss")
```

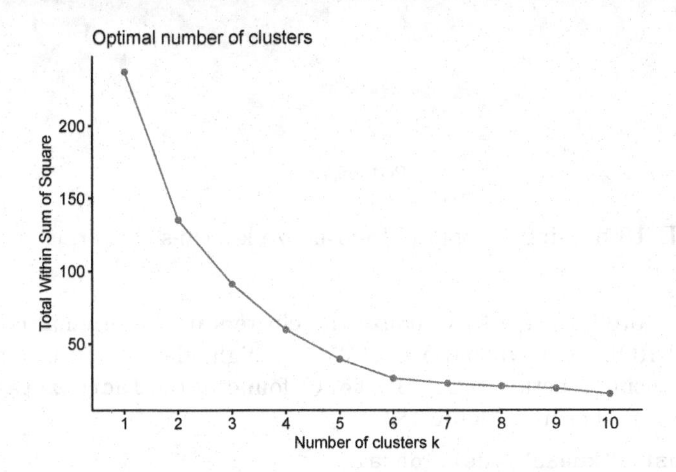

FIGURE 18.7 Derermining and Visualizing the Optimal Number of Clusters using package factoextra

Looking at Figure 18.7, we see that 3 is indeed a good trade-off: going from 3 to 2 would imply a large increase in WSS, whereas going from 3 to 4 would bring a much smaller reduction in WSS.

Function **kmeans()** of the **stats** package does the actual work of K-means clustering:

```
kmean.model = kmeans(data, 3)
```

Like in PCA, K-means clustering has determined the dimension that explains most of the dispersion (PC1). Like for PCA, we use function **autoplot()** of **ggfortify** that we saw on page 282. The result is Figure 18.8.

```
library("ggfortify")
autoplot(kmean.model, data = data)
```

We can see that one direction clearly captures most of the variance and separates Cluster 1 from the two others, and the second direction separates Cluster 2 from Cluster 3.

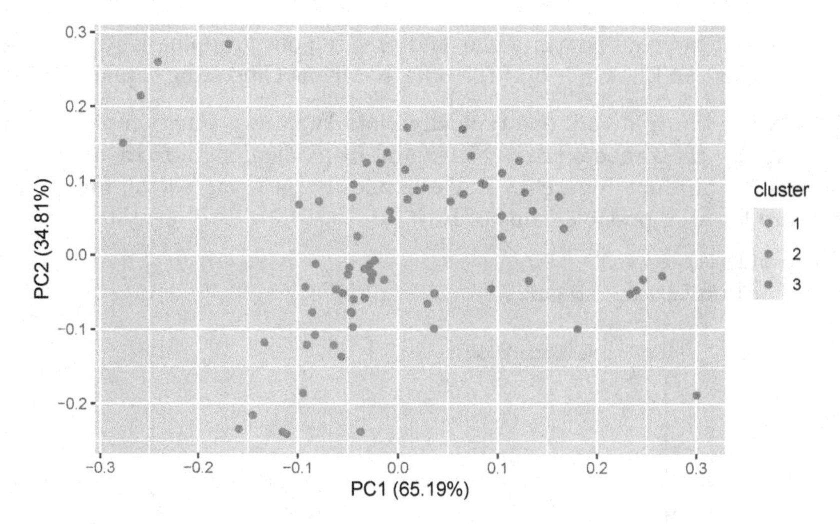

FIGURE 18.8 Using autoplot() to visualize k-means clustering

But it is more intuitive to visualize the clusters in the original coordinate system of ROE vs. Dividend Yield. We can highlight the clusters found by **kmeans()** using function **fviz_cluster()** found in the **factoextra** package. The output is shown in Figure 18.9.

```
fviz_cluster(kmean.model, data)
```

We can verify that K-means clustering found similarities in the dividend yields and ROEs of most of the IT stocks in the top left plot and of most of the Energy stocks at the bottom of the graph.

As a technical side-bar, we should note that two applications of K-means can result in two significantly different outcomes. That's because the number of attempts is only 10 by default in R. Setting it to something like 50, for example, typically solves that problem. You can do it using the **nstart=** option as illustrated below:

```
kmean.model = kmeans(data, centers = 3, nstart=50)
```

18.7 Hierarchical Clustering

Hierarchical clustering is related to KNN in that we are interested in finding "close" elements according to some definition of distance. A typical dataset contains different values across columns for different records (or entries, or elements) across rows. If we take two of these columns, then the distance

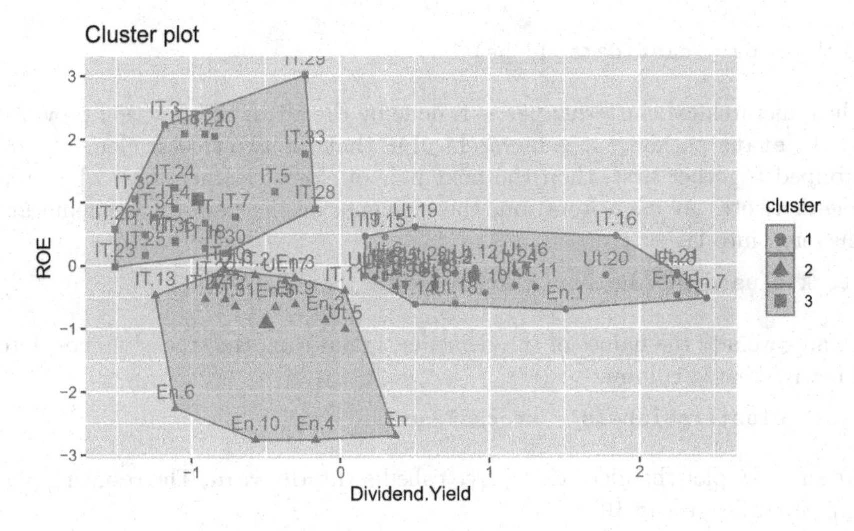

FIGURE 18.9 Visualizing clusters using factoextra

between these elements is the Euclidean distance we saw on page 348 and whose formula appears in Equation (18.4).

Let's see how that works on a concrete example – again, our **ACWI** index:

```
stocks = read.csv("ACWI.csv")
```

To make our graphs legible, we will limit ourselves to only a few constituents – say 30:

```
stocks = head(stocks, 30)
```

As you'll remember, some of the columns in the file are read as character strings, so we convert them to numerical values:

```
stocks$Dividend.Yield      = as.numeric(stocks$Dividend.Yield)
stocks$Price.to.Cash.Flow = as.numeric(stocks$Price.to.Cash.Flow)
stocks$Price.to.Sales      = as.numeric(stocks$Price.to.Sales)
stocks$ROE                 = as.numeric(stocks$ROE)
```

We are almost there! We clean up our data frame by removing `NA`'s, then take only two of the columns; in this case, dividend yield and the price-to-sales ratio:

```
stocks = na.omit(stocks)
data_dy_ps = data.frame(x=stocks$Dividend.Yield,
                        y=stocks$Price.to.Sales)
```

As said earlier, hierarchical clustering relies on calculating distances, which can be done using the `dist()` function:

```
dist_dy_ps = dist(data_dy_ps)
```

Then, hierarchical clustering per se is done by the **hclust()** function provided by the **stats** package. It is hierarchical in that the two closest elements are grouped together first, then the next pair of closest elements are clustered together, etc. At each iteration, the elements or the pairs of elements are clustered into larger pairs.

```
stock_clusters = hclust(dist_dy_ps)
```

Finally, we add the names of the elements (in our case, the stocks' names) into a newly created column:

```
stock_clusters$labels = stocks$Name
```

We can now plot the hierarchical tree called a **dendogram**. The resulting plot appears in Figure 18.10.

```
plot(stock_clusters)
```

As we can see, J&J and Procter & Gamble are clustered close to each other, and so are Bank of America and JPMorgan Chase. Likewise, ASML, Apple and TSMC, which all design and manufacture microchips, are clustered together.

One may wonder whether this intuitive grouping is a result of our choosing dividend yield and price-to-sale. Exercise 18.9.2 asks you to verify that hierarchical clustering on price-to-cash-flow and return on equity would produce a similar result.

18.8 Chapter-End Summary

We discussed several machine learning algorithms producing categorical outputs, **classification algorithms**, that decide to which possible discrete category (which **class**) an input data point corresponds to. We saw five such classification methods: KNN, logistic regression, decision trees, regression trees and K-means clustering.

We saw that **supervised** algorithms learn from **train data** (also called **in-sample** data) before being vetted on **test data**. KNN, logistic regressions and decision trees are examples of supervised algorithms. We measure the efficiency of supervised algorithms that produce two classes (thus, excluding regression trees) using confusion matrices, and we saw a few of the statistics that we can calculate on these confusion matrices: specificity, sensitivity and accuracy.

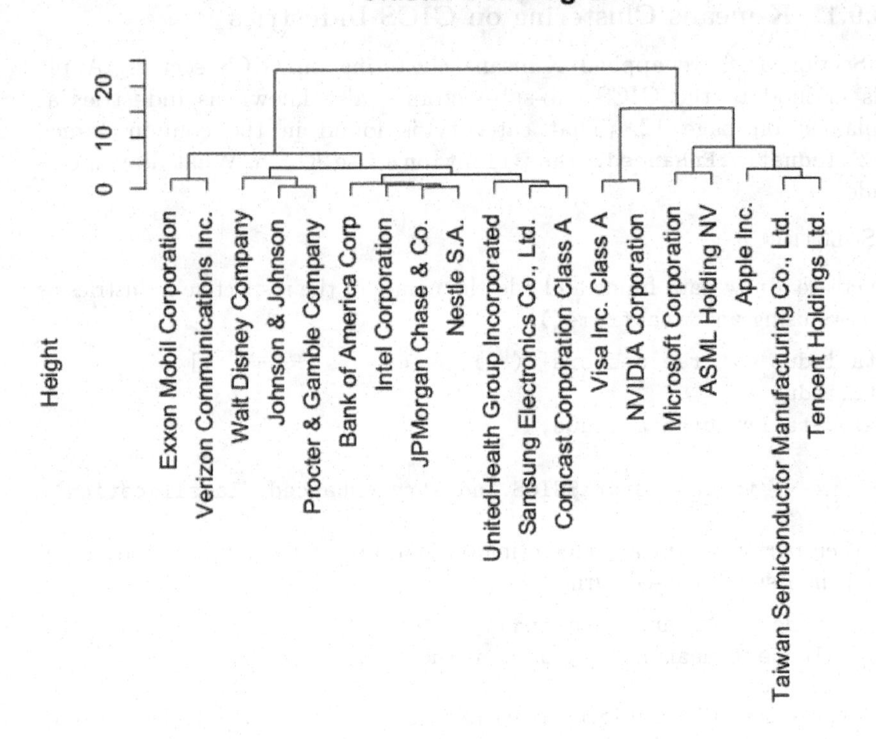

FIGURE 18.10 Dendogram of Hierarchical Clustering based on Dividend Yield and P/Sales

In contrast, **unsupervised** algorithms detect patterns or similarities –for example, based on "closeness"– and continuously improve their pattern detection as they are fed more data. K-means clustering and hierarchical clustering are two examples of this family of algorithms.

18.9 Exercises

18.9.1 K-means Clustering on GICS Industries

In Section 18.6 we applied K-means clustering on GICS sectors. Apply this method to the GICS sub-sub-sectors – also known as industries as explained on page 12. (That category is found in the column named `GICS.Industry.Enhanced.Classification`.) Use $K = 5$. What do you conclude?

■ SOLUTION

We create a new data frame and add the names of the respective industries as its rows using `as.data.frame()`:

```
data.indus = three.sectors[c("Dividend.Yield","ROE")]
data.indus =
  as.data.frame(data.indus,
        row.names =
          three.sectors$GICS.Industry.Enhanced.Classification)
```

We then perform K-means clustering on 5 groups, with enough iterations to avoid the issue discussed earlier.

```
kmean.model = kmeans(data.indus, centers = 5, nstart=50)
fviz_cluster(kmean.model, data.indus)
```

As you can see in Figure 18.11, K-means clustering was quite efficient at separating Energy Equipment companies in the bottom cluster from Oil and Gas in the middle-right clusters, and at separating Hardware and Semiconductors firms in the top left cluster from IT Services companies in the middle-left cluster.

18.9.2 Hierarchical Clustering on P/CF and ROE

Perform hierarchical clustering on price-to-cash-flow and return on equity. What do you conclude?

■ SOLUTION

```
data_pcf_roe =
  data.frame(x=stocks$Price.to.Cash.Flow, y=stocks$ROE)
```

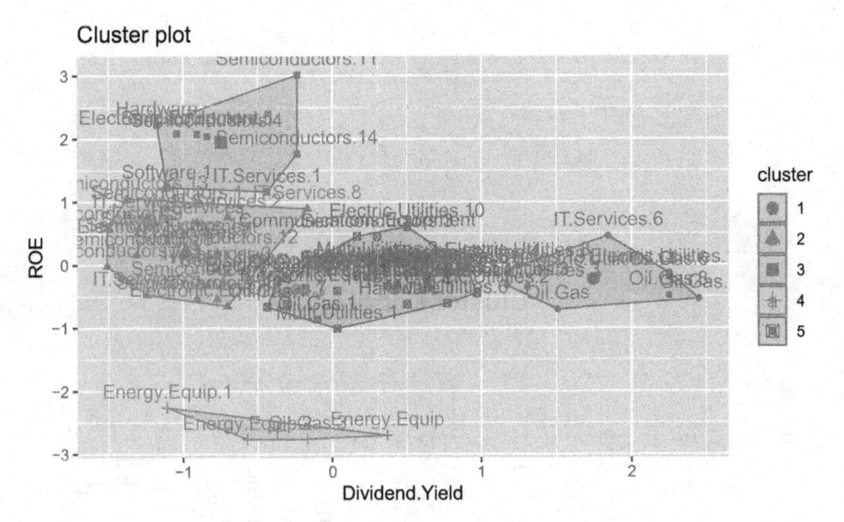

FIGURE 18.11 Testing if GICS Industry categories match the KNN clusters

```
dist_pcf_roe = dist(data_pcf_roe)
stock_clusters2 = hclust(dist_pcf_roe)
stock_clusters2$labels = stocks$Name
```

The dendogram resulting from this clustering appears in Figure 18.12.

```
plot(stock_clusters2)
```

Here again, we see some natural clustering of companies in similar industries: Bank of America and JPMorgan Chase are again clustered closely, and NVIDIA and ASML are also close and so are P&G and J&J.

■ END OF SOLUTION

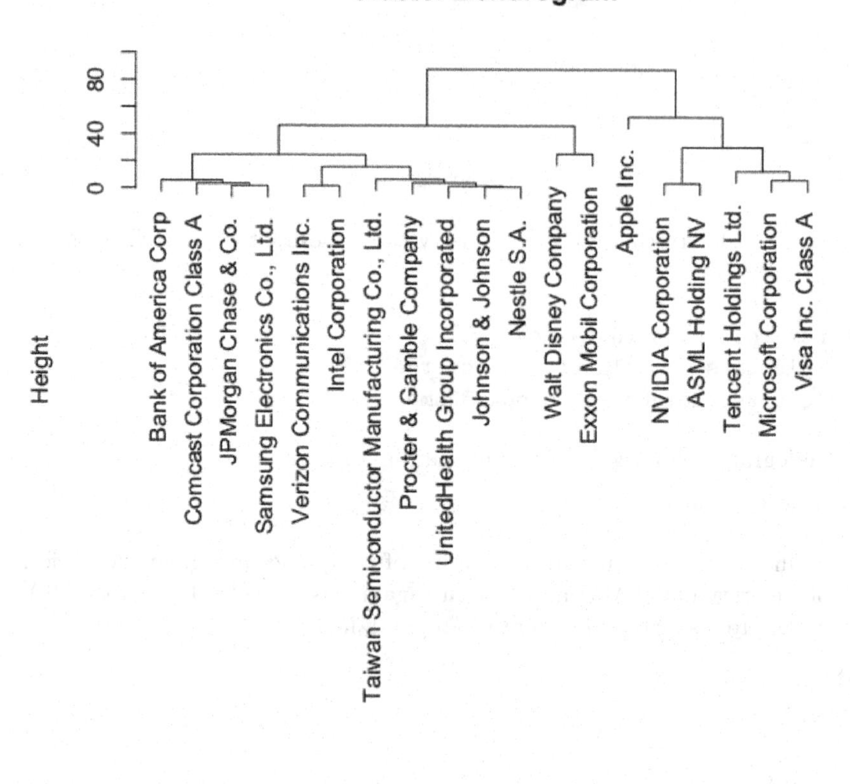

FIGURE 18.12 Dendogram of Hierarchical Clustering based on P/CF and ROE

19

Presenting the Results of Your Analyses

Analyzing data with the hope of gaining insights, sometimes called exploratory data analysis, certainly is an exhilarating endeavor. However, you'll typically need to explain your analysis and the conclusion you are reached to another person: a boss, a client, a coworker, etc. Moreover, that other person may want to understand how you reached your conclusion – and possibly see the code that you crafted, or the steps you took in your analysis. On your side, you may want to show exactly what you've done and – critically – to have all the computations recalculated and the graphs updated automatically each time you re-run your analysis on a new data set.

I call slide presentations and writing memos the "20th century" way of presenting the results of data analysis. 20th century tools still are pervasive of course, but it would be a shame not to introduce you to more modern ways – the 21st century ways of conveying information. In the same ways as downloading real-time data directly from the internet is more efficient than the 20th-century technology of reading static data files, using these methods is more efficient and allow for fast refresh of your analyses. These two methods are **Markdown** and **Shiny**.

19.1 Markdown Documents

We discussed notebooks in Section 1.2 and saw how they allow you to blend your notes with your code and see the results of executing pieces of your code. The next logical step is to have an entire document assembling your thoughts, your code itself, and the result of executing this code – all in one professional-looking document generated for you automatically. That document can be just one page; it can be an article or a corporate memo or a slide presentation; or it can be an entire book. In fact, this textbook is entirely produced by Markdown within RStudio, including its index, the table of content, page references, etc.

Markdown can produce documents in different formats: **HTML**, **Microsoft Word** or **PDF**. PDF files are generated through the use of one of the many free versions of the **Latex** software (RStudio recommends MiKTeX on Windows and MacTeX on Macs, but TeXShop has worked well for me on a MacBook).

DOI: 10.1201/9781003320555-19

More specifically, Markdown generates Latex code, then calls Latex to generate a PDF. This entire textbook was generated that way.

The first step in creating your first Markdown document to open a New File and to select the "R Markdown..." option from the drop-down menu. Then, provide a document title and select which document format you prefer. If you are not familiar with Latex, we recommend you select the Word file type. If you are familiar with Latex, then congratulations! The output document will look much better, as you already know.

But you might still be wondering: How are these Markdown documents useful, or any different than typing a document in any word processor? The answer is twofold: First, Markdown makes certain your code snippets *work*. That's because what the reader sees is exactly the output of the very code that is shown in the document. In particular, since this textbook is generated in Markdown, I (and you!) can be certain that the code examples work as they are shown. Second, and this is immensely useful in a professional setting, the numbers in your document are recalculated each time you run Markdown. It means that up-to-date data can be retrieved from the Internet each time you run Markdown (unless you explicitly say otherwise), and that calculations are updated each time – ensuring that all the figures and results in the document you produce apply to these refreshed data.

How to Use Markdown

The main tip to keep in mind with Markdown is to pay attention to colors and to let colors guide you. If part of what you type appears in a different color, or on a different background, there is a good reason for that. You'll quickly learn what these visual clues mean, and once you know them, you'll know when to *expect* them – and suspect something is wrong when you don't see them.

YAML

The top part of a markdown consists of a few likes that look like the code below:

```
---
title: "Here's the Title of My Document"
author: "Your name here"
date: "March 1, 2022"
output: word_document
---
```

These lines are written in a language called **YAML** (which stands for Yet Another Mark-up Language). YAML is beyond the scope of this book, but know that it allows you to define most properties of your document. For our purposes, the main properties we need are the title, name of the author, date, and the filetype of the document that Markdown will generate.

Code Chunks and Their Options

A piece of code, called a code **chunk**, starts with a line consisting of three back-quotes followed by `{r}` and ends with a line consisting of three back-quotes again. For example, imagine you want to run the `head()` command on the `stock` dataframe, as follows:

```
head(stocks, 2)
```

```
  Ticker                  Name            GICS.Sector
1   AAPL            Apple Inc. Information Technology
2   MSFT Microsoft Corporation Information Technology
        Country Ending.Weight Market.Cap.
1 United States          3.47   2,193,582
2 United States          2.81   1,899,924
  Dividend.Yield Price.to.Cash.Flow Price.to.Book
1           0.62              30.73         33.40
2           0.85              31.17         14.13
  Price.to.Sales   ROE ReturnPast52Weeks   Buy
1           8.41 73.69             78.98  TRUE
2          13.55 40.14             40.72 FALSE
```

The above was produced using the following Markdown text:

```
9897  as follows:
9898
9899 - ```{r}
9900  head(stocks, 2)
9901 - ```
9902
9903  The above was produced using the following Markdown text:
```

Everything between lines 9899 and 9901 that delineate the code chunk is considered as executable code by Markdown. Also, note that by default Markdown prints this content in courier font with a gray background, passes this content to the relevant programming language (which is the meaning of the `r` between the curly braces – Markdown works also with languages like Python), waits for R to come back with the output of these commands, and inserts that output below the code chunk. Of course, that output can be a graph, and almost all the figures in this textbook were generated directly from a Markdown file. For example, Figure 19.1 shows how Figure 3.1 was produced:

Note that the code chunk can be executed without showing its content using the `echo=FALSE` option within the curly braces. For example, we could plot

```
3
4 ```{r fig.cap="Simple example of plotting the closing prices of a stock"}
5 plot(Apple.closing.prices)
6 ```
7
```

FIGURE 19.1 How Figure 3.1 was produced

Apple's price without showing the R code using the Markdown specification shown in Figure 19.2.

```
9569
9570 ```{r echo=FALSE}
9571 plot(Apple.closing.prices)
9572 ```
9573
```

FIGURE 19.2 How to execute a command without showing it

Finally, note that you can embed R calculations within your text. For example, you can confidently write that the area of a disc with diameter 42 is 1385.4424 because you can let R calculate it for you within Markdown. You just have to start your R expression with a single back-quote followed by `r` and close the expression with a single back-quote. This paragraph for example looks like this in the Markdown file that generated this book:

```
28  note that you can embed R  calculations within your text. For example,
    you can confidently write that the area of a disc with diameter 42 is
    `r pi*(42/2)^2` because you can let R calculate it for you within
    Markdown. You just have to start your R expression with a single
```

FIGURE 19.3 Embedding R calculations within your text

The embedded R expressions can include formatting, for example to limit the number of significant digits. Embedded R expressions without formatting were used on page 159 when calculating the $R_{t,k}$'s, which explains why so many decimals were calculated and shown.

Sections and Subsections

Finally, you can organize your memos using sections and subsections, as usual. The syntax used by Markdown to indicate chapters, sections and subsections is to insert an empty line followed by one, two or three pound signs. For example, the header of this subsection was produced using the Markdown code shown in Figure 19.4.

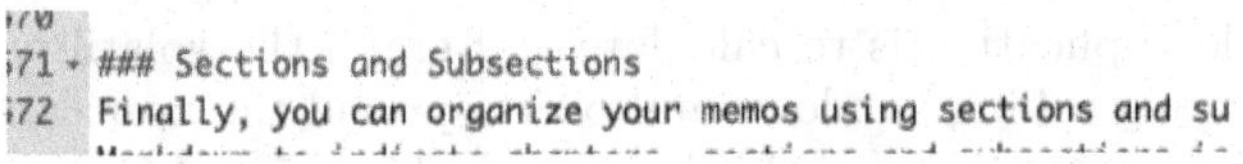

```
71 ▾ ### Sections and Subsections
72   Finally, you can organize your memos using sections and su
```

FIGURE 19.4 How to sepecify a section header in Markdown

19.2 Shiny

Producing insight-creating visualization is a critical part of data analysis and then of this textbook. But how do you share that? Of course, you can send documents by email, but they are static: you can't change a graph once it is sent, you can't refresh it with new data, and your documents are not interactive – so your audience cannot change any parameter of your analysis.

Fortunately, RStudio is integrated with a powerful notebook called **Shiny**. Shiny lets you create graphs that are interactive, in the sense hinted earlier that users can modify the parameters of your visualization. But an even more powerful feature of Shiny is that you can create mini-webpages that any user can access immediately, like any web page. There are no fees to pay, and only a one-time registration to **shinyapps.io** is necessary[1].

The easiest way to see how Shiny works is to start with an example, and RStudio offers a built-in example. To access it, create a new file, as usual – but instead of creating an R Script or a Markdown file, request a Shiny Web App. This opens a ready-made file and, because that file comes with RStudio, it should work immediately. That example code graphs a histogram of eruptions of a geyser called Old Faithful. We will not study that code but will use its skeleton to develop a simple example related to finance.

The Shiny Web Apps have a very simple three-pronged structure: the code that displays the output to the user and interacts with them, the code that performs the calculations you want to perform, and a little piece of code that glues the first two pieces together. The first part is the **user interface** (UI) side, and the second is the **server** side. Each time the user requests a modification to the output (a change in parameter value, for example), the UI side sends the request to the server code, which crunches the numbers again according to the new request, and sends back the output to the UI side.

The UI is created using the `shinyUI()` function, which can be as simple as a single `fluidPage()` function as shown below. As in the Old Faithful example, that function provides two information: The title of the web app, and the desired layout using the `sidebarLayout()` function.

[1] A complete and up-to-date tutorial on this service can be found at `https://docs.rstudio.com/shinyapps.io`.

Our example application is to calculate and graph the volatility of Apple's stock on varying rolling windows and over a period the user can also select using slide bars. Note that we added a second **sliderInput()**, separated by a comma. The two variables adjusted through the sliders are called **window** and **history**, respectively.

The rest is very similar to the out-of-the-box Old Faithful example provided by RStudio. Note that we gave more personal names to the two functions created on the UI and server sides, respectively.

```
my.ui = fluidPage(
   titlePanel("Rolling Volatility of Apple's Stock"),
   sidebarLayout(
      sidebarPanel(
         sliderInput("window",
                     "Window for Vol calculation (days)",
                     min = 20,
                     max = 100,
                     value = 30),
         sliderInput("history",
                     "Length of Historical Data, in days:",
                     min = 100,
                     max = 1000,
                     value = 250),
      ),
      mainPanel(
         plotOutput("distPlot")
      )
   )
)
```

The layout here consists of just two panels: the main panel, declared using the **mainPanel()** function, and a sidebar panel. The main panel simply outputs whatever plot is contained in an object called "**distPlot**." That object is not defined on the UI side: It will be up to the server side to define (and produce the content of) that object.

The sidebar simply consists of a slider, specified using the **sliderInput()** function. This slider allows the user to change the number of bins in the histogram constructed by the server side and sent back to the UI side to be displayed by **mainPanel()**.

Server Side

The server side is where the action really takes place and is defined using the **shinyServer()** function. The output of that function has to contain what is promised to the UI side however, and in our case we promised something called **distPlot**. That **distPlot** is contained in the output of the function. What it receives as an input is called **input** that contains **window** and **history**. Both are defined in the UI code, so reading either value is done by writing **input$window** or **input$history**.

The code first gets today's date using **Sys.Date()**, then calculates the **start.date** by subtracting the number of days specified in the **history** slider. We then pull historical data on AAPL thanks to **getSymbols()**, calculate the returns using **periodReturn()**, and graph the volatility using **chart.RollingPerformance()**:

```
my.server = function(input, output) {
    output$distPlot = renderPlot({
        today = Sys.Date()
        start.date = today - input$history
        getSymbols("AAPL", from = start.date)
        aapl.rets = periodReturn(AAPL, period = "daily")
        chart.RollingPerformance(R=aapl.rets,
                                 width=input$window,
                                 FUN="sd.annualized",
                                 scale=252,
                                 main="Rolling volatility")

    })
}
```

Note that the calculation done by the server is a **renderPlot()**. **renderPlot()** produces a plot, which is provided as the last command in the block of code. That plot is assigned to what is called "distPlot" in **plotOutput()** on the UI side.

We are almost done. In addition to **shiny**, we need the **quantmod** and **PerformanceAnalytics** libraries:

```
library(shiny)
library(quantmod)
library(PerformanceAnalytics)
```

The last step is to bind the two sides together using the **shinyApp()** command. In contrast to the default example provided in RStudio, we don't bother with the named argument **ui** and **server**:

```
shinyApp(my.ui, my.server)
```

You can then run this application locally. But we can do even better and make a mini-webpage out of it.

Web Applications

The easiest way to create a web application is to deploy or "publish," in RStudio parlance, the interactive window you just created on the **shinyapps.io** website. As of Spring 2022, this website offers to host 5 applications for free. You can monitor which of your old codes are still taking one of these 5 slots using ShinyApps.io's dashboard, such as the one shown below.

Once your account is created, go back to **RStudio** and click on the button that offers to "publish." You will be asked to select which platform you want to publish to, and select Shinyapps. As of early 2022, RStudio will ask you to provide a "token" or "key" that you can find on the `Shinyapps.io` website, typically in the section that details your account. Once you provided that token to RStudio, hit the Publish button and, after a minute or two, a window should appear in your default browser. Copy the URL there and send it to anyone, and ask them to check it out!

The result of our simple application appears in Figure 19.5. As indicated in the URL banner at the top of the browser, anyone can play with this dynamic graph by connecting to the indicated URL. (Note that this webpage might not be functional by the time you read these pages.)

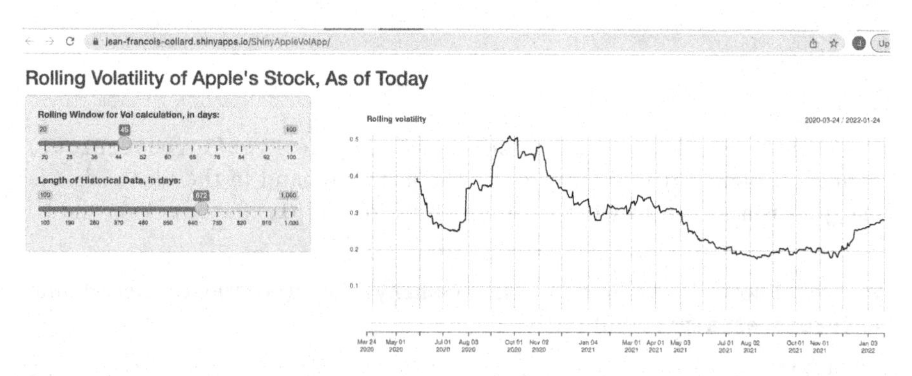

FIGURE 19.5 Final result of our small website

20

Appendix: Main Packages Seen in this Book

Package Name	Page(s)	Main Uses
gtsummary	106	Produce clean-looking tables
tidyr	62	Data wrangling and reorganization
gtsummary	106	Clean-looking tables
xts	182	Manipulate time series
stats	365, 368	Statistics, hierarchical clustering
normtest	188	Statistics
tseries	188	Statistics
quadprod	212	Quadratic programming, optimization
ggplot2	127	Visualization
ggfortify	365	Visualization
ggrepel	146	Visualization
factoextra	365	Visualization
corrplot	250	Visualization of correlation matrices
PerformanceAnalytics	5, 14, 180	Financial analysis
PortfolioAnalytics	173, 178	Construction of financial portfolios
QuantMod	17, 161	Quantitative finance
tidyquant	17	Quantitative finance
RQuantLib	5	Quantitative finance
Quandl	25, 121	Quantitative finance
class	348	Classifier (machine learning), KNN
rpart	356	Decision tree (machine learning)
tseries	334	Time series analysis

DOI: 10.1201/9781003320555-20

Index

Printed in the United States
by Baker & Taylor Publisher Services

Printed in the United States
by Baker & Taylor Publisher Services